Logical Empiricism and the Physical Sciences

This volume has two primary aims: to trace the traditions and changes in methods, concepts, and ideas that brought forth the logical empiricists' philosophy of physics and to present and analyze the logical empiricists' various and occasionally contrary ideas about the physical sciences and their philosophical relevance. These original chapters discuss these developments in their original contexts and social and institutional environments, thus showing the various fruitful conceptions and philosophies behind the history of 20th-century philosophy of science.

Logical Empiricism and the Natural Sciences is divided into three thematic sections. Part I surveys the influences on logical empiricism's philosophy of science and physics. It features chapters on Maxwell's role in the worldview of logical empiricism, on Reichenbach's account of objectivity, on the impact of Poincaré on Neurath's early views on scientific method, Frank's exchanges with Einstein about philosophy of physics, and on the forgotten role of Kurt Grelling. Part II focuses on specific physical theories, including Carnap's and Reichenbach's positions on Einstein's theory of general relativity, Reichenbach's critique of unified field theory, and the logical empiricists' reactions to quantum mechanics. The third and final group of chapters widens the scope to philosophy of science and physics in general. It includes contributions on von Mises' frequentism; Frank's account of concept formation and confirmation; and the interrelations between Nagel's, Feigl's, and Hempel's versions of logical empiricism.

This book offers a comprehensive account of the logical empiricists' philosophy of physics. It is a valuable resource for researchers interested in the history and philosophy of science, philosophy of physics, and the history of analytic philosophy.

Sebastian Lutz is senior lecturer of theoretical philosophy at Uppsala University. He works on philosophy of science, philosophical methodology, and the history of logical empiricism.

Adam Tamas Tuboly is postdoctoral researcher at the Institute of Philosophy, Eötvös Loránd Research Network, MTA BTK Lendület Morals and Science Research Group, and a research fellow at the Institute of Transdisciplinary Discoveries, Medical School, University of Pécs. He works on the history of logical empiricism and philosophy of science.

Routledge Studies in the Philosophy of Science

Logical Empiricism and the Physical Sciences

From Philosophy of Nature to Philosophy of Physics

Edited by Sebastian Lutz and
Adam Tamas Tuboly

Routledge
Taylor & Francis Group

NEW YORK AND LONDON

First published 2021
by Routledge
52 Vanderbilt Avenue, New York, NY 10017

and by Routledge
2 Park Square, Milton Park, Abingdon, Oxon, OX14 4RN

Routledge is an imprint of the Taylor & Francis Group, an informa business

Library of Congress Cataloging-in-Publication Data
A catalog record for this book has been requested

ISBN: 978-1-138-36735-7 (hbk)
ISBN: 978-0-367-76821-8 (pbk)
ISBN: 978-0-429-42983-5 (ebk)

Typeset in Sabon
by Apex CoVantage, LLC

Dedicated to the memory of Thomas Oberdan (1949–2018)

Contents

1 Introduction

From Philosophy of Nature to Philosophy of Physics

Sebastian Lutz and Adam Tamas Tuboly

1.1. Introduction

Naturphilosophie, or its English counterpart, philosophy of nature, has a long and fascinating history and has undergone significant changes over time.[1] During the 19th and 20th centuries, many forms of philosophy of nature emerged, from monographs of philosophical remarks by physicists – often written as diversions and philosophically naïve – and the treatises of the German idealists – often ambiguous and scientifically naïve – to the philosophically motivated writings of revolutionary scientists.

The logical empiricists' philosophy of nature and science, however, stands out for its ability to synthesize various forms of inquiries: while its roots reach deep into the neo-Kantian "Wissenschaftslehre" of its day, it was framed in the empiricist traditions of English philosophers and scientists of the modern era. It developed both traditions further with the help of Frege's and Russell's logic and amalgamated them with contemporary French thinking about the conventionalist theory and practice of science. (Katherine Dunlop discusses conventionalism in her contribution to this volume, especially with reference to Poincaré and Neurath.) As Hans Reichenbach (1931/1978, 383) claimed in his programmatic pamphlet on philosophy of nature,

> philosophy of nature constitutes a great triumph of rationalism. . . . However, modern philosophy of nature must also be regarded as a triumph of empiricism. . . . Modern empiricism does not contradict rationalism because knowledge is conceived as a system constructed by reason upon which experience exerts a regulative and selective influence.

The logical empiricists' philosophy has become a focus of research, both as a historical object of study and as a systematic philosophical position. However, there has been comparably little research on the logical empiricists' philosophy of the natural sciences, especially that of physics. This is an unfortunate gap in the history of philosophy of science, for one because the logical empiricists' views on physics heavily influenced their views on philosophy of science and philosophy in general. Furthermore, their analyses of scientific theories were arguably clearest when it came to physics. But the lack of research on the logical empiricists' philosophy of physics is also unfortunate for systematic philosophy, as their views can provide new input for both the philosophy of physics and the philosophy of science and for philosophy in general.

This volume accordingly has two overarching themes: first, it traces the traditions and changes in methods, concepts, and ideas that brought forth the logical empiricists' philosophy (or philosophies) of physics, and, second, it presents and analyzes the logical empiricists' various and occasionally contradictory ideas about the physical sciences and their philosophical relevance. The logical empiricists' ideas are discussed in their original contexts, including their social and institutional environments, to show the fertile ground on which the history of 20th-century philosophy of science was built. In the remainder of this chapter, we will briefly introduce the protagonists of this volume (Section 1.2) as well as some deuteragonists (Section 1.3) and circumscribe the extent of their involvement in the philosophy of nature (Section 1.4). This discussion tamely (but in keeping with the logical empiricists' methodology) concludes with the suggestion that the extent of their involvement in the philosophy of nature depends on how "philosophy of nature" is understood.

1.2. The Protagonists

Moritz Schlick (1915/1979, 1917/1979) gave the first philosophically informed meta-scientific presentations of the special and general theories of relativity that were also accepted by Einstein. Schlick's 1917 book on space-time (*Space and Time in Contemporary Physics*) was quickly translated into English, and several editions were published. After working out the epistemological consequences of relativity in his magnum opus, *Allgemeine Erkenntnislehre*, Schlick's attention turned towards other issues in the philosophy of mind, language, ethics, and science in general. Nonetheless, as Herbert Feigl (1982, 61) noted, "[s]ince Wilhelm Ostwald's rehabilitation of the concept of 'philosophy of nature', there can surely have been no more dedicated 'philosopher of nature' than Schlick".

In 1921, Rudolf Carnap wrote his doctoral dissertation in Jena (after some back-and-forth between the physics and philosophy departments) under the supervision of Bruno Bauch, with Max Wien on the committee. It was published one year later as a separate volume of the *Kant-Studien* (1922/2019). The dissertation concerned the meanings of 'space' across the sciences (mathematics, geometry, and philosophy) and the challenge presented by the theories of relativity. Soon thereafter, he published the short monograph *Physikalische Begriffsbildung* (*Physical Concept Formation*, Carnap 1926/2019) on measurement and abstract concept formation in physics.[2] An investigation of the notion of entropy in statistical physics from the 1950s remained unpublished until after his death (Carnap 1977). (Carnap's philosophy of physics and science is the subject of the chapters by Sebastian Lutz, Jordi Cat and Robert DiSalle.)

After defending his doctoral dissertation in 1906, Philipp Frank, as the successor of Einstein in Prague, published many important – though mostly forgotten – papers on the simplification of the special theory of relativity (see Frank 1932/1998, 290–296 for a list) and collaborated with the Austrian physicists and engineer Hermann Rothe: together, they derived the Lorentz transformation based on group theory, without the postulate of the constancy of the speed of light. Though their achievement was recognized now and again, Frank did not influence the mainstream of physics.[3] As the director of the Physics Department at the German University of Prague for almost 25 years, before his emigration to the United States, Frank was a perfect candidate to write the *Foundations of Physics* monograph for Neurath's *International Encyclopedia of Unified Science*.[4] (Frank is discussed in Don Howard's and Adam Tamas Tuboly's chapters.)

Of all the logical empiricists, Hans Reichenbach was the most productive and well-known investigator and popularizer of the physical sciences. After a dissertation on the interpretation of probability at Erlangen (1915/2008), Reichenbach published numerous books on the philosophy (1920/1965) and axiomatization (1924/1969) of the theory of relativity. As a student in Albert Einstein's famous seminar on relativity in Berlin during the 1918–1919 winter semester, Reichenbach had up-to-date and detailed knowledge of the debates and concerns surrounding the theory. Later, when he focused his attention on a systematic presentation of probability theory and on quantum mechanics (1944/1965), he achieved widespread recognition with his three-valued logic and scientific realist interpretations. (Reichenbach's work is central to the chapters by Alan Richardson, Marco Giovanelli, and Flavia Padovani).

If not in physics itself, many logical empiricists had earned degrees with dissertations on the philosophy of physics or the philosophy of nature and produced new insights through their early works in this field. One of the best examples is Herbert Feigl, who defended his dissertation *Zufall und Gesetz* (*Chance and Law*) under Schlick in 1927. (The entire dissertation was later published in Haller and Binder 1999, 1–191.) Feigl had studied physics in Vienna with Schlick and Hans Thirring until Edgar Zilsel's first monograph, on the problem of applying mathematical-numerical descriptions to the natural world, had convinced him to pursue philosophy instead. His dissertation focused on probability and induction – an interest that Feigl never abandoned throughout his career. This work also enabled him to write a short monograph about theory and experience in physics (Feigl 1929).[5] Though Feigl later became known mainly for his writings on the philosophy of mind and as the leader of the Minnesota Center for the Philosophy of Science, he kept publishing papers on induction and probability for years.[6] (Matthias Neuber's chapter examines Feigl's mature views.)

1.3. Some Deuteragonists

There were at least two other students of Schlick's whose dissertations concerned issues in physics. The first was Marcel Natkin, who wrote a dissertation on simplicity, causality, and induction (see Haller and Binder 1999, 193–301). After leaving academia, Natkin worked as a photographer and never returned to the philosophy of nature (although he occasionally wrote about photography). His dissertation is an interesting document of the inner tensions within the Circle: while Feigl (1929) reserved some role for explanation, in contrast to the tradition of Mach, Kirchhoff, and Duhem (though, according to Feigl, this was mainly just a terminological issue in the face of the new philosophy), Natkin followed Mach and Duhem, stating that science shall aspire to the production of the most economical descriptions of the natural world.

The third student of Schlick's who wrote a dissertation on the physical sciences was Tscha Hung, who discussed causality in the new physics (see Haller and Binder 1999, 303–353). He studied physics, mathematics, and philosophy, first with Reichenbach in Berlin and from 1928 on with Schlick and others in Vienna. He was a regular member of the Circle meetings but left for China after Schlick's death. Although he published many papers in German and English about the Vienna Circle in the 1980s (when he held visiting fellowships at Oxford and Cambridge), they were mainly concerned with the protocol-sentence debate. Nonetheless, his dissertation under Schlick again testifies to the breadth and direction of the latter's interests in physical theories during the early 1930s.

Though his dissertation was on moral philosophy, Béla Juhos published several articles and books about the epistemological dimension of physical theories and on general philosophy of science. Juhos had always worked on the periphery of the Circle, but as one of the few original members who stayed in Vienna during and after the Second World War, he tried to preserve the spirit of logical empiricism in Austria (with limited success). Those works of his that were published in English under the title *Selected Papers on Epistemology and Physics* (Juhos 1976) mainly concern questions of causality, but in the 1960s, Juhos also published two books on the epistemic-logical foundations of classical and modern physics (Juhos and Schleichert 1963; Juhos 1967).

1.4. Doubts

As even this rather short and cursory overview shows, the logical empiricists (those in the center as well as those working on the periphery) were occupied with questions of physics, natural sciences, and the philosophy of nature,[7] perhaps more so than with any other topic. Nevertheless, there are doubts about this positive conclusion. (These are also discussed in Clark Glymour's chapter.) Thomas Ryckman (2007, 193) notes that the term "'philosophy of physics' was little used by the logical empiricists themselves, and that, with notable exceptions, they produced little of what is currently understood by that name, *viz.*, detailed investigations into particular aspects or interpretations of physical theories". While many scholars agree that logical empiricism played a major (if not the most important) role in institutionalizing philosophy of physics and bringing it into the mainstream in the second half of the 20th century, "it is something of an anachronism", concludes Ryckman (ibid.), "to speak of logical empiricism's 'philosophy of physics'".

For instance, none of the members of the Vienna Circle published any detailed work on the theories of relativity during what Friedrich Stadler (2001/2015) calls the constitutive non-public (1924–1929) and the public (1929–1936) phases of the Circle. From the mid-1920s on (following his textbook article on Naturphilosophie, which was presumably commissioned years before), when Schlick started to organize the informal Thursday discussions in the Mathematical Institute's library, he did not publish anything on the nature and analysis of physical theories.

Although Carnap's papers in the early 1920s on space-time and physics (now published in Carnap 2019) related, in varying degrees, to the logico-mathematical questions of the physical sciences, they were mainly produced before his permanent move to Vienna and thus still during his neo-Kantian period.[8] (Interestingly, Carnap's dissertation on space was intended to be "a contribution to the theory of science", i.e. "Wissenschaftslehre", as its subtitle says, and not to philosophy of nature.) And in contrast to Reichenbach, Schlick, and Frank, Carnap did not keep up with the latest developments in physics by himself. Having devoted most of his time during his student and doctoral years to relativity, he needed help following the latest debates and results in quantum mechanics. In Prague, Frank presumably provided this assistance, and otherwise, as Carnap (1963, 14–15) mentions in his autobiography, Reichenbach was always at hand in correspondence.[9] Especially during its American period, Carnap's main research was in the general philosophy of science, albeit with applications in the philosophy of the natural sciences.

As Frank had already left Vienna for Prague in 1912 and rarely participated in the discussions, his work, while being of primary importance for understanding the logical empiricists' philosophy of physics (or at least one version thereof), took place mainly on the periphery. (Frank's late *Relativity: A Richer Truth* is mainly concerned with the socio-cultural bearing of relativity in general, not just in physics.)

Reichenbach could be considered an exception to this lack of focus on the philosophy of physics, and he himself often claimed that what distinguished the Berlin Group of logical empiricists from the Vienna branch was the former's close adherence to the actual state and problem nexus of the natural sciences. (The Berlin branch of logical empiricism is taken up in Nikolay Milkov's chapter.) However, Stöltzner (2013) in particular raises serious doubts about Reichenbach's self-confident and proudly presented role in the philosophy of natural sciences. While Reichenbach's initial investigations into relativity theory had been prompted by the most recent ideas of physicists, there "was a major difference between the evolution of Reichenbach's agenda and the thinking of most physicists between 1916 and 1936" (Stöltzner 2013, 125). Though he was in close contact with physicists after his move to the United States, and some of his later considerations came quite close to those of the physics community, his motivations and moves differed radically.

Thus, from the mid-1920s on, the attention of the logical empiricists (both in Vienna and Berlin) turned towards more general issues in science. Even when Einstein's theories were discussed in the Circle, they did not necessarily play a substantial role.[10] As Viktor Kraft (1973, 16), a peripheral member of the Circle, stated later in an interview with Heinrich Rutte, "actually the theory of relativity played almost no role, at best it was an example of modern natural science".

Besides relativity, it was quantum mechanics that most excited professional philosophers and educated laymen in the first half of the century. Nevertheless, it was only the interpretational challenges of Bohr, Heisenberg, Schrödinger, and others that sparked discussions among the logical empiricists. Due to the several debates concerning the notions of probability and especially of causality, Frank, Schlick, Reichenbach, and Richard von Mises again started publishing works on the philosophy of physical theories, that is, on the philosophy of nature. (von Mises' involvement in the debates concerning physics and probability is the topic of Maria Carla Galavotti's chapter.) Though their works focused less on the technical details of the primary physical theories (as they did before with relativity), they tried to show with their own (varied) methods and from their own viewpoints how the new theoretical physics might influence philosophical conceptions (or not). (Quantum mechanics is the topic of Jan Faye and Rasmus Jaksland's chapter and, with respect to Schlick, Richard Dawid's chapter.)

The 1936 Unity of Science Congress in Copenhagen was devoted explicitly to the discussion of causality in the physical and biological sciences (the event was opened in Niels Bohr's home). There was thus apparently a brief return to the philosophy of nature, but it did not last long. After the death of Schlick, and with the forced emigration of Frank and Carnap (Reichenbach had left Germany already in 1933), the remaining physicists-turned-philosophers of the Circle left for the United States, where they encountered a new atmosphere that would in due course also change their perspectives and approaches.

1.5. What Was Philosophy of Nature Anyway?

As noted previously, "Naturphilosophie" had various meanings and denoted various forms of inquiries, often not even linked by any common topic. The same might also be said for individual logical empiricists' use of the word.

Carnap's textbook *An Introduction to the Philosophy of Science*, published towards the end of his career, contains a short section, inserted in the chapter on causality, outlining the task of philosophers of science and how the field is related to the sciences. Carnap (1974, 187) starts from the observation that for the ancient Greeks, philosophy of nature was both an "empirical investigation of nature and the philosophical clarification of such knowledge", while during some periods of philosophy (the German Naturphilosophie of the 19th century, presumably), these empirical considerations were replaced by metaphysical speculations about nature. But this latter "old philosophy of nature", Carnap claims (1974, 188), "has been replaced by the philosophy of science". This new philosophy

> is not concerned with the discovery of facts and laws (the task of the empirical scientist), nor with the formulation of a metaphysics about the world. Instead, it turns its attention toward science itself, studying the concepts employed, methods used, possible results, forms of statements, and types of logic that are applicable.
>
> (Carnap 1974, 188)

Thus, Carnap's replacement of the old philosophy eschews the Greeks' empirical investigations and the metaphysical speculations of the German Naturphilosophie but remains closely connected to the Greeks' clarification of empirical investigations.

In his own textbook entry on the topic, "Outlines of the Philosophy of Nature", Schlick (1925/1979, 2) still defended a certain weak form of a foundationalist approach to philosophy of nature. He explained the relation as follows:

> As general philosophy inquires into the foundations of all human knowledge, and investigates the principles whereby the overall framework of knowledge is constructed and held together, so the task of the philosophy of nature is to do the same thing for the special area of natural science.

While this indeed sounds like the old program of securing the foundations of the sciences, Schlick's statement already indicates that for him, philosophy (of nature) and (natural) science go hand in hand. As "all sciences once sprang from the womb of philosophy" (ibid.), they contain presuppositions and make use of principles from philosophy and are thus philosophical at their core.

A few years later, by the time he held a public lecture at the University of Vienna in 1929, Schlick's conception of philosophy and its relation to science had changed radically under the influence of Ludwig Wittgenstein. Philosophy no longer had any foundational task; instead, it became an activity that "consists in the acts of giving or finding meaning, which lend significance to all the words occurring in our propositions" (Schlick 1934/1979, 142). By searching for clear meanings and demonstrating the relations between words, propositions and concepts, the aim was to scrutinize the world in new ways; "[science] provides a picture of the world, [philosophy] uses it . . . in order to construct a world-view" (1934/1979, 139). Philosophy of nature was pursued by those who could grasp scientific theories and practice philosophy at the same time.

> The scientist must be a philosopher if he is to understand and build further upon the basic concepts of his science; and the philosopher can arrive at his world-view no otherwise than by starting from the world-picture of the sciences.
>
> (1934/1979, 153)

This conception of Schlick's remained rather stable under Wittgenstein's ever-growing influence. Schlick's posthumously published lecture "Philosophy of Nature" again contained the idea that, first, philosophy is not a foundational enterprise to secure the natural sciences and, second,

> [t]he task of a philosophy of nature is . . . to interpret the meaning of the propositions of natural science; and therefore the philosophy of nature is not itself a science, but an activity which is directed to the consideration of the meaning of the laws of nature.
>
> (1949, 3)

While Schlick claimed that insightful science, by collecting the true propositions about nature, is deeply philosophical in its essence (a word often used by Schlick in the 1920s), he still upheld a strict division between science and philosophy.

For Reichenbach (1931/1978, 360), there was no such division, as he argued that "we cannot accept the distinction between philosophy of the natural sciences and philosophy of nature". Reichenbach thought that, as science advances, philosophical reasoning would also have to evolve as it tries to keep up with the latest developments, because

> [i]t is not true that the general frame of theoretical reasoning is given once and for all, and that the progress of science consists merely in gradually filling this frame with more content. Reason is not a

system of ready-made compartments in which the scientist deposits more and more subject-matter; this system of compartments is itself subject to historical change, and will constantly be transformed in such a way that it becomes progressively adjusted to the findings of experience.

(1931/1978, 361)

As both measurement results (1929/1978a, 159) and the general categories that are used to depict them are subject to change, Reichenbach argued, philosophy of nature *has to be* pursued along the lines of philosophy of the natural sciences. In fact, the two pursuits should be one and the same, since modern science, after 2000 years of development, provides the sharpest and most comprehensive analytical tools for both. Philosophy of nature (that is, for Reichenbach, the philosophy of the natural sciences) is a theoretical enterprise in its own socio-historical context, so that

> the answers that we must find today can only be discovered by the logical tools of our time, as they have mainly been elaborated in the mathematical and scientific methods. An approach geared to the present is the way, a reorientation of theoretical philosophy of nature.
> (Reichenbach 1931/1978, 387)

Reichenbach was not at all worried about using the term "Naturphilosophie"; the titles of many of his papers in the late 1920s and early 1930s even explicitly refer to it. In the title of his report on the 1929 Prague Conference on the Epistemology of the Exact Sciences, for instance, Reichenbach called the new philosophical approach "Die neue Naturphilosophie", "The New Philosophy of Nature" (Reichenbach 1929/1978b). He was adamant that this new scientific philosophy was a genuine descendant of Descartes, Leibniz, and Kant, because

> what is new in this kind of philosophy is its method rather than its aims. Its aim is the solution of certain fundamental epistemological questions, some of which have played a role in the older philosophy, some of which, however, were only discovered in our times.
> (1931/1978, 359)

Neurath aggressively opposed any form or mention of philosophy of nature and in a letter to Reichenbach tried to dissuade him from using the term "Naturphilosophie", because according to Neurath, "Philosophie" conveyed a certain old-fashioned, dusty, internalist conception about knowledge, which was exactly the opposite of what "the movement" tried to achieve.[11] In this debate, Carnap took Reichenbach's side, expressing his happiness "about the actual proposals for systems" in Reichenbach's forthcoming writings. While Carnap took Reichenbach's stance to be the opposite of Schlick and Wittgenstein's activity-based approach, he saw that it was also in tension with Neurath's position.[12]

As already noted, in his later works, Reichenbach did not try to solve the same problems as the physics community. Rather, he found a special task for philosophers in the interpretation of quantum mechanics. When scientists themselves tried to account for their steps and commitments, Reichenbach (1944/1965, vii) claimed, they either used the ambiguous terminology of the old 19th-century philosophy (subject/object, picture/reality) or tried to evade interpretations at all costs as "unnecessary ballast" in favor of empirical operation manuals. Reichenbach, as a philosopher of physics, did not avoid these interpretational challenges and aimed at developing a "philosophical interpretation of quantum physics which is free from metaphysics, and yet allows us to consider quantum mechanical results as statements about an atomic world as real as the ordinary physical world" (1944/1965, vii), as he stated in the preface to his *Philosophic Foundations of Quantum Mechanics*. In the preface, Reichenbach did not talk about philosophy of nature as such but limited his attention to specific physical theories and questions. He still maintained his belief that the philosophy of the natural sciences (or now the philosophy of physics) was about the physical world, albeit through the prism of scientific theories. "All that is intended in this book is clarification of concepts. . . . Whereas physics consists in the analysis of the physical world, philosophy consists in the analysis of our knowledge of the physical world" (Reichenbach 1944/1965, vii).

Thus, it seems that Reichenbach took a position similar to Carnap, who eschewed the Greeks' empirical investigation of nature as part of the new philosophy of science but remained close to their idea of philosophy of nature as a clarification of the knowledge obtained by empirical investigation. What, then, about Reichenbach's earlier refusal to "accept the distinction between philosophy of the natural sciences and philosophy of nature"?

A possible way of retaining a continuity in practice is given by Carnap:

> Although the work of the empirical scientist and the work of the philosopher of science must always be distinguished, in practice the two fields usually intermingle. A working physicist is constantly coming upon methodological questions. What sort of concepts should he use? What rules govern these concepts? By what logical method can he define his concepts? How can he put his concepts together into statements and the statements into a logically connected system or theory? All these questions he must answer as a philosopher of science; clearly, they cannot be answered by empirical procedures.
>
> (1974, 188)

Thus the philosopher of science has all the justification needed to get involved in the endeavors of the natural scientist, since much of what is usually considered science is actually philosophy. And in this sense, then, the philosophy of science is a continuation and indeed part of the philosophy of nature.

1.6. The Chapters in This Volume

The first part of the volume casts its view widely, examining the background and rise of philosophy of physics, with special attention to particular connections and parallels between philosophy and science. JORDI CAT challenges the predominant role in the rise of logical empiricism that is accorded to relativity theory, and secondarily quantum mechanics, by virtue of their being modern or revolutionary. He argues instead that the scientific practices and results of electromagnetic theory exerted as much an influence.

NIKOLAY MILKOV shows that Kurt Grelling, the "third man" of the Berlin Group besides Reichenbach and Walter Dubislav, could grasp the academic trends of the time faster and was better informed about logic and philosophy of nature than his two prominent colleagues. Thus he could better delineate, although tentatively, central threads of research of the Berlin Group.

KATHERINE DUNLOP argues that Henri Poincaré's account of the conventionality of geometry, a central theme in the philosophy of physics, exemplifies a broad view of the relationship of theory to data more familiar from the writings of Otto Neurath. Like Neurath, Poincaré holds that experience cannot confront theory until it has undergone a selective regimentation, which makes it impossible to base theory directly on experience. Poincaré and Neurath both developed Mach's view in the same novel direction, although there is not sufficient evidence to conclude that Neurath was influenced by Poincaré.

DON HOWARD surveys the nearly 50-year-long friendship and collaboration between Albert Einstein and Philipp Frank, exploring their detailed discussions of the nature and role of conventions in physics and Frank's argument that Duhem pointed the way to synthesizing the conventionalism of Poincaré and the empiricism of Mach, a synthesis that Frank thought to be central to Einstein's philosophy of science.

The first part of the volume is closed with a transitory chapter by ALAN RICHARDSON, who offers a framework for thinking about the significance of Hans Reichenbach's project in scientific philosophy. He argues for a reappraisal of the main purpose of Reichenbach's 1920 book, *Relativitätstheorie und Erkenntnis A Priori,* that both connects it more closely to the details of scientific neo-Kantianism and lays out more explicitly its attempt to provide an empiricist alternative to neo-Kantianism. The tensions in this attempt lead to some of the central themes in Reichenbach's mature work in philosophy of science.

The second part of the volume narrows its focus on the treatment of specific physical theories in logical empiricism. ROBERT DISALLE offers a critical analysis of the Carnapian picture of science as an account of the empirical, synthetic character of physics, as distinct from the analytic character of its mathematical formalism. By considering the case of space-time theory, in particular general relativity, DiSalle shows the contrast between Carnap's picture and the way in which the theory actually determines its characteristic theoretical magnitudes, such as the curvature of space-time.

THOMAS RYCKMAN argues that the logical empiricists' reading of Einstein's "Geometry and Experience" as championing their "scientific world view" deliberately stripped the text of its context, the ongoing attempts by Weyl, Kaluza, and Einstein himself to extend general relativity via a highly non-empirical method of mathematical speculation to a "unified field theory" encompassing electromagnetism. Logical empiricism kept its disagreements with Einstein under wraps, as doing so was crucial to its self-portrayal.

MARCO GIOVANELLI provides an overview of the Appendix to Reichenbach's 1928 *Philosophie der Raum-Zeit-Lehre*, which presents an intentionally artificial example of geometrization of the electromagnetic field by allowing spacetime to have torsion in addition to curvature. While Reichenbach's geometrization attempt was doomed to fail, the philosophical message of the Appendix is an integral part of the book's argument that general relativity was not the beginning of the new era of the "geometrization of physics" but the culmination of a historical process of the "physicalization of geometry".

After these discussions of general relativity and the project of a unified field theory, the volume turns to quantum mechanics. JAN FAYE and RASMUS JAKSLAND argue that, despite frequent claims to the contrary, logical positivism did not influence physicists' view of quantum mechanics until well into the 1930s, if at all. Bohr in particular believed that understanding quantum mechanics provided a general philosophical lesson to which the logical positivist would be responsive and might compel them to join him in the attempt to extend his ideas of complementarity and contextualism beyond quantum mechanics.

RICHARD DAWID argues that the direction of influence is rather the reverse: Moritz Schlick's interpretation of the causality principle is based on his understanding of quantum mechanics. Furthermore, the difference between Schlick's logical empiricism and Bas van Fraassen's constructive empiricism reflects a difference between the understanding of quantum mechanics endorsed by Schlick and the understanding that had been established by the time of van Fraassen's writing.

The second part of the volume concludes with a critical chapter by CLARK GLYMOUR, who investigates why there were so few contributions of the logical empiricists that influenced fundamental issues surrounding developments in physics or other sciences in the first half of the 20th century. He argues that late 20th- and 21st-century philosophy of science has largely followed Carnap's more passive vision of philosophy as metascientific logical reconstruction rather than Reichenbach's vision of an active contribution to scientific problems. Some latter-day philosophical and scientific work, however, more or less in Reichenbach's spirit, has addressed many of the methodological problems the logical empiricists avoided.

The third part of the volume widens its focus again to the general philosophy of physics. It is opened by three investigations of the logical empiricists' view on the relation between theory and experience. MARIA CARLA GALAVOTTI maps how Richard von Mises developed a perspicacious version of the frequency theory of probability with a deeply empiricist and operationalist flavor. The leading idea in von Mises' program was that the theory of probability is a natural science rooted in experience, and he was convinced that statistics provides the means for dealing with natural science's problem of specifying the connection between experimental data and scientific laws and theories.

Focusing on Reichenbach's early writings, FLAVIA PADOVANI considers the role of conventionalism in Reichenbach's approach to the way we define some fundamental terms of a theory in relation to the application of theory to experience. It is here where his notion of "coordination" plays an essential role.

Carnap's view on the relation of theory and experience is taken up by SEBASTIAN LUTZ. He argues that, contrary to the received view on Carnap's philosophy of science, Carnap held from the very beginning that all observational terms can be explicitly defined in theoretical terms, but not vice versa. Lutz further describes how, from the very beginning, Carnap suggested formalizing physical theories as restrictions in mathematical spaces, as in the state-space conception of scientific theories favored by van Fraassen.

MATTHIAS NEUBER compares Ernest Nagel's, Herbert Feigl's, and Carl G. Hempel's respective conceptions of the cognitive status of physical theories. All three of these conceptions stood in the logical empiricist tradition. However, whereas Feigl favored a semantic point of view and sought a realist interpretation of the logical empiricist program, Nagel and Hempel favored a syntactic point of view and were more cautious regarding realism. Ultimately, Feigl's account of the cognitive status of physical theories remained at the periphery of the logical empiricist movement.

Finally, ADAM TAMAS TUBOLY details how, besides his work on theoretical physics, Philipp Frank remained interested in the philosophical questions of physics throughout his entire career. While this interest surfaced in different forms in different periods, Frank's late approach constituted a sophisticated and complex inquiry into what he called the "humanistic background of science", which involved two core issues: (1) interpreting metaphysics in a way that would make it a meaningful enterprise for empiricists and (2) investigating to what extent metaphysics, in this interpretation, can contribute to a better understanding of the natural sciences. Using the results of the latter, Frank tried to bridge the gap between the natural and social sciences.

1.7. Dedication

THOMAS OBERDAN was working on a contribution that would have explored the structuralism in Schlick's early thought as it evolved from his pre-Vienna thinking through his "form and content" philosophy. The contribution would have revealed a profound continuity linking Schlick's scientific philosophy with the earliest form of his logical empiricism. Tom did not live to finish his investigation. This volume is dedicated to him.

1.8. Acknowledgments

During the editorial process, Adam Tamas Tuboly was supported by the MTA Premium Postdoctoral Scholarship and the MTA Lendület Morals and Science Research Project. We are indebted to our authors for their participation and for the stimulating discussions in this volume and during our correspondence. We are also grateful to our reviewers and our editors at Routledge for their kind work, help, and patience. We are also indebted to the archives for permission to quote their materials: RC = Archives of Scientific Philosophy, Rudolf Carnap Papers, 1905–1970, ASP.1974.01, Archives & Special Collections, University of Pittsburgh Library; MSN = Moritz Schlick Nachlass, Wiener Kreis Archiv, Rijksarchief in Noord-Holland, Haarlem, The Netherlands. Sebastian thanks Alana and Hannah. Ádám thanks Gabó, Sára, and Róza Lilla.

Notes

1. This history is still to be told. For a start, see Grant (2007) and the chapters in Part III of Buchwald and Fox (2013).
2. The first candidate for writing the *Physikalische Begriffsbildung* had been Hans Reichenbach, who had been unable to accept the invitation due to time constraints (see Ungerer to Carnap, 2 January 1925, RC 110–05–43). Carnap was thus only the second choice after Reichenbach, as he was for his academic positions in Vienna (1925) and Prague (1931).
3. Although Einstein held Frank in high regard, Arnold Sommerfeld wrote to Schlick that "despite all his perspicacity, [Frank] has never touched a physical problem as a positivist thinker" (Sommerfeld to Schlick, 17 October 1932, MSN).
4. One perfect candidate: before inviting Frank, Neurath had asked Reichenbach to write about physics, in particular space-time and relativity; while Reichenbach seems to have accepted the invitation, he never produced any lines for the *Encyclopedia*.

5. Chapter 3 of Feigl's book also appeared in English (Feigl 1929/1981). Originally, Carnap had been invited to write the monograph, after his *Physikalische Begriffsbildung* had appeared in the same series a few years earlier. Carnap accepted the invitation, and for two years, the volume was slated to appear under the title 'Der Aufbau der physikalischen Theorien', until Carnap withdrew from his contract and suggested Feigl in his stead (see E. Ungerer to Carnap, 17 October 1926, RC 110–05–36; Carnap to Ungerer, 18 December 1926, RC 110–05–34; and Carnap to Ungerer, 23 July 1927, RC 110–05–27). Carnap heavily influenced the content of Feigl's book, however (Carus and Friedman 2019, xli, n. 24).
6. Though Feigl's *Festschrift, Mind, Matter, and Method* contains a section on "Philosophy of the Physical Sciences", this is significantly shorter than the parts dedicated to the philosophy of mind, induction, and method.
7. Felix Kaufmann could also be mentioned here, as he wrote many papers about induction and probability in the 1940s. He came closest, however, to the philosophy of physics in his contribution to the *Library of Living Philosophers* volume dedicated to Ernst Cassirer. See Kaufmann (1949).
8. These papers are analyzed in Runggaldier (1984) and in the editorial introduction to Carnap (2019).
9. The same conclusion is supported by a letter of Carnap's to Wolfgang Yourgrau, 3 October 1958; RC 027–42–03.
10. Frank (writing on Einstein's relation to positivism), Reichenbach (writing on the philosophical significance of relativity), and Karl Menger (writing on geometry and relativity) contributed papers about physics in the *Library of Living Philosophers* volume dedicated to Einstein. Many other contributors were conceptually linked to logical empiricism, such as Percy Bridgman, Victor Lenzen, and Kurt Gödel.
11. See Carnap's diary entry, 15 July 1929 (RC 025–73–03); Neurath to Reichenbach, 22 July 1929 (RC 029–15–15) and Neurath to Carnap, 9 October 1932 (RC 029–12–24). Cf. Neurath to Reichenbach and Carnap, 17 January 1935 (RC 029–09–97).
12. Carnap to Reichenbach, 1 February 1935 (RC 102–64–05).

References

Buchwald, J. Z. and Fox, R. (eds.) (2013), *The Oxford Handbook of the History of Physics*. Oxford: Oxford University Press.

Carnap, R. (1922/2019), 'Der Raum/Space', in Carnap (2019), pp. 21–171.

——— (1926/2019), 'Physikalische Begriffsbildung/Physical Concept Formation', in Carnap (2019), pp. 339–427.

——— (1963), 'Intellectual Autobiography', in P. A. Schilpp (ed.), *The Philosophy of Rudolf Carnap*. La Salle, IL: Open Court, pp. 3–84.

——— (1974), *An Introduction to the Philosophy of Science*. New York: Dover Publications.

——— (1977), *Two Essays on Entropy*. Edited and with an Introduction by A. Shimony. Berkeley and Los Angeles: University of California Press.

——— (2019), *Early Writings: The Collected Works of Rudolf Carnap, Volume 1*. Edited by A. W. Carus et al. Oxford: Oxford University Press.

Carus, A. W. and Friedman, M. (2019), 'Introduction', in Carnap (2019), pp. xxiii–xli.

Feigl, H. (1929), *Theorie und Erfahrung*. Karlsruhe: Verlag G. Braun.

——— (1929/1981), 'Meaning and Validity of Physical Theories', in R. S. Cohen (ed.), *Herbert Feigl: Inquiries and Provocations: Selected Writings 1929–1974.* Dordrecht: D. Reidel, pp. 116–144.

——— (1982), 'Moritz Schlick, a Memoir', in E. T. Gadol (ed.), *Rationality and Science: A Memorial Volume for Moritz Schlick in Celebration of the Centennial of His Birth.* Wien and New York: Springer, pp. 55–82.

Frank, Ph. (1932/1998), *The Law of Causality and its Limits.* Dordrecht: Springer.

Grant, E. (2007), *A History of Natural Philosophy.* Cambridge: Cambridge University Press.

Haller, R. and Binder, Th. (eds.) (1999), *Zufall und Gesetz: Drei Dissertationen unter Schlick.* Amsterdam and Atlanta: Rodopi.

Juhos, B. (1967), *Die Erkenntnislogischen Grundlagen der modernen Physik.* Berlin: Duncker & Humblot.

——— (1976), *Selected Papers on Epistemology and Physics.* Edited by G. Frey. Dordrecht: D. Reidel.

Juhos, B. and Schleichert, H. (1963), *Die erkenntnislogischen Grundlagen der klassischen Physik.* Berlin: Duncker & Humblot.

Kaufmann, F. (1949), 'Cassirer's Theory of Scientific Knowledge', in P. A. Schilpp (ed.), *The Philosophy of Ernst Cassirer.* Evanston, IL: The Library of Living Philosophers, Inc., pp. 183–213.

Kraft, V. (1973), 'Gespräch mit Viktor Kraft', *Conceptus*, 7 (21–22): 9–25.

Reichenbach, H. (1915/2008), *The Concept of Probability in the Mathematical Representation of Reality.* Translated and Introduced by F. Eberhardt and C. Glymour. Chicago and LaSalle, IL: Open Court.

——— (1920/1965), *The Theory of Relativity and A Priori Knowledge.* Berkeley and Los Angeles: University of California Press.

——— (1924/1969), *The Axiomatization of the Theory of Relativity.* Berkeley and Los Angeles: University of California Press.

——— (1929/1978a), 'The Aims and Methods of Physical Knowledge', in M. Reichenbach and R. S. Cohen (eds.), *Hans Reichenbach: Selected Writings 1909–1953, Volume Two.* Dordrecht: D. Reidel, pp. 120–225.

——— (1929/1978b), 'The New Philosophy of Science', in M. Reichenbach and R. S. Cohen (eds.), *Hans Reichenbach: Selected Writings 1909–1953, Volume One.* Dordrecht: D. Reidel, pp. 258–260.

——— (1931/1978), 'Aims and Methods of Modern Philosophy of Nature', in M. Reichenbach and R. S. Cohen (eds.), *Hans Reichenbach: Selected Writings 1909–1953, Volume One.* Dordrecht: D. Reidel, pp. 359–388.

——— (1944/1965), *Philosophic Foundations of Quantum Mechanics.* Berkeley and Los Angeles: University of California Press.

Runggaldier, E. (1984), *Carnap's Early Conventionalism.* Amsterdam: Rodopi.

Ryckman, Th. (2007), 'Logical Empiricism and the Philosophy of Physics', in A. Richardson and Th. Uebel (eds.), *The Cambridge Companion to Logical Empiricism.* Cambridge: Cambridge University Press, pp. 193–227.

Schlick, M. (1915/1979), 'The Philosophical Significance of the Principle of Relativity', in H. L. Mulder and B. F. B. van de Velde-Schlick (eds.), *Moritz Schlick: Philosophical Papers, Volume I (1909–1922).* Dordrecht: D. Reidel, pp. 153–189.

——— (1917/1979), 'Space and Time in Contemporary Physics', in H. L. Mulder and B. F. B. van de Velde-Schlick (eds.), *Moritz Schlick: Philosophical Papers, Volume I (1909–1922).* Dordrecht: D. Reidel, pp. 207–269.

—— (1925/1979), 'Outlines of the Philosophy of Nature', in H. L. Mulder and B. F. B. van de Velde-Schlick (eds.), *Moritz Schlick: Philosophical Papers, Volume II (1925–1936)*. Dordrecht: D. Reidel, pp. 1–90.

—— (1934/1979), 'Philosophy and Natural Science', in H. L. Mulder and B. F. B. van de Velde-Schlick (eds.), *Moritz Schlick: Philosophical Papers, Volume II (1925–1936)*. Dordrecht: D. Reidel, pp. 139–153.

—— (1949), *Philosophy of Nature*. Translated by A. von Zeppelin. New York: Philosophical Library.

Stadler, F. (2001/2015), *The Vienna Circle*. 2nd ed. Dordrecht: Springer.

Stöltzner, M. (2013), 'Did Reichenbach Anticipate Quantum Mechanical Indeterminism?', in N. Milkov and V. Peckhaus (eds.), *The Berlin Group and the Philosophy of Logical Empiricism*. Dordrecht: Springer, pp. 123–149.

Part 1
The Rise of Philosophy of Physics

2 The Electromagnetic Way to the Scientific World-Conception

Maxwell's Equations at the Service of Logical Empiricism

Jordi Cat

2.1. Introduction

Writing on the role of physics in the rise of logical empiricism, Thomas Ryckman has expressed the prevalent view: "Relativity theory and quantum theory, the two revolutionary developments of twentieth-century physics, happily coincided with the rise and consolidation of logical empiricism" (2007, 193; see also Ryckman 2005). And he adds: "Logical empiricism was conceived under the guiding star of Einstein's two theories of relativity" (2007, 194). Early works by Schlick, Reichenbach and Carnap, he continues, "were principally concerned to show how the philosophy of natural science should be necessarily transformed in [relativity theory's] wake" (2007, 195). This prevalent view, with a historiographic commitment to the category of revolution, is notoriously also at the center of Friedman's *Dynamics of Reason* (Friedman 2001).

Crucial to this view is the assumption of a particular historiographical relation between history, science and philosophy: a time-indexed historical epistemology that identifies scientific developments, often considered revolutionary, in whose relatively proximate aftermath historical figures considered them evidence in relation to established or controversial philosophical views in support of a new philosophical view, which, in turn, is also considered a significant event in the history of philosophy. This kind of historical argument often expresses a similar one put forward by the historical figures themselves, in a case of participant history.

This picture of the relation of philosophy to science and of the path to logical empiricism indeed does some justice to the importance of relativity theory. But I want to suggest that it also leaves out the way electromagnetic theory, with its historical formulations, applications and developments, played a role, even in relation to relativity theory. In particular, it leaves out the role it played in the paths different thinkers took to the logical empiricist movement, both in their shared and their differentiating problems and positions. Electromagnetic theory, through its own historical evolution, tracks the development of their respective thinking. The widespread picture leaves out, in particular, perhaps more surprisingly, Neurath's own engagement of physical theory and applications. The revised picture must include a revised historiographical assumption, rooted also in the protagonists' arguments, which includes a broader historical scope of relevant scientific practices or products that stand in a relation to 'modern' science not of error, but of exemplary value, or else the source of changes that would lead to an exemplary 'modern' consequence.

The revised picture should also take into account the way in which relevant practices and products such as electromagnetic theory also received philosophical description and evaluation. In fact, prior to the works of Carnap, Frank, Schlick and Neurath, Maxwell's electromagnetic theory had also been the subject of philosophical consideration and criticism in the hands and minds of a number of influential scientist-philosophers such as Mach, Hertz, Boltzmann, Duhem and Poincaré. Their appeal to it by members of the movement of logical empiricism reflected and extended the theory's dual life and its evolving philosophical and scientific significance.

2.2. Scientific and Philosophical History of Maxwell's Theory

2.2.1. *Maxwell*

I begin with historical background, with an emphasis on its dual theoretical and epistemological dimensions – or methodological, in narrower terms, and philosophical, in broader terms.

Maxwell's equations were introduced by James Clerk Maxwell between 1856 and in 1865 to provide a unified mathematical field theory of electrical and magnetic phenomena (see Maxwell 1856/1890, 1861/1890, 1862/1890, 1865/1890). Reflecting prior conceptions of electrical and magnetic fields forces and their mutual relations of induction through charges and currents, they have the form of differential equations connecting measurable electric and magnetic magnitudes (to bring out their more intuitive and empirical, geometrical meaning, Maxwell often presented mathematical equations with integrals along lines and across surfaces or else verbal descriptions of those relations).[1]

In 1861, he also identified light with electromagnetic waves. The result was an even more significant turning point in the history of physics. With it, Maxwell's theory was opening the door to the possibility of reducing optics to electromagnetism. In his development and presentation of the theory, Maxwell also linked different mathematical representations of electric and magnetic phenomena to particular mechanical models of the ether and thereby to spatial, temporal and causal properties of the different mechanical phenomena. For instance, in the model that decomposed the ether into connected molecular vortices, magnetic force corresponds to the centrifugal force of the vortices and electric displacement corresponds to the elastic yielding of the connecting mechanism (see Siegel 1991; Darrigol 2000 for discussion and references). Maxwell called this cognitive and methodological heuristic the method of physical analogy (see Cat 2001). In this way, Maxwell's equations had entered, then, both the history of physics and the history of scientific epistemology and methodology, in fact raising controversies in both.

2.2.2. From Helmholtz to Hertz

Maxwell's theory competed against Continental alternatives based on direct action at a distance between electric particles or currents. In the face of competing theories, the physicist and physiologist Hermann Helmholtz devised in 1870 a framework within which different theories could be classified and compared, constituting a theoretical condition for an *experimentum crucis* (Heidelberger 1998; Darrigol 2000, Ch. 6). He had already adopted a generalizing strategy, although with a unifying purpose, with the formulation and justification of the principle of conservation of energy. In the so-called parametric generalized formulation, theories were identified by different features of physical situations corresponding to different numerical values of the relevant parameters. The particular case of Maxwell's theory corresponded to the null value of a parameter for currents in conductors. From that standpoint, Maxwell's theory was the form of Helmholtz's general theory in special limiting conditions. Experimental results suggested the choice of Maxwell's theory, at least on grounds of simplicity of the equations.

In the 1880s, Helmholtz's student Heinrich Hertz provided new experimental evidence supporting the existence of Maxwell's electromagnetic waves and their contiguous propagation of action. Hertz declared his difficulty forming a consistent physical conception of Maxwell's ideas: what, he asked, was Maxwell's theory? Testing Maxwell's theory alongside Continental alternatives required following Helmholtz's approach and comparing only their mathematical equations and characteristic parameters.

Like Helmholtz, Hertz embraced the framework of energetics and, prompted also by a unifying principle based on the fundamental character of electrostatic force, he derived the so-called Maxwell-Hertz equations, a symmetric system of equations equivalent to Maxwell's, connecting electric and magnetic quantities but without sources or discussion of the nature of charge, current or electric displacement (Darrigol 2000, 253–254). His experimental detection in 1888 of invisible 'air waves' induced by oscillating charges and propagating at the speed of light supported Maxwell's theory with higher probability over its alternatives.

The research process prompted Hertz to engage in general and prescriptive considerations of epistemic and methodological nature. Underlying Helmholtz's theory and Hertz's experimental approach was the decoupling of Maxwell's theory from its supporting mechanical models, the "gay garments which we use to clothe nature" (Hertz 1892/1893, 28). Instead, he declared, "Maxwell's theory is Maxwell's system of equations" (Hertz 1892/1893, 21). The view expressed a dual significance, as a matter of intelligibility and empirical testability, and both, I will argue, would be recognized by different members of the Vienna Circle.

For Hertz, these claims were also part of a specific epistemic standard of electrodynamic research that allowed him to circumvent a number of methodological challenges. His account distinguished between a physical theory, its presentation (*Darstellung*) and its representation (*Vorstellung*) – or its meaning (*Bedeutung*) (Heidelberger 1998). The theory expressed the regular relation between observed properties or phenomena; the presentation or expression helped communicate the theory symbolically and intuitively and the representation contributed a mental conception of the ultimate unobservable causes of the phenomena. As Heidelberger has noted, Hertz's claim about the identity of Maxwell's theory in fact expressed a methodological application of his epistemic views; the theory should be freed from any problematic presentation and be established only prior to the formulation and experimental testing of its valuable physical representation. Needless to say, I will show that other scientists and philosophers adopted a different understanding and use of his famous prescription.

Hertz (1894/1899, 26) had also adopted a mechanical ideal and subsequently persuaded many with the intuitive notion he attributed to Maxwell that electromagnetic forces are due to the motion of hidden masses. Also, the new mechanics was for Hertz the vehicle for a new epistemology. Hertz proposed a notion of mechanics (and physics) as an image or picture (*Bild*), which in this case was of the mechanical behavior of material points. A picture or symbol is a constructed image of the natural world that includes an axiomatic structure, which Hertz used to identify proper physics. But it is not a theoretical representation of hidden kinds of causes in the sense applied to electrodynamic theory. The theoretical symbol is not a sign either in a sense introduced by Helmholtz, namely of unobservable external reality, or the sensation associated with an unknown stimulus, or of one among different kinds. Instead, it was also a picture in its similarity to the sensations it represented. Still, as constructed theoretical structures, theoretical pictures are not determined by experience; yet, as symbolic structures, they present the following features: such symbolic systems are fact fitting, or correct, in their observable predictions, logically consistent and appropriate in the avoidance of arbitrary terms or symbols such as 'force' denoting nothing (see Hertz 1894, introduction; see also Mach 1883/1907, Appendix, sect. XXI; Barker 1980, 246–249; Kjaergaard 2002, 130).

The empirical correctness of the pictures was embedded and extended within the logical structure so that necessary relations in thought would track necessary relations in physical reality: that is, the necessary consequences of given ideas in thought would agree with the consequences of necessary consequences in nature. In particular, the conditions of the possibility of the mechanical picture were restricted by the concepts of space, time and mass.

The standard of correctness, and its application to the rejection from mechanics of the concepts of absolute motion, force and least action,[2] shared much with Mach's phenomenology and its emphasis on a distinction between the acceptable symbols and unintelligible or empty metaphysics – the shared attitude might have prompted Hertz to mention in the introduction to the *Principles* his debt to Ernst Mach's book *Contributions to the Analysis of Sensations* (Mach 1885/1894; Hertz 1894/1899, xxiv). This overlap allowed many philosophically or methodologically sensitive scientists to accept Hertz's perspective and project, or else loosely to consider both projects similar enough and otherwise also express sympathy towards Mach's. But Hertz's pictures are not Machian theories, abbreviated symbols for complexes of elements of sensations with symbolic relations describing the relations between those elements. Instead, they are constructions whose coordination with facts of observations is more holistic and serves testing purposes. Hertz's picture theory of physical representation, modified further by the physicist Ludwig Boltzmann, also played a role in Wittgenstein's development of his early philosophy of language.[3] The independent influence of Hertz, Mach and Boltzmann among scientists and scientific philosophers would support the new approach to Maxwell's equations as an epistemic standard.

2.2.3. Mach

Towering over other scientists, especially philosopher-scientists, was the physicist Ernst Mach with his phenomenological doctrines – while he used the label "phenomenological", his follower Hans Kleinpeter introduced the often-used label "phenomenalist" (Kleinpeter 1913). His views have subsequently been classified and examined alongside others developed at the time under the same labels: Husserl's phenomenological philosophy, Stumpf's phenomenological psychology and Mach's phenomenological science (Spielberger 1982; Berg 2016).

Mach's phenomenalism is an ontological and epistemological monistic doctrine intended to be distinctively neutral, that is, neither objective nor subjective, and its framework was explicitly psychological. (Kleinpeter 1913, Berg 2016; Banks 2003). In addition, Mach's psychophysiological account was an epistemological and methodological phenomenalism of thought economy. The cultivation of science involved for Mach obtaining a picture of the world in an economic way (Mach

1896/1986, Ch. XXVI, 361): Economy of thought, he declared, is the basis of science (Mach 1883/1907, esp., Ch. IV, Sect. IV). It aids in science's reproduction of facts in the domain of so-called uncompleted experience, reaching beyond familiar instances but tied to them – for confirmation and refutation. In general, cognition is the useful and economic reproduction of facts in thought.

In physics, Mach opposed a phenomenological physics to mechanical physics; while the former was illustrated by his commitment to anti-atomism, the latter was illustrated by Boltzmann's commitment to atomism (Mach 1896/1986, Ch. 22). Also, physics, he claimed, is experience arranged in economical order. Mach's anti-atomism sits between the doctrines of pure thermodynamics and energetics. The first posited general laws derived from experiment (without constitutive atomistic hypotheses, only constructive ones); the second posited a unified phenomenological description of all physical phenomena (without external explanations or justifications).

Especially relevant to the growing scientific attitude to theorizing and Maxwell's equations was Mach's defense of anti-atomism linked to his preference for differential equations. Differential equations represent the continuous characteristic of the manifold of our experience and constitute the proper aim of physical description (Mach 1883/1907, Appendix, Sect. XXI, 551). The idea of an atom is not a direct description of experience, nor does it involve properties of observed bodies, violating the principle of continuity. It can be at best a provisional "mathematical model for facilitating the mental reproduction of facts" (Mach 1883/1907, Ch. V, 492). The idea, he added, is no different from the idea of fourth spatial dimension or non-Euclidean geometry. But Mach found them artificial and arbitrary and lacking in connecting, unifying power across theories and disciplines outside physics. By such a standard, he considered Boltzmann's statistical mechanics epistemologically deficient, as it didn't meet the phenomenalist standard for a theory of matter and was ontologically deficient as ideas as images of and analogies to sensations, not identities.

In relation to the use of atomic models in physics, Mach made no mention of Maxwell. Yet he praised Maxwell's mathematical formulation for eliminating with contact action the unnecessary complication of Laplacian atomic models connected by forces acting at a distance (Mach 1883/1907, 534–535). This preferred formulation bears analogies to theories in other parts of physics and, as a result, has the additional merit of contributing to a simpler, more economical, homogeneous physics (Mach 1883/1907, 498). In that regard, Mach also praised Hertz's formulation of electromagnetic theory for its use of differential equations in the description of electric phenomena (Mach 1883/1907, 583). Yet, while rejecting Boltzmann's atomistic pictures in thermodynamics and mechanical hypotheses in electromagnetic theory, Mach, unlike Hertz, specifically praised Maxwell's use of the method of analogies as an important instance of the heuristic value of comparison and analogies in scientific thinking, which also constitute examples of continuity and proper theoretical, indirect description. Ironically, Mach also noted, as Boltzmann did, the similarity of Maxwell's view to Hertz's:

> Without giving up intuition, Maxwell thus succeeds in preserving his open mind and conceptual purity, combining the advantages of hypothesis and mathematical formulation. His picture is such that its mental consequences are pictures of the consequences of facts, to adapt a phrase of Hertz. Thus, Maxwell comes close to an ideal method of scientific inquiry: hence his uncommon success.
>
> (1905/1976, Ch. XXIV, 169)[4]

It is little surprising, then, that Kleinpeter (1913, 215–217) argued in Mach's footsteps not only that Maxwell was as great a thinker as he was a physicist. He also declared that Maxwell's method of mechanical analogies and intuitive illustrations established him as a foremost phenomenalist. And next, Kleinpeter (1913, 217–219) extended the same evaluation to Hertz's picture theory and surprisingly, little Maxwellian interpretation of Maxwell's equations.

One significant difference from Hertz's picture theory and phenomenological approach consisted in Mach's requirement that each symbol, and not just composite and interconnected symbolic systems, be properly linked to elements of sensations, as abbreviated expressions of statements of sense data (Barker 1980, 246). In his discussion of Hertz's *Principles*, Mach nevertheless identified Hertz's requirement of appropriateness, or minimum arbitrariness, with his own criterion of economy (Mach 1883/1907, Appendix, Sect. XXI, 549), noting Hertz's elimination of the need for force, however at the expense of a theory of hidden masses less simple than his own. He also noted the roots of Hertz's mechanical theory of connections in Maxwell's theory of contiguous electromagnetic action. As a form of action, Mach (1883/1907, 535, 553–554) considered the latter, as matter of experience, more intelligible than action at a distance (Mach also noted the similarity Hertz's program bore to Descartes'). And, again, he also noted the similarities between Hertz's picture theory and Maxwell's method of analogies. Yet Mach (1883/1907, 555) also criticized some of Hertz's objections to Newtonian mechanics and the teleological character of energy principles such as Lagrange and Hamilton's principle of least action and found Hertz's system, also by Hertz's own admission, less simple than his own.

2.2.4. *Boltzmann*

Not only for Hertz and Mach, also for Boltzmann, Maxwell's theories of electromagnetism and gases were the focus of both scientific and philosophical discussion, even in debate with the others. Boltzmann was yet another leading light in the rising German-speaking world of scientific philosophy.

Boltzmann's admiration of Maxwell's research dated back to the late 1860s in the wake of Maxwell's first kinetic-molecular theory of gases in 1867 (Darrigol 2018). Soon after his initial investigations of Maxwell's molecular theory of gases, Boltzmann also studied Maxwell's theory of electromagnetism and in 1873 presented its main resources and results to the Graz science society. Boltzmann praised as a distinguishing feature of Maxwell's theory that it reached beyond known experimental phenomena and, with numerous graphic illustrations, that its consideration of molecular vortices in the ether helped represent lines of magnetic force, predict the propagation of electromagnetic waves and suggest the electromagnetic nature of light (Boltzmann 1892/1974, 11).

Subsequently Boltzmann published a detailed set of lectures on Maxwell's electromagnetic theory (Boltzmann 1891/1893), the first in Germany. In 1890 in France, Henri Poincaré had just published his own lectures on Maxwell's theory (Poincaré 1890) and Hertz his experimental evidence for it and in 1892 his collection of investigations on the propagation of electric force (Hertz 1892/1893). Through the prism of kinetic theory, Boltzmann picked up on Maxwell's combination of an energy framework and concrete mechanical analogies, especially molecular ones. He even contributed mechanical models of electromagnetic contiguous action of his own in lecture and, even more unusual in Germany, in print.

While agreeing with Hertz about the primacy of the electromagnetic field and the obscurity of Maxwell's derivative notions of charge and current, Boltzmann criticized the nude formal relations and operational definitions of electromagnetic quantities without experimental realizability and mechanical illustrations and declared electric motion a "thing of thought" for the purposes of "picturing" the integration of electromagnetic equations (Darrigol 1993, 2000, 258–259).

Boltzmann framed his Maxwellianism, much as Hertz did, within an epistemic or methodological viewpoint that gave it extended expression and significance. It featured prominently in his late, philosophical period. First in 1892, in the midst of his Maxwellian lectures and investigations on electromagnetism, Boltzmann wrote about the broader significance of Maxwell's method of analogies and use of models on the occasion of the publication of a catalog of mathematical and physical models at the university (Boltzmann 1892/1974, 5–12). Maxwell's examples helped show what Boltzmann called the practical and epistemological value of models in theoretical physics. Specifically, they provided "illustrative and tangible representations" for concepts used in mathematical analysis and its application to physical theory (see also his 1902 *Encyclopedia Britannica* article "Model" [1902/1974, 213–220]). In 1899, Boltzmann also declared that Maxwell's theory was for Maxwell himself a picture of nature he had called a mechanical analogy that offered the most uniform and comprehensive description of the totality of electromagnetic phenomena in a way that was distinctively provisional and non-unique (Boltzmann 1899/1974a, 83). Moreover, Boltzmann had drawn special attention to the fact that Maxwell's mechanical analogies – "mechanical models [that] existed only in thought" or "dynamical illustrations in imagination" – contributed not only understanding but, as Boltzmann (1892/1974, 10) emphasized, also the very equations of the theory. And that, claimed Boltzmann, was the epistemic and methodological weakness of Hertz's exclusive focus on bare equations.

Another limitation of Hertz's late notion of a picture theory was its reliance for appropriateness on a priori laws of thought at the expense of the standard of correctness from empirical adequacy (Boltzmann 1899/1974b, 104–105); that is, in his deductive axiomatic approach, Hertz sacrificed empirical justification to a priori necessity rather than relying on sufficiently clear and consistent, yet arbitrary, construction (Boltzmann 1899/1974b, 108).[5] In spite of this attitude, Boltzmann claimed in 1899, albeit with reservations, that Hertz's picture theory in fact completed Maxwell's epistemology, but it is Boltzmann who had the last word:

> In his book on mechanics Hertz has given a certain completion not only to Kirchhoff's mathematico-physical ideas but also to Maxwell's epistemological ones. . . . [Maxwell's] own account he called mere pictures of phenomena. Following on from there, Hertz makes physicists aware of something philosophers had no doubt long since then started, namely that no theory can be objective, actually coinciding with nature, but rather that each theory is only a mental picture of phenomena, related to them as sign is to designatum.
>
> From this it follows that it cannot be our task to find an absolutely correct theory but rather a picture that is, as simple as possible and that represents phenomena as accurately as possible.
>
> (1899/1974a, 90–91)

Thus, for both Hertz and Boltzmann, Maxwell's physics was a catalyst not just for scientific research but also for philosophical analysis and participation in a debate that also included Mach. Boltzmann was explicit about being concerned with Maxwell's research "in two respects, the epistemological and the specifically physical" (Boltzmann 1899/1974a, 83). Maxwell's mechanical analogies and molecular models, Boltzmann believed, supported the value of this approach and, specifically, of atomistic pictures (not explanatory hypotheses).

His Maxwellian view was that theories are mental (internal) pictures – also models, or signs – for the sake of predicting appearances (Boltzmann 1897/1974, 41–53). For the purpose of prediction or general representation, pictures cannot be free from arbitrary elements, undetermined by particular facts of experience, but must include as few as possible.

This epistemic standard was at the center of Boltzmann's attitude towards Mach and Hertz's epistemological and methodological positions. He offered a critique of what he called general phenomenology, mathematical phenomenology and energetic phenomenology (Boltzmann 1899/1974a, 97). Contra Mach and Hertz, he declared theoretical pictures constructions free from axioms of phenomenalism and monism; he meant it only in the spirit of pluralism about possible approximations and resulting partial analogies and truths whose measure of success relies in their resulting empirical descriptive and predictive value, not sources. He added a qualification: Hertz's Euclidean, axiomatic, deductive model of theory construction indeed shared this condition of empirical evaluation, but it prioritized a priori formal constraints in the formulation of the theoretical pictures according to laws of thought. In addition, Boltzmann pointed out, Hertz's nude and astringent conception of Maxwellian electromagnetic theory is an atomistic theory only with minimal arbitrariness.

2.2.5. Hilbert

Hilbert's work on geometry established the program and standards of axiomatics as the deductive ideal of scientific theory, especially in the foundations of mathematics and the physical sciences. Hilbert's distinctive contribution first in geometry, as formulated famously in *Die Grundlagen der Geometrie* (1899), included the central role of formal criteria in the construction of systems of axioms such as independence and completeness and the resulting role for implicit definitions – emblematic of the rise of formalism in mathematics alongside logicism. In the years prior to the formulation of the system of geometry and subsequent application of the standard to physics, Hilbert had become acquainted, through Minkowski, with Hertz's book on the foundations of mechanics. The book might have provided a key catalyst in his adoption of the axiomatic treatment first and foremost of geometry and then, by 1905, of physics, with specific methodological criteria such as simplicity, consistency, completeness and especially independence of interrelated axioms about basic facts of a science (for this suggestion, see Corry 2006 142, 144, 148). Hilbert also pointed to Hertz in the formulation of the ideas as they appeared by 1899 in his lectures on geometry and recent lectures on mechanics (Corry 2006, 145).

2.2.6. Poincaré and Duhem

Maxwell's theory received an extension of its dual scientific and methodological life in the hands of, among others, French conventionalists such as the physicists Henri Poincaré and Pierre Duhem. Poincaré led the way with a technical study of Maxwell's theory in 1890 that prompted a more critical one by Duhem in 1902. The critique continued in their more philosophical works.

Poincaré had discussed Maxwell's theory in the first volume of his published lectures (Poincaré 1890) and Helmholtz's theory and Hertz's experiments in the second volume (Poincaré 1891). He later examined it in the context of his philosophical reflections, insisting on a commitment to the factual truth of structural features of mathematical equations in physics, insofar as they enabled intersubjectivity and manifested invariance over time (Poincaré 1902/1905, 114). In the case of geometry, he famously concluded that of all formal constructions, the choice of Euclidean axioms was merely a convenient matter of convention, while arithmetic deserved synthetic a priori status in its manifestation of the formal constructive powers of the mind. Poincaré (1902/1905, 149) opposed French standards of simplicity, precision and logic to Maxwell's theory as a disunified collection of provisional and independent, even contradictory, constructs. Maxwell's last development and formulation of his theory, however, was for Poincaré mechanically general, acknowledging the infinity of possible specific mechanical representations or explanations, but, unlike in previous accounts, the theory avoided any mention of any particular mechanical model of matter or ether, for instance, of connected molecules (as in the kinetic models of gases he still endorsed, explored and developed). The theory was now "a form almost devoid of matter" (1902/1905, 155). And he acknowledged Hertz's merit providing experimental proof of the non-instantaneous character of electromagnetic induction – of closed currents or displacement currents – propagating, as Maxwell predicted, at the speed of light (1902/1905, 166).

Duhem endorsed the autonomy of physics from metaphysics and, accordingly, of representation or classification of phenomena from their explanation. One epistemic virtue of such autonomy was the preservation of consensus in physical theory in the face of multiple contradictory possible metaphysical systems. Another was the coordination of reason and experience (experiment). Theories contain rules for connecting mathematical equations and quantitative experimental results, but their expression take the form of symbolic laws and symbolic measurement results. As a result, Duhem claimed laws are neither true nor false, but approximations (Poincaré had declared them to be neither true not false, only conventions), and he would endorse a positivist, conventionalist attitude, sometimes alongside a form of provisional realism about explanations such as atomic hypotheses (Duhem 1906). To bridge the gap between physical theory and the ideal, true representation of the world he called natural classification, he also required a minimal uniqueness condition of historical continuity.

Duhem set his notorious criticism of Maxwell's theory at the heart of his arguments and did so partly prompted by Poincaré's own criticisms and in nationalistic terms: as part of a broader critique of the so-called English school of physics, represented by Faraday, Thomson and Maxwell and his followers (Duhem 1902, 1906). Distinctive of their work was model-building, which incorporated in electromagnetic theory a multiplicity of – often inconsistent – mechanical analogies grounded neither in the exercise of reason nor in additional results of experiment, only in imagination (Duhem 1902, 1906; Ariew and Barker 1986).

Such a practice conflicted most grievously with the epistemic strictures and methodological standards he endorsed, in particular with the theoretical standards of logical clarity and coherence and historical continuity. Not surprisingly, Duhem also acknowledged Hertz's separation of Maxwell's equations from Maxwell's models (e.g., Duhem 1902, 221–222). The interpretation helped Maxwell's electromagnetic theory meet Duhem's logical standards, transforming Maxwell's "imprudent audacity into prophetic divination" (1902, 8). But these were not Duhem's only requirements. Besides believing that Hertz was identifying theories too algebraically, too narrowly with symbolic equations (1902, 223), Duhem thought Helmholtz's comprehensive – classificatory – theory, also incorporating Maxwell's equations, met more adequately the criterion of historical continuity, namely with earlier French and German mathematical theories of electricity and magnetism (1902, 225).

The tangle of the philosophical dialogue between Mach, Hertz, Boltzmann, Duhem and Poincaré is inseparable from the role of Maxwell's physics in their views. Maxwell's theories featured prominently and recurringly in their scientific and philosophical contributions. The two aspects, I claim, set the stage and the resources for further attention to Maxwell by, among others, Neurath, Frank, Schlick and Carnap.

2.2.7. *Electron Theory and the Electromagnetic World-Picture*

By the end of the century, electromagnetic theory had reached new heights of theoretical and methodological significance. In 1895, Hendrik Antoon Lorentz made a crucial contribution by introducing the so-called Lorentz-Maxwell equations for the electrodynamics and optics of moving bodies. With the equations taken axiomatically, Lorentz's theory broke away from the mechanical world-picture and provided a broader unification of electromagnetism and optics with a new electron theory. It also sought to explain Michaelson and Morley's perplexing experimental results about the invariant speed of light in any direction independently of the speed of the source. For that purpose, Lorentz postulated new relations between locations, times and motions and a hypothesis of contraction of electrons when moving through the ether. The growing success of Lorentz's theory explaining new phenomena prompted him in 1900 to speculate on the possibility of reducing gravitation to electromagnetism. As a result, at a conference in Lorentz's honor of the same year, Wilhelm Wien officially announced the project of an electromagnetic world-picture unifying all matter and forces. The proposal was promptly and widely endorsed and enlisted new theorists and experimenters (Miller 1998, 122–124; Wien 1900).

Any credibility left for the mechanical world-picture was further shattered by Planck's statistical proposal of the quantum of radiation, also in 1900 (Planck 1900). The quantum, Einstein noticed, also challenged the electromagnetic project (Einstein 1905b). To deepen the crisis, Einstein also showed that the electromagnetic behavior of moving bodies and the kinematic behavior of light conflicted with the laws of mechanics (Einstein 1905a). Following the example of Mach, Hertz and Lorentz, he sought safety and generality in positivism and axiomatics. Specifically, with new axioms, he could protect electromagnetic theory from the conflict with mechanics and the burden of Lorentz's hypothesis of physical contraction. It is worth noting that unlike the earlier axiomatic systems, Einstein's were not the field equations. One was the axiomatic principle of relativity, or the invariance of the equations for all inertially moving bodies and observers (that is, independence from uniform motion), and the other was the principle of the absolute constancy of the velocity of light. From them followed a new space-time structure for any theory of matter and radiation, that is, any form of energy.

Along the way, Einstein's electromagnetic path to the theory of relativity ended up placing Maxwell's equations further at the core of modern physics, only now alongside a new image of space and time whose job was to protect them (see Darrigol 2000).

2.3. Logical Empiricism

Where do logical empiricists enter the picture? The older scientifically trained generation of logical empiricists included Philipp Frank, Otto Neurath and Moritz Schlick.[6] Maxwell's equations and electromagnetic theory were the focus of Schlick's doctoral thesis of 1904 under Planck in Berlin and Frank's habilitation of 1908 in Vienna. Frank had studied with Boltzmann in Vienna and with Klein and Hilbert in Göttingen and received a doctorate with a thesis on dynamics.

2.3.1. Schlick

In 1904, Schlick received his doctorate in physics with a dissertation supervised by Planck on optics and the theoretical application of Maxwell's equations to inhomogeneous media without ether models. He soon turned to philosophical research on questions in areas as diverse as ethics, aesthetics and epistemology. While writing on the philosophical significance of relativity theory – in parallel to Hans Reichenbach – Schlick articulated a general view in the book *General Theory of Knowledge* (1918, 1925/1974).

There he developed a sophisticated epistemology of scientific knowledge around the idea that it is not intuition but only knowledge of relations and ordering of concepts that help coordinate judgments to domains of phenomena. Following Poincaré and subsequently Reichenbach's neo-Kantian analysis of relativity theory, he also pointed to the role of conventions. His account opposed both Kantian idealism and Machian phenomenalism.

A recurring illustration in the book is the case of electromagnetic theory in the stringent formal interpretation he associated with Hertz. The physicist, he noted, for instance, can coordinate relevant propositions to the entire domain of electromagnetic phenomena with the help of Maxwell's equations (Schlick 1925/1974, 79). In relation to the optical phenomena, in particular, Schlick also stated that laws for electromagnetic waves were noticed to fit optical phenomena better than laws for mechanical waves, and, he added, this was the result of developments facilitated by Hertz's double contribution, discovering electromagnetic waves and setting forth their laws "in rigorous mathematical form" (Schlick 1925/1974, 10).

The epistemic status of Maxwell's equations, in Hertz's sense of Maxwell's theory, is empirically distinctive and irreplaceable. As Schlick further noted more generally, they provide the only knowledge, the only essence of electromagnetic phenomena, just as Einstein's equations alone provide the knowledge of the essence of gravitation (Schlick 1925/1974, 242; the discussion appeared first in 1917 in the lecture published as "Appearance and Essence" and next in the first edition of *General Theory of Knowledge*, Schlick 1918).

Schlick followed Hertz and Hilbert's formal axiomatics as a model of scientific knowledge in general and in particular the implicit definition of concepts coordinated with experience through quantitative determinations. The affinity between the interpretation of the physical and mathematical theories is no coincidence, as Hilbert himself had to some extent followed Hertz (Corry 2006). In this sense, Schlick illustrated the way in which formal axioms connect concepts and provide definitions of associated terms with the history of the term 'electricity': he contrasted the earliest (explicit) definition, in relation to the observable effect of rubbing a piece of amber, with the contemporary definition "at the highest level of theoretical physics" as given by the fundamental equations of electrodynamics, from which known phenomena are then deduced as consequences (Schlick 1925/1974, 47). Only the formal nature of the equations enabled such logical relations. In the first edition of the book, he had written "the so-called Maxwell equations" instead of "fundamental equations of electrodynamics" (Schlick 1918, 45).

Schlick's scientistic critique of Kantianism targeted, besides idealist epistemology, the concepts of space, time, substance and causality. During the same period, Schlick considered electromagnetism in his analysis of space, time and substance in the theories of relativity as exemplars of contemporary physics (Schlick 1917, 1919/1920).

Schlick, partly following Mach and Poincaré, drew a distinction between physical and intuitive, especially visual, space. And physical space, now a theoretical construct and coordinated with, but not determined by, intuition, is the space of real physical objects – another construct identifying and representing physical, not metaphysical, reality. Schlick associated with the distinction between the two spaces the distinction between two corresponding kinds of concepts, psychological qualities and physical quantities: "Physics does not know colour as a property of the object with which it is associated, but only frequencies of the vibrations of electrons" (Schlick 1919/1920, 78). This correspondence, and this example, would play a central role in the subsequent discussions of empiricism and physicalism, and psychology, within the Vienna Circle. Also in line with the Hertzian prescription, he noted that in the special theory of relativity, Einstein considered electromagnetic fields – the physical nature of light signals traveling at an absolute velocity – an independent physical system and no longer properties of a substance, the electromagnetic ether (1919/1920, 20).

Schlick also placed Maxwell's equations at the center of his investigation of the role of causality in physics (Schlick 1920/1979, 1925/1979). With clear Kantian overtones, he declared the principle of causality of 'the greatest importance for the shaping of our picture of the world' (1920/1979, 320), especially to the extent that its application defines inductive cognition and lawfulness and is also the necessary condition for our knowledge of nature (1920/1979, 309).

According to Schlick, the causal relation is a deterministic relation between physical states manifested solely by processes in the temporal direction (pace the central role for discontinuity in Planck's and Bohr's quantum theory [Schlick 1920/1979, 297]). It also presupposes a relation of separateness that reflects the notions of space and time as forms (not content) – "following Kant's terminology" (1920/1979, 307) – and whose mathematical expression takes the form of independence from absolute space and time, or spatio-temporal homogeneity or relativity; in other words, independence from particular time and position coordinates (1920/1979, 308). In fact, Schlick credits Maxwell for the formulation and defense of this relational (relativistic) formulation of the causal principle – in *Matter and Motion* (Maxwell 1876): "The difference between one event and another does not depend on the mere difference of the times or the places at which they occur, but only on differences in the nature, configuration, or motion of the bodies concerned" (Schlick 1920/1979, 308). After Mach, Schlick noted, it is Einstein who gave the formulation central status in the principle of relativity.[7]

In terms of physical theories, the causal principle presumes the rejection of causal action at a distance, in space or time, in the classical field physics of matter and electromagnetic forces. Therefore, he concluded, its formulation must be the following: "the natural processes occurring at a point in space are thus completely determined by those in its immediate neighbourhood, and it is only indirectly, that is, by means of the latter, that they also depend on more distant processes" (Schlick 1920/1979, 296–297). And its mathematical expression must be in terms of differential equations. Schlick called such causal laws micro-laws, so that any macro-laws involving macroscopic differences in the representation of physical action are reducible to or deducible from – by integration, for instance – from causal micro-laws.

From this causal-mathematical standpoint, the significance of Einstein's theory of general relativity lies, according to Schlick, in extending the domain of causal explanation in nature to the case of gravitation (see also Ryckman 2007, 198–199). In a subsequent and more ambitious essay in the footsteps of the *General Theory of Knowledge*, "Outlines of the Philosophy of Nature" (Schlick 1925/1979), Schlick repeated the point about the significance of field physics with explicit reference to the role of differential equations. Euler's equations of hydrodynamics represent the causal relation between contiguous motions and, supplementing the case of gravity, Maxwell's equations of electrodynamics "allow us to derive the changes in the electrical and magnetic forces at a point from the values of these quantities in the vicinity of that point" (Schlick 1925/1979, 33). Coulomb's law of electrostatic action over macroscopic distance is only a macro-law derived from Maxwell's equations, which are, formally speaking, more fundamental and genuinely causal (Schlick 1925/1979, 34). This formal emphasis reflects the epistemology in *General Theory of Knowledge*, whose second edition was published the same year (Schlick 1925/1979), and his interpretation therein of Maxwell's equations as the grounds for the modern definition of electricity (whose original Maxwellian models Helmholtz, Hertz and others had objected to) and as the sole form of the knowledge or essence of electromagnetic forces.

Ultimately, Schlick placed both Maxwell's and Einstein's theories on the same causal footing, as classical approximations to future theories that would resolve the conflict posed by the role of discontinuity in quantum theory, in the form, for instance, of a "deepened theory of continuity" (Schlick 1925/1979, 62). He would change his mind in 1931, in the aftermath of his acquaintance with the new quantum mechanics, Wittgenstein's doctrines and the Vienna Circle manifesto.

He now considered laws of nature, with Wittgenstein, prescriptions for making particular assertions – that is, inference tickets – and, connectedly, considered the causality principle neither a tautology nor an empirical proposition but a methodological prescription (Schlick 1931/1979, 193). He added that the lesson from quantum mechanics, to the extent that it was adequate in its current form, was that the rule was useless; it conflicted with the fact that the theory placed a principled limitation on determinism and the exactness of prediction (1931/1979, 197). Exact prediction, and determinism, had become impossible. Along the way, Schlick criticized the status of Maxwell's causal criterion as too narrow: its validity, he now pointed out, is limited to actually known laws of nature (except for quantum mechanics) but not to all possible – imaginable – ones (1931/1979, 183).

The critique is significant insofar as it illustrates the historical character of some philosophical evidence. The modal gap between the possible and the imagined, sign of a failure of conceptual imagination in philosophy, is marked by the introduction of conceptual schemes such as quantum mechanics, which were logically possible, but unconstrained by sensory intuition, and previously unimagined (Schlick 1931/1979, 176–177). Schlick gave his recognition of the factual success and fruitfulness of the application of quantum mechanics – superior to that of relativity – a historical character, indexed as novel and contemporary. It represents a measure of progress that, for Schlick, gave distinctive epistemic authority to 'contemporary physics'. The equivalent authority of his recurring appeals to electromagnetic theory, alongside Einstein's relativity theory, must rest on similar grounds; in this case, they are the contemporary validity and foundational dimension of its Maxwellian core, theoretical extensions – such as the Maxwell-Lorentz equations – unification and application – electron theory and the electromagnetic worldview. In both cases, however, Schlick still qualified their epistemic status: it was restricted by their approximate character, marked, for instance, by their conflict with the principled discontinuity (and indeterminism) characterizing quantum mechanics, and quantum mechanics itself was restricted by the provisional status of its adequacy. Philosophy draws lessons from, one might say, the best available science, but the best status and the value of the lessons are historically relative, relative to both past and future.

The distinction between formal statements (analytic and true by construction), factual statements (synthetic) and prescriptions was distinctive of the new logical empiricism. In order to analyze the objectivity of scientific knowledge in the footsteps of Carnap's *Aufbau* and the Vienna Circle's manifesto, Schlick cast in such terms his realistic epistemological views in *General Theory of Knowledge*. Key ideas were the related distinction between form and content (Schlick 1932/1979), between physicalism and affirmations (Schlick 1934/1979, 1935/1979a) and between physical and psychological concepts (Schlick 1935/1979b). Schlick supported and illustrated them also by appeal to the equations of electromagnetic theory.

Science, defended Schlick, is a system of explanatory knowledge; specifically, a system of descriptions by formal and general laws organized in a logical structure derived from the simplest set of such laws, or axioms (Schlick 1932/1979, 327–328). The ideal case, and also the one that allows for philosophical analysis, takes a mathematical form defended by

Hilbert and the Italian school in the relations expressed by symbolic formulas and the relations between them. By that standard, Schlick declared that the most advanced knowledge of nature through the application of mathematics is found in theoretical physics. There, the measurements of physical systems provide the content, meaning or interpretation of the formal structure and specify of what reality it provides knowledge (Schlick 1932/1979, 331).

From an epistemological standpoint, Schlick agreed with Carnap – in the *Aufbau* – that it is the formal structure that bears the objectivity of scientific knowledge. But also in physics, he observed, "the same structure is used to describe many essentially different physical processes" (Schlick 1932/1979, 332). His example is the application of the wave equation. The substitution of measurable signs, and private experience, for abstract mathematical variables is precisely the substitution of Hertzian Maxwell's equations for empty Hilbertian axioms:

> the word 'electric force', for instance, will have no definite meaning, but will signify any entity that fulfills certain axioms (these axioms, in the classical theory, will be Maxwell's fundamental equations), and there may be innumerable such entities; which of these is really meant?
>
> (Schlick 1932/1979, 332)

Formal axioms and measurement conditions – linked to both personal experience and physical reality – constituted the two poles, theoretical and empirical, of Schlick's post-Kantian, scientific epistemology. Regarding the empirical, his analysis sought to reconcile positivism and realism (without any of the talk of transcendental reality of metaphysics). He articulated his position in the linguistic context of the so-called protocol-sentence debate, first emphasizing the phenomenological private element of experience and subsequently in terms of Carnap's physicalism (see subsequently).

The language of physics, echoed Schlick (1935/1979b, 424), was the only language meeting the scientific requirements of universality and objective validity – intersubjective and inter-sensual. This characteristic was the result of the distinctive structural form of the method of physical measurement he had long analyzed in terms of (spatio-temporal) coincidences (1935/1979b, 424–425). Accordingly, he reduced the problem of understanding experience from the standpoint of scientific epistemology to the problem of understanding the relation between psychological and physical concepts or languages. He chose two examples that illustrated and supported his view of the relationship between the two kinds of concepts, color and grief. With the case of color, Schlick could address the domain of experience central to (realist) empiricism, and, as Neurath and Carnap also did (see below), he introduced the only available physical concept, from electromagnetic theory: "Physically, a colour is defined by a frequency, a number of vibrations per second" (1935/1979b, 428). In terms of measurement, he added that "one observes the coincidence of a spectral line or of an interference fringe with certain marks on the measuring apparatus" (1935/1979b, 428). Only the problem of ambiguity in the relation between psychological and physical color concepts would require additional physical concepts, describing physiological or behavioral states.

It is worth noting that Herbert Feigl adopted his teacher Schlick's attention to the philosophical significance of electromagnetic theory and the historical evolution of its application and appreciation. In *Theorie und Erfahrung in der Physik* (*Theory and Experience in Physics*) Feigl (1929) also adopted – from Mach, Hertz, Boltzmann, Poincaré, Hilbert, Duhem, Wittgenstein and Schlick – the notion of a theory as a construction with arbitrary elements to be organized in the form of a system of axioms, such as Hilbert's system of geometry (Feigl 1929, 96). While the arbitrariness could be reduced as a pragmatic matter of simplicity, it could also be considered – following Wittgenstein – a heuristic in the construction of working hypotheses. An example of such a mental and methodological aid is Maxwell's use of mechanical analogies, fictions or models (Feigl 1929, 99; in preparation, he wrote to his friend Kurt Gödel asking for help understanding Maxwell's theory).[8] Such intuitive representatives of the mathematical theory of electromagnetism constitute one concrete representation of the theory in addition to the empirical observations derived by deductions (with the aid of auxiliary hypotheses). But Hertz, observed Feigl, had drawn attention to the dispensability of such heuristic analogies in addition to the formal, exact, general laws of electromagnetic phenomena. The same development and contrast could be found, he added, between Bohr's atomic model and Heisenberg's "completely abstract quantum theory" (Feigl 1929, 99). Appeals to electromagnetic and electron theory also played a central role in subsequent work on existential hypotheses and the mind-body problem.

2.3.2. Frank

Philipp Frank, Otto Neurath and the mathematician Hans Hahn, among others, met regularly in Vienna between 1907 and 1912 to discuss current issues in science and philosophy. Their readings included recent foundational work in logic, mathematics and physics and the philosophy of science of Mach and French conventionalists such as Duhem, Poincaré and Abel Rey. As Frank reported, much of their discussions revolved around the crisis in physics associated with the failure of the atomistic mechanical worldview and the proliferation of alternative systems such as the electromagnetic worldview around electron theory and energetics. In a similar spirit, they paid attention also to the foundational significance of

Hilbert's new geometrical paradigm in the axiomatic systematization of mathematics. The group's reception of French new positivism and conventionalism cannot be separated from the scientific and methodological attention Mach, Hertz, Boltzmann, Hilbert and the French had given to Maxwell's equations and electromagnetic theory.

Frank had studied with Boltzmann in Vienna and with Felix Klein and Hilbert in Göttingen. Eventually in 1906, he wrote a doctoral dissertation on dynamics – "On the Criteria of Stability of the Motion of a Material Point and Its Relation to the Principle of Least Action" – and in 1909 also a habilitation at the University of Vienna on the principle of relativity in mechanics and electrodynamics (Stadler 2001, 631). If Einstein's special theory of relativity is inseparable from the scientific value and validity of electromagnetic theory and from its theoretical conflict with mechanics, Frank's presentation of positivism is inseparable from (1) historical arguments involving considerations of the philosophical significance of Maxwell's method and electromagnetic theory, (2) philosophical conflicts between scientists and (3) philosophical conflicts between, on the one hand, modern science and scientific worldview and, on the other, its criticisms.

In one of his earliest publications (1907/1941), Frank considered electromagnetic theory in relation to the world-picture crisis. He was already working on his habilitation thesis in the wake of relativity theory. Lorentz and others' electromagnetic world-picture, Frank reported, takes as fundamental physical quantities electric charge and electric and magnetic field intensities. Extending Mach's doctrines with theses of conventionalism, he seized on Poincaré's conventionalist interpretation of general physical principles such as inertia and conservation of energy (Poincaré 1902/1905, Chs. 6 and 8). Experience alone, concluded Frank, cannot decide between the mechanical and electromagnetic pictures (Frank 1907/1941, 25); both are more or less complicated expressions of the same empirical world (1907/1941, 26).

Prompted by Mach's death, in 1917, Frank penned a defense of some of Mach's views from criticisms by scientists such as Planck and the geometer Eduard Study (Frank 1917/1941). Mach had himself argued that the aim of science is the formulation of theories for the economic representation of connections between phenomena or complexes of perceptions. According to Mach, the role of auxiliary concepts such as quanta, atoms and electrons has to be evaluated by the aims of economic abstraction and interdisciplinary connections. The ensuing controversy focused primarily on the value of this philosophical view as a tool for a physicist's scientific research, but Frank declared its main value "defending the edifice of physics against attacks from outside" (1917/1941, 37), for instance, from objections of skepticism and from metaphysical doctrines.

Frank did not hesitate, however, to defend the value in research. In the footsteps of Einstein's recent publication on the general theory of relativity, Frank pointed to it as evidence: "Einstein's general theory of relativity and gravitation grew immediately out of the positivistic doctrine of space and motion, as Einstein himself has discussed in detail in his reference to Mach" (1917/1941, 38). More surprising, perhaps, was a much longer reference to the case of electromagnetism. The philosophical role of electromagnetism in the paper is not again the role of a historically recent expression such as the electromagnetic world-picture but of an earlier one, Maxwell's own work.

Clearly following Kleinpeter, Frank placed Maxwell within a Machian framework and declared that Maxwell had adopted a "phenomenalistic standpoint" (Frank 1917/1941, 31) and "doubtless thought positivistically" (1917/1941, 42).[9] Distinguishing between descriptive and constructive theorizing, Frank noted that phenomenalism would be a doctrine favored by physicists with purely descriptive rather than constructive work – such as Maxwell's and Planck's (1917/1941, 31–22). Still, Maxwell's phenomenalistic conception of theorizing enabled the assistance of free imagination (1917/1941, 31) and, beyond the mechanical models of electromagnetic theory, his positivistic thinking didn't prevent him from setting the basis for a molecular theory of gases (1917/1941, 42).

Frank considered electromagnetic theory again following the constitution of the Vienna Circle and the publication of the manifesto. The occasion was a lecture at the 2nd Conference on the Epistemology of the Exact Sciences in Königsberg on September 5–7, 1930 (Stadler 2001, 350); it was part of the public presentation of the Vienna Circle and the Berlin group (Frank 1941, 10). He now placed the emphasis on the conflict between scientific conception and school philosophy – the scholastic standard of metaphysics.

As a matter of epistemology of the exact sciences, Frank adopted the following criterion: that any unsolved problem whose solution involves achieving knowledge involves a conception of the experience indicative of the specific knowledge solution. Whether such an experience can be described (conceived, anticipated) determines a different kind of problem (Frank 1930/1941, 59–60). When it cannot, Frank used the criterion in the Machian spirit of the manifesto and suggested that a physicist might consider many such problems philosophical, often presuming an eternal truth, in the tradition of the unified world conception he calls "school philosophy". The alternative possibility was a new scientific world conception, in which problem-solving and cognition are tied to empirical data. The application of the standard was illustrated, according to Frank, by relativity theory, without untestable commitments to absolute length or velocity (1930/1941, 61–63), and by quantum mechanics, without untestable commitments to a precise position and velocity of an electron (1930/1941, 65).

But he also considered the case of electromagnetism, now with a new historical and philosophical twist in line with Hilbert and Schlick's views, stating that "Hertz [not Maxwell!], as it is often said, elucidated the nature of light" (1930/1941, 60). The elucidation, in the Hertzian formal approach, takes the form of a quantitative difference in wavelength between light and other electromagnetic waves while satisfying the same mathematical equations (1930/1941, 60). But the nature of electricity – electromagnetic forces – is still unclear, he added, an unsolvable philosophical problem (1930/1941, 60, 67). By contrast, physical knowledge of "the identity of light and electricity" has a perfectly scientific, empirical meaning in terms of statements of concrete experiences (experimental results) involving light sources and electrical transmitters (1930/1941, 67). Next he introduced explicitly the Hertzian interpretation reducing Maxwell's theory to formal mathematical relations between quantities expressed by symbols and linked to measurement results:

> Definite rules assign to the electromagnetic experiences symbols, the field quantities, among which there exist formal relations, the field equations. From given combinations of symbols one can then, using mathematical operations, derive new combinations with the help of the equations. These combinations may be translated into experiences again with the help of the same rules of assignment.

Then, he concluded, the "identity of light and electricity then means an identity of mathematical relations between symbols" (Frank 1930/1941, 67).

The Hertzian interpretation of Maxwell's theory allowed Frank to use it – and the electromagnetic world picture – in illustration of the logical empiricist scientific world conception put forward in the manifesto, reducing sources and kinds of cognition to (a priori) formal symbolic relations by construction and (a posteriori) empirical descriptions prompted by experiences and experimental results, also in terms of his problem-solving criterion: "the solution to a problem means the assignment to experiences, among which there exist relations which can be stated" (Frank 1930/1941, 68).

In addition, echoing Schlick's proposals,[10] Frank introduced a univocality (uniqueness) condition: the symbol system constitutes true cognition of electrical phenomena and physical phenomena, more generally, whenever the system – based on rules for assigning symbols and the relations among them – can assign the symbols uniquely (Frank 1930/1941, 80–81). The equations of electromagnetism satisfy the condition: they can provide a univocal measure of the distribution of electricity on the surface of a sphere by assigning a single constant to the charge density as a function of position (1930/1941, 80). Much classical physics, linked to experiment and technological application, holds up as sound scientific theoretical knowledge.

Classical physics, however, when expressed by general projects such as the mechanical and electromagnetic world pictures, failed, according to

Frank, to establish fixed, complete truths about the "real world", especially in relation to atomic elements of matter in terms of space, time and causality (Frank 1930/1941, 101) or when projects such energetics and the electromagnetic world picture were misused scientifically in order to support grand idealistic doctrines about the end of materialistic physics (1930/1941, 126). When the assignation for individual electrons, for instance, is not unique or precise, but the law precisely connects probabilities or statistical averages, a theory such as quantum mechanics is statistical (1930/1941, 99). Frank concluded that the progress of science goes hand in hand with progress in observation and its unique symbolic representation and control.

Frank presented the next historical argument in response to a historical critique of the evolution of science towards a psychical, subjective or idealist foundation (Frank 1934/1941). The presentation was part of a lecture at the Preliminary Conference of the International Congresses for the Unity of Science in Prague (August 31–September 2, 1934).[11] He was responding to both defenders and critics of the development. The former was represented mainly the French philosopher Raymond Ruyer, whose article prompted Frank's response (Ruyer 1933; Frank 1934/1941, 105), the latter by Russian materialist thinkers such as Lenin (1934/1941, 109–110). In this new kind of historical argument, the philosophical significance of the past cases is also newly distinctive: it is relevant to the philosophical significance of the present to the extent that they are parts of a connecting historical trend. Frank presented a counterargument of the same historical kind, with a corresponding new kind of significance of electromagnetic theory.

The historical critique targets relativity and quantum theories as recent revolutions in physics (Frank 1934/1941, 105). In Frank's general version, the more recent kind of argument, such as Ruyer's, seeks to support a general world conception of a fundamentally anti-mechanical, organismic, subjective, ideal, spiritual or even Platonistic character (1934/1941, 106). The arguments contrast an older, mechanical world conception with a new, mathematical one (ibid.) that identifies the theory with abstract mathematical formulas presenting mathematical symbols and relations and reduces the intuitive content of the theory to its geometrical features or connections to experimental observations. From that conception, the subjectivist, or idealist, philosophical interpretation of relativity and quantum theories points to the fundamentally mental aspect of mathematics and especially to the role of observation (1934/1941, 108–113).

Frank's reply is twofold, and each part seeks to challenge the alleged contrast between classical and modern physical theories. First, he pointed out that, in fact, the metaphysical project of a spiritualist, animistic, subjectivist or idealist philosophy of reality in relation to physical theory preceded the twentieth century and therefore does not track scientific progress (Frank 1934/1941, 110, 126). Second, he added, both in classical and modern physics – whether relativity or quantum theory – the instruments of observation provide the relevant facts (Einstein's coincidences), and the role of observation is the same and only one, namely the intersubjective determination of an experimental or measurement result (1934/1941, 112). Observation is regarded here, just as in classical physics, as something "intersubjective" (1934/1941, 113), and, therefore, "there is nothing psychological, at any rate any more than in classical physics" (1934/1941, 111–112).

This alternative, strictly scientific interpretation includes the empirical elements of his earlier characterization of electromagnetic research:

> What one can learn from the relativity and quantum theories in this connection is only what is also given by a historically consistent presentation of classical physics: every physical principle is, in the final analysis, a summary of statements concerning observations, or, if one wishes to speak in a particular physical way, concerning pointer readings.
>
> (Frank 1934/1941, 111–112)

In a subsequent paper, Frank placed this aspect of electromagnetic research in the context of the anti-metaphysical positivism of the Vienna Circle (Frank 1935/1941).

He noted that relations between physical quantities and observations, in the positivistic view, involve definite values of physical quantities such as electric field strength in Maxwell's electrodynamics that correspond to, and predict, definite observations (1935/1941, 128, 130). The application of physical quantities connects scientific theory, or thought, to experience; that is, it connects "the world of physical quantities" (thought) to "the world of the senses" (experience) (1935/1941, 129). Frank was echoing the protocol-sentence debate, with its linguistic and unifying standards, and the introduction of physicalism in the doctrine of a scientific kind of positivism. Attempting to present an integrated view, he simply declared that a unified language of science in the form of Neurath's unified science through physicalism expresses the kind of scientific radical positivism (such as Schlick's) that prevents science from getting lost in the pursuit (denounced by Carnap's) of metaphysical pseudo-problems (1935/1941, 134).

Frank finally returned to the historical argument of 1934 with a far more detailed treatment that merits closer attention (Frank 1937, 1937/1949). The target was the same, the distinction between the mechanical and mathematical conceptions and the question of the alleged anti-materialist philosophical revolution based on a lesson that both idealists and materialist drew – the ones announcing, the others denouncing – from new physical theories – relativity and quantum mechanics: that "the world is no longer a machine but a mathematical formula" (Frank 1937, 41). The new discussion integrates a more detailed account of Maxwell's contribution to physics and epistemology introduced in the papers on positivism of 1917, 1930 and 1935.

Frank's historical analysis is guided by one question:

> Is it really correct to see in the physics of the 20th century a transition from a "mechanical" to a "mathematical" explanation, from a materialistic palpable or intuitive perceptibility (*Anschaulichkeit*) to a mathematically abstract intellectual spirituality? (*Geistigkeit*). Or has the rebirth of idealism entirely different sources, and is simply dragging along with it whatever material comes to hand?
>
> (1937, 49)

As the basis for his criticism, Frank denied a strict theoretical and historical opposition between mechanical and mathematical theories and between their mathematical and intuitively perceptible ("*anschaulich*") character that would justify the idealist interpretation or criticism of new twentieth-century physics. He claimed, instead, that "the transition from mechanistic to mathematical physics was the work of the positivistic doctrine in science" and that "it had nothing to do with the [alleged] 20th-century tendency towards idealism and metaphysics" (Frank 1937, 55).

Frank's historical counterargument began with Newton's case and quickly focused on Maxwell. Newtonian theory, he observed, was widely considered, including by Huygens and Leibniz, mathematical rather than mechanical, missing mechanical hypotheses that Newton himself claimed not to feign (Frank 1937, 50). Frank introduced two distinctions. He distinguished first between two virtues of mechanical theories, the unity of physics – or the whole of science – and intuitiveness. By the end of the nineteenth century, he noted, the superior success of comprehensive energetic and electromagnetic representations had prompted a shift between the two mechanical virtues, from an explanatory interest in reductive unity to a mere representational interest in intuitiveness – also linked to the prevention of animism (1937, 50). He further distinguished between two concepts of intuitiveness. The kind of intuitiveness associated with mechanical theories is characteristic of the representation of phenomena in terms of most familiar (mechanical) objects, properties and behaviors (1937, 51). The other, broader, kind is an empiricist criterion of meaning as verifiability, to derive from physical laws "an experimental result that is directly observable" (1937, 51). Frank acknowledged that such observations involve, in the end, the same "gross mechanical events" that help verify mechanical theories (1937, 51).

Both kinds of intuitiveness, according to Frank, were illustrated by different perspectives on the field equations of Maxwell's theory of electromagnetism. He insisted that mechanical intuitiveness can no longer be an ideal of fundamental scientific explanation, not only because of the failures to provide a unified explanation of all physical phenomena, but also because of the conflict with the criterion of empirical verification – that is, they fail to be verified independently of the non-mechanical theory. Yet the ideal was part of the development and understanding of Maxwell's theory (Frank 1937, 52).

At this point, he credited Hertz with cutting the 'Gordian knot' that tangled electromagnetic theory up with the ideal of intuitive mechanical explanation and with introducing, alongside Mach, Duhem, Boltzmann and others, a new positivistic conception of physics that replaces that ideal of intuitive mechanical explanation but is not incompatible with considerations of atomic models of matter (Frank 1937, 51, 54). He quoted Hertz:

> Maxwell's Theory is nothing else than Maxwell's equations. That is, the question is not whether these equations are *anschaulich*, i.e. can be interpreted mechanistically, but only whether *anschaulich*, interpretable conclusions can be derived from them by means of gross mechanical experiments
>
> (1937, 52)

What about Maxwell himself? As part of his historical argument, Frank defended both the soundness of a positivistic interpretation of Maxwell's theory and Maxwell's own epistemology – besides Hertz's prescription, closer in spirit to the Circle's official positions. In earlier papers, Frank had credited Maxwell with a positivistic, or phenomenalistic, conception of physics. Here Frank took a more indirect route and began by rejecting the alleged conflict between atomism in Maxwell and Boltzmann and positivism in Hertz, Mach and Kirchhoff (Frank 1937, 52). Next he defended that Maxwell had himself rejected a commitment to a physical (mechanical) explanatory theory of electromagnetism and to a conflict between a "purely mathematical formula" and a "physical hypothesis" (1937, 53). Mathematical formulas alone allow the formal derivation of consequences but do not present the phenomena to be explained or any extended physical connections; a physical hypothesis does what the formula cannot, but it's only a partial explanation that can be blind to facts and to the uncritical assumption of its uniqueness or necessity. Instead, Frank quoted Maxwell defending a more general kind of physical theory that integrated the benefits of each and none of the liabilities (1937, 53–54). This is Maxwell's method of physical analogies. It supplements the formal and positivistic virtues of the mathematical theory with partial analogies between two kinds of phenomena, guided by their formal similarities; they provide possible and intuitive representations without committing to an explanatory mechanical theory.[12]

Centered on both Maxwell's theory and epistemology, the new historical argument allowed Frank to make relevant the philosophical history of Maxwell's theory to the new philosophical interpretation of twentieth-century physics. Frank concluded that twentieth-century physics did not represent a commitment either to a new, distinctively mathematical theory or to a positivist epistemology prompted newly exclusively by it. What about the connection to logical empiricism? As a twentieth-century contribution to this longer history of positivism in physics, Frank drew attention to the "neo-positivistic" proposal of the Vienna Circle and pointed out the distinctive linguistic character of the their "logically constructed empiricism". The emphasis on the intersubjectivity of language protects the intuitive character of mathematical theories from radically subjective interpretations, already inaccurately associated with Mach (Frank 1937, 57–58). Therefore, Frank then concluded,

> the professional philosophers did not have to wait for the relativity and quantum theories of the 20th century to misconstrue the transition from mechanistic to mathematical theory. They succeeded in doing this with the physics of the 19th and earlier centuries.
>
> (1937, 57)

More generally, Frank argued that classical physics also illustrated the "plan connecting existing mathematical concepts with new physical theories", for instance, the central role of Maxwell's potential functions in his theory of electromagnetism and Einstein's curvature tensors in his general theory of relativity (1937, 71–72).

2.3.3. Neurath

Neurath's association with electromagnetism is more surprising than Schlick's and Frank's, discussed so far, and Carnap's, discussed in the next section. Identifying his particular electromagnetic path to the scientific world-conception will show that neither his interest in political economy nor his contributions to logical empiricism can be fully understood in isolation from considerations of physics and technology. Regarding his acquaintance with the science and technology of electromagnetism, the Vienna discussion group, which he joined around 1907 on his return from doctoral work in Berlin and his military service, only added to his earlier science education.

One context for the relevance of physics and its applications to economics is the history of political economy itself. The so-called Industrial Revolution had taken place on the basis of a new organization of labor and the use of machines and engines. Marx and others promptly analyzed its significance for economic theory.

Neurath even considered valuable the economic role for machines supplementing labor shortages in war economies (Neurath 1910/2004, 169). In the same spirit, he didn't omit references to the economic relevance of the steam engine, but he drew special attention to the social and political significance of the use of electricity (already noted by Lenin), especially noting its superior socializing effect and its power to set up networks of production. For instance, he declared that "through the introduction of electricity the Russian peasants were brought closer to socialism in their attitudes" and, more generally, that "electricity has a much more 'socializing' effect than steam", for example, by enabling an "interlinking network of places of production" (Neurath 1925/2004, 449). It is also worth noting in this regard that Walther Rathenau, whose social planning Neurath repeatedly praised, had been consulted by Josef Simon and expressed approval of Neurath's plan for a socialized economy for the short-lived Bavarian Republic of 1919. By then, Rathenau had become chairman of AEG, General Electric Company, founded by his father. Finally, in a telling anecdote, when during his voluntary military service in 1906, the commanding officer required an essay on steam at the service of the army, Neurath wrote that the more adequate subject should "no longer be steam, but electricity in the service of the army" (Neurath 1973, 8).

In the history of economics physical as well as organic analogies were also dangerously common. Throughout the 1910s, Neurath also began introducing mechanical, thermodynamic and engineering analogies to illustrate the distinctive holistic, modal and constructive features of his own model of planned administrative economy. Engines illustrated the challenge of tracking the causal complexity of quality of life. Engineering illustrated the modal consideration of possible designs and the active and constructive dimension of planning: utopias, declared Neurath, are the business of social engineers (Neurath 1919/1973, 151).

The consideration of possibilities was for Neurath even part of the historical methodology of political economy. His view resulted from engaging two different but related debates over unity still ongoing at that time: one over the relation of the historical and cultural sciences, including economics, to the natural sciences, and the other over the different economic methods and perspectives. During the same period, Neurath extended his critical investigation to the methodological unity and cooperation in history and chose the case of the history of optics.

He wrote two overlapping papers, "On the Foundations of the Theory of Optics" (1915/1973) and "On the Classification of Systems of Hypotheses" (1916/1983). It is in these works, especially in the second, that he argued that electromagnetic theory and Maxwell's equations play a role. To begin with, Maxwell's theory put an end to the isolation of optics. By the mid-nineteenth century, observed Neurath, optics entered a unification with electrical theory introduced by Maxwell to order phenomena of electricity and magnetism (1916/1983, 16–17).

Neurath proposed an objective, unifying method of classification in history of science that he modeled after the method of analysis and synthesis in physical theory, as well as chemistry, even after the algebraic logic of the political economist and mathematician Stanley Jevons – as well as work on algebraic logic by Ernst Schröder (Cat 2019). Jevons' technique offered a combinatorial mechanical approach to composition applied to duals of conceptual components and their negations. As a result, it could systematically explore and classify realized and unrealized possible combinations. Neurath adopted as elementary notions periodicity, polarization, interference and diffraction.

But what was the required form of the analyzed theories? Neurath introduced a weighted criterion of physical theory as a system of hypotheses. One criterion, which he attributed to modern physicists, including Duhem and Poincaré, gave almost exclusive priority to mathematical form, with a role in logical argument. The alternative granted superior educational and methodological value to the role of imagery and analogy (Neurath 1915/1973, 102–103; 1916/1983, 25). For the methodological purpose of looking to actual science, Neurath endorsed the second, which was also Maxwell's own.

Without explicitly endorsing Maxwell's method of analogies or dismissing Duhem's criticism, he declared the heuristic value of analogies to present, guide and extend the imaginable systems of relations, and this, he added, must be done 'purely logically' and by deducing further consequences (Neurath 1916/1983, 25). Here Neurath provided several explicitly Maxwellian examples: (1) mechanical analogies for electric and magnetic phenomena (1916/1983, 26–7) and (2) analogies between the large and the small such as the application of Maxwell's equations for electrical fields to the field of electrons (1916/1983, 27) and analogies between different fields and their kinds of phenomena such as the ones that led to the successive unification of light, electricity, magnetism and radiating heat, the very achievement attributed to Maxwell's theory (1916/1983, 27).

For Neurath, the significance of the formal criterion of theory was historical: it tracked changes in the history of science, and not just in the twentieth century, as he attributed to Duhem and Poincaré and he used to discuss with Frank and others. Accordingly, he also mentioned Hertz explicitly on two accounts: as having developed Maxwell's theory of light (Neurath 1916/1983, 17) and as having identified the theory with its mathematical field equations. On this occasion, Neurath also made sure to note that Hertz had justified the identification in Neurath's own Duhemian historical way on the basis of the convergence and continuity of results (1916/1983, 29). He quoted Hertz accordingly, beyond the famous identity statement:

> To the question, "What is Maxwell's theory?" I know of no shorter or more definite answer than the following: – Maxwell's theory is Maxwell's systems of equations. Every theory which leads to the same system of equations, and therefore comprises the same possible

phenomena, I would consider as being a form of or special case of Maxwell's theory.

(Neurath 1916/1983, 29)

The collective efforts of the Vienna Circle manifesto would give expression to the goal of unity. The manifesto emphasized a rigorous linguistic framework prominently featuring the axiomatic method and logical analysis and an emphasis on intersubjective, neutral constructed systems of formula with precise symbolic relations (Carnap, Neurath and Hahn 1929/1973, 306). Hertz's interpretation of Maxwell's theory met the new standard and thereby gave it new philosophical significance. Frank made just this point explicit the following year at the first international presentation of the Circle and its movement at the Königsberg Congress on the Epistemology of the Exact Sciences.

In the wake of the manifesto, Neurath would still make occasional reference to electromagnetic theory, but now to serve the purposes of illustrating and supporting his own views and marking out differences from the manifesto's ideals.

His well-known proposal was an anti-metaphysical, materialist account of unified language characterized by the interconnected doctrines of syntacticism and physicalism (Neurath 1931/1983a, 1931/1983b).

The unified language of empirical science would have to be intersubjective and, from the empirical standpoint, inter-sensory. And such features depended, according to Neurath, on relations of order (1931/1983b, 62). For instance, in reports of spatio-temporal data, that is, of spatio-temporal order – or "space-time linkages" (Neurath 1931/1983a, 49) – protocol statements would consider only material things or events in space and time. Neurath sought to enforce the social and scientific requirement of objectivity and to challenge Carnap's reliance on subjective experience in the epistemology of the *Aufbau*. His brand of physicalism also provided a new solution to his old problem of unifying the natural and the human sciences.

To illustrate and support his physicalist doctrine of empiricism, in "Sociology in the Framework of Physicalism" (Neurath 1931/1983b), Neurath considered the use of the everyday term "blue" to report an experience. One way to provide a physicalist, inter-sensory and intersubjective formulation, Neurath suggested, was to have recourse to electromagnetic theory, namely the physical concept of "the number of oscillations of electromagnetic waves" (Neurath 1931/1983b, 63). Carnap (1923/2019) had introduced the same correspondence earlier. The appropriateness of the choice was obviously based on Maxwell's theory's reduction to electromagnetic theory of optics and the associated concepts for qualities such as color. The statement "here is a blue cube" could then be replaced, according to Neurath (1931/1983b, 63), by "a physical formula in which place is defined by coordinates".

But Neurath insisted that the substitution was neither required nor unavoidable, since behavioral descriptions, for instance, could similarly provide acceptable physicalist alternatives (1931/1983b, 63). In the purified form, electromagnetic theory could now contribute to the physicalist project of empiricism. But, unlike Carnap, Neurath's use of the example expressed also his anti-reductionistic approach to unity that preserved the value of the physicalistic but otherwise autonomous human sciences.

Also, one later appeal to electromagnetic theory in 1936, in "Individual Sciences, Unified Science, Pseudorationalism", illustrated his anti-reductionistic approach to unity (Neurath 1936/1983). Now, it did so from an extended Duhemian standpoint of methodological holism across different disciplines, not just different hypotheses. In addition to the familiar example of the forest fire, the fighting of which requires the application of multiple sciences at once, here Neurath mentioned electromagnetic theory much in the way Einstein had introduced it in 1905. The theory, stated Neurath, cannot be empirically "controlled in isolation", without predictions integrating statements of different disciplinary sources:

> The theory speaks of electric currents that originate when closed conductors and magnetic fields move relative to each other in a certain way whereas a prediction has to speak of a dynamo in a certain laboratory and of the behavior of an experimenter.
>
> (1936/1983, 133)

I want to conclude my discussion of Neurath's philosophical use of Maxwell's electromagnetic theory by citing two references to Maxwell's work in defense of the value of considering actual scientific practice, for instance, of significant historical figures, in philosophical debate. Both appear in the wartime correspondence with Carnap.

The first occurs in a rambling, disparaging letter of December 22, 1942, prompted by disagreements over Carnap's commitment to semantics. Neurath charged against Tarski's talk of truth and Popper absolutism about falsification on the grounds that one ignores scientific practice and the other trivializes it. Instead, countered Neurath,

> I test all these ideas by looking into the sciences, I am reading carefully Maxwell's letters, speeches, etc. and I found fine things. I am looking, how Lord Kelvin argued, Faraday, Marx, Max Weber, and then I try to find out, where we could sharpen our doubts.
>
> (Letter to Carnap, 22 December 1942;
> Cat and Tuboly 2019, 567)

The second reference occurs the following year in a letter of September 25, 1943. Neurath again criticized the absolutist, intolerant spirit of Popper's commitment to one system and one method, much against the tolerant attitude in groundbreaking actual scientific research (which in Maxwell's case, Duhem had also criticized). Neurath reported as follows:

> I look at the procedures of scientists as follow[s]: (I re-read in the last two years and particularly during the last months many authors, Maxwell, Darwin, Newton, Kepler, Malthus, Marx, Smith, etc. and many single papers in various subjects) we start from certain observation statements, which may be dropped sometimes, and try to catch as many of them as possible by means of theoretical tools. Should we find holes for our pegs, we are very happy as research workers, and do not bother too much about the pegs without holes and the holes without pegs, feeling it a progress compared with a situation, without pegs wh[ich] fit into holes.
>
> (Letter to Carnap, 25 September 1943;
> Cat and Tuboly 2019, 594)

But that was precisely part of Duhem's criticism of Maxwell's theory. Here and in the papers on theory of optics, then, Neurath was also defending Maxwell's scientific practice.

2.3.4. Carnap

Carnap intended and presented his pre-*Aufbau* works as contributions to the theory of science. In particular, he applied recent formalist, axiomatic and psychological perspectives. The goal was to investigate the sources of physical knowledge in terms of the construction and organizations of concepts, and, derivatively, the evaluation of theories.

In his doctoral dissertation, *Space (Der Raum)*, Carnap (1922/2019) followed in the footsteps of Helmholtz, Mach, Poincaré and others in engaging Kant's problem of the a priori conditions of the possibility of experience. Like them, he considered the formation of intuitive cognition of space and its relation to recent formal theories of space. In his case, Carnap followed Schlick and Reichenbach in turning to Einstein's theories of relativity. But he promptly extended the project to the world of physics, which required a theoretical representation, again, in the form of an axiomatic system, of the properties and laws that constitute matter, forces and causality – that is, constitutive of their concepts.

In "On the Task of Physics" ("Über die Aufgabe der Physik"), for instance, Carnap (1923/2019) investigated the decisions he considered involved in evaluating and selecting physical theories according to principles. Extending now the scope of the conventionalism he encountered in Poincaré and Dingler, he argued that the relevant decisions concerned, first, three stipulations: a system of space, a system of time and an action law fixing the dynamics and the description of the state of the world (Carnap 1923/2019, 211, 239). Second, decisions were required to choose to which of the three stipulations one must apply the conventionalist principle of maximal simplicity and, accordingly, what specific form of the principle should be applied. He then introduced the elements of what he called the ideal physical system or completed construction of physics (1923/2019, 221–233). The first is an axiom system that includes the space and time postulates and the action law. The second is an empirical dictionary that translates the descriptions of qualities in the domain of perception and the descriptions of the objects of physical theories associated with the choice of axiom system. The third element, also fixed by the axiom system, is the description of the corresponding physical state of the world at any two points in time.

It is worth noting that Carnap required the three elements to accommodate the concepts and laws of electromagnetic theory. Why? Without them, the elements of Carnap's ideal of physics and thereby his own account lacked a necessary credible scientific image of the physical world, one with the epistemic authority of actual science. In the case of the axiom system, he considered three possible kinds (Carnap 1923/2019, 223–227). All included Maxwell's equations either as axioms or required theorems, including those systems with Einstein's space-time equations. The first is characterized by a base with the axioms of Euclidean geometry, Newton's laws of motion and Newton's law of gravity. In such a system, the reduction of electromagnetic theory, that is its desired derivation from the base, is only a logical possibility. Carnap considered the variant system with a base extended axiom with Maxwell's equations, with the possible replacement of the law of gravity by a statistical-kinetic theory of material atoms. The second kind has a base that includes the equations of electron theory and a quantum postulate. Only the third kind is a family of variants of Einstein's equations of general relativity, Mie-Hilbert's, Weyl's and Kaluza's, and they all include irreducible electromagnetic magnitudes. This is not surprising; recall that Einstein's special theory provides the space-time structure to the Lorentz-Maxwell equations of electron theory and the general theory a more symmetric structure that succeeds in geometrizing away gravitation, not electromagnetism.

Moreover, note that the deductive nature of the axiom system required the strictly formal nature of any of the laws either in the axioms or theorems. The implicit choice, then, was the Hertzian version of Maxwell's equations or the subsequent Lorentz-Maxwell equations for the motion of electrons, the core of electron theory and the electromagnetic worldview that marked the physics of the turn of the century.

If in the first element of the ideal of completed construction of physics, the equations of electromagnetic theory illustrated and grounded the formal structure of the unified ideal of physics, in the second element, they also illustrated and grounded the empirical, phenomenological dimension. Carnap pointed to the case of colors, which would be recognized only within an ordered color system – he mentioned the example of Ostwald. The corresponding physical object or process is electromagnetic, but it would vary according to the chosen axiom system for physical theory. Thus, for the case of blue, also Neurath's choice, he observed that the color would correspond in the second kind of system to a periodical movement of electrons denoted by the frequency of oscillation (Carnap 1923/2019, 227). Similarly with smells – despite the caveat, he noted, of the lack of a clear classificatory system – and sensations of warmth; Carnap associated them, within a system of the second kind, for example, with different properties of electron complexes (1923/2019, 227).

In the more systematic discussion in *Physical Concept Formation* (*Physikalische Begriffsbildung*), Carnap (1926/2019) introduced further details and also considered the pervasive issue of unity in the distinction between the natural and the cultural sciences, the latter distinguished by the aim of understanding without general laws. On that issue, he added that, as a matter of theory formation, the unity of physics depended on the theory of electromagnetism, echoing the widespread support of the electromagnetic world-picture. Not only had optics and magnetism become parts of the theory of electricity, he noted; in the new atomic model of the physical world (including the quantum postulate), he added, all physical and chemical appearances, with the exception of gravitation, had been either reduced or declared in principle reducible to electromagnetism (1926/2019, 405). Without it, Carnap insisted in a grander tone, the most important result in the development of physics in the last hundred years, the development of electromagnetic theory in the form of electron theory, would not have been achieved, or with it, a new unified theory of physics (1926/2019, 407). With this result, he concluded, the theory of electricity had "revolutionary impact on physics" (1926/2019, 409). If considerations of revolution in physics are relevant to philosophical change, this example is neither about relativity nor about quantum mechanics.

Now, where in the *Aufbau* (Carnap 1928) is the theory of electrons and electromagnetic fields? They featured more discreetly in the set of choices of a physical basis. Carnap listed only a selection from the examples of axiom systems for natural laws that he had introduced in "On the Task of Physics": electron theory, where the acceleration of electrons and protons and the atoms they make up helps construct the electromagnetic fields and gravitation; Weyl's integration of electromagnetic potentials, defined over the points of Einstein's curved space-time and Einstein's tensor field for gravitation (curvature) and the same integration defined over the points on Minkowski's world-lines (Carnap 1928, 84, art. 62). Electromagnetic theory, along with gravitation, underwrites the possibility of unity of the sciences. The narrower set of available projects still illustrated and also supported the conventional nature of the required choice.

The second role, in the epistemological or experiential system, allows for the objects of perception constructed out of experiences in the autopsychological basis to be used in the construction of physical objects. Carnap referred to the explication of the physical-qualitative correlation he had offered in the earlier essays I have presented (Carnap 1928, 182, art. 136).

The new manifesto had already pointed to the linguistic nature and unity of science. In *The Unity of Science* (originally published as an article, "*Physikalische Sprache als Universalsprache der Wissenschaft*"), Carnap (1931/1995) heeded Neurath's call for physicalism in the proper scientific account of linguistic empiricism. Yet Carnap's choice of a physical language that would capture the role of experience, or protocol language, was much closer to physics than was Neurath's. The context was relatively new, but his proposal in part wasn't. It included the terms for sensory qualities that were either characterized or characterizable in terms of the numerical determinations of physics, of "a definite value or range of values of a coefficient of physical state" attached to "a specific set of co-ordinates (three space, one time co-ordinates)" (Carnap 1931/1995, 52–53). Any acceptable alternatives would have to be reducible accordingly. Among the qualities he considered, he again paid more attention to the visual and the case of color. Since Maxwell's identification of light with electromagnetic waves, the correlation, as he had discussed it in his pre-*Aufbau* essays, required measures of wave oscillations. He was now explicit that the application of science required the mathematical formulation of general laws of nature. Here he explicitly pointed to Maxwell's second equation, linking the spatial distribution of the electric field in the infinitesimal neighborhood of a point and the rate of change of the magnetic field at the same point (1931/1995, 56).

More importantly, he argued, with Neurath, that the mathematical determination allowed by the equation had the virtue of being both inter-subjective and inter-sensory, independent of color perception and visual perception altogether. In fact, the technological arrangement that would make the cross-modality possible involved the use of electricity, so that, by a further application of Maxwell's theory (or a modern development), the information about the set of frequencies associated with a certain color could have its ordering or structural property expressed in the motion of a palpable pointer or the audible frequency of acoustic waves (Carnap 1931/1995, 60).

He also pointed out that the formal sameness of content of qualitative and physical representations or propositions was independent of the images and conceptions associated with them (Carnap 1931/1995, 91). Now, notice that the rejection of associated images or conceptions and the emphasis on the common numerical determination constitute precisely, as I have already mentioned, the sort of epistemic decoupling that Neurath had identified in Hertz's restrictive conception of Maxwell's theory and that Duhem had noted and approved.

Finally, I want to conclude this brief examination of the enduring and significant role of electromagnetic theory in Carnap's philosophical evolution with a reference to *The Logical Syntax of Language* (1934/1937). Carnap was now walking in lockstep with Neurath's syntacticism as well as physicalism. In the new theory of scientific knowledge, Carnap (1934/1937, 316–318) characterized the logical syntax of science in terms of sets of transformation rules and concluded with an examination of physical language. Testing physical P-sentences and introducing primitive physical P-terms through the derivation of protocol terms was, following Duhem, a holistic affair, not a single-file chain of logical derivations. To argue the point, he drew new attention to the case of Maxwell's equations (1934/1937, 319).

The argument also illustrated the kind of foundational investigation that should characterize what he called non-metaphysical philosophy, the logical analysis of science. The task he now declared syntactical was the analysis of scientific statements, of so-called language-forms, expressed by formal statements about other statements, that is, in the formal mode. Now he called these syntactical statements. The others he now called descriptive. However, he also warned against assuming that the distinction between the logical or syntactical analysis of science and the specific sciences rests on the distinction between syntactical and descriptive statements (Carnap 1934/1937, 331). He offered a detailed example from the empirical sciences, namely the analysis of Einstein's discussion of Maxwell's equations for moving bodies and the propagation of light. Beginning with Einstein's opening statement, he offered a paraphrase that allowed for the identification of descriptive and syntactical statements. Carnap's parsed quote is as follows:

> That Maxwell's electro-dynamics/lead to asymmetries in their application to bodies in motion/which do not appertain to the phenomena/ is well known.

Carnap paraphrased the breakdown as follows:

> In the laws which are sequences of the Maxwell equations/certain asymmetries are shown/which do not occur in the appertaining protocol-sentences./Contemporary physics knows that.

The first three sentences Carnap identified as purely syntactical statements about laws and protocol-sentences, while the last one was a historical descriptive sentence (1934/1937, 329). As descriptive primitive P-rules, rather than L-rules, in this later analysis, Maxwell's equations were replaceable, consistent with his early conventionalism. The account differs from both Hertz's and Neurath's.

The evolving significance and diverse use and interpretation of Maxwell's equations and electromagnetic theory reappear here now in the syntactic analysis, also one of the recent significant new places in Einstein's argument for his special theory of relativity. Attention to relativity, in fact, requires attention to electromagnetic theory. In the earlier discussions, among examples of actual projects of axiom systems, he had also recognized the place of Maxwell's equations in every unified theory of physics based on the space-time structure of general relativity.

2.4. Conclusion

To conclude, each round of examples involving electromagnetic theory and phenomena illustrated and supported interventions in philosophical evolution. And, in Carnap's case, Maxwell's equations track an evolution different from Neurath's, Frank's and Schlick's, and also their interaction, surrounded by affinities and marked contrasts. In each case, considerations of electromagnetism tracked their respective disciplinary choices, their epistemological priorities, their methodological standards and, in particular, their conceptions of unity. Also in each case, the role of attention to electromagnetism is embedded in different considerations of the historical character of evidence from scientific practice, whether as a matter of the authority of most recent physics, of its revolutionary character, of historical change or as part of the widespread historical domain of different scientific practices.

Acknowledgments

The chapter develops the content of a paper presented at the conference *Wege der Wissenschaftlichen Weltauffassung Rudolf Carnap and Otto Neurath*, at the University of Graz on September 26, 2019. Portions of Section 2.3 are published in the conference proceedings and reproduced here with permission of the editors, Johannes Friedl and Christian Damböck.

Notes

1. Of all the relations he called laws, four have been considered fundamental:

$$\text{div } \mathbf{D} = 4\pi\rho$$
$$\text{curl } \mathbf{H} = 4\pi\mathbf{J}$$
$$\mathbf{curl\ E} = -\partial\mathbf{B}/\partial t$$
$$\text{div } \mathbf{B} = 0$$

   Div and **curl** are differential operators that represent local, infinitesimal variations in space, either along each dimension – div – or in a circular or vortical arrangement around a line vector passing through its center and in its particular direction – *curl* (the bold type representing its vector character).

In the first law – known as Coulomb's law in reference to Coulomb's laws of electrostatics – **D** is the electric displacement in a non-conducting medium and r is the charge density. In the second law – known as Ampère's law in reference to Ampère's laws of electric currents – **H** is magnetic field intensity and **J** is the electric current density that induces the field around it (as discovered by Oersted). In the third law – known as Faraday's law in reference to Faraday's discovery of the induction of electric fields and currents by changing magnetic fields – **E** is the electric field and **B** is the quantity of magnetic induction (**H** and **B**, and **D** and **E**, are related by a constant depending on the medium). The fourth equation, by contrast with the first, represents the absence of magnetic charges or poles.

2. The concepts were meant to be rejected except as auxiliary concepts or symbols for, in the case of forces, the result of hidden connections between visible or hidden masses.
3. See for instance, Janik and Toulmin (1973), Barker (1980), Wilson (1989), Kjaergaard (2002).
4. In an earlier, shorter essay, on comparison in physics, he mentioned the examples of mechanical and mathematical analogies in relation to the representation of phenomena of electricity and magnetism but omitted any reference to Maxwell or his method of analogy (Mach 1895, 249–250).
5. See also his critique of Hertz in the first section of *Principles of Mechanics* (Boltzmann 1897).
6. It is equally worth drawing attention to Hans Reichenbach's different interests in electromagnetic theory, including his early interest in radio engineering and subsequently in its use in pedagogical communication, in the role of light signals in relativity theory, in the project of unification of gravitation and electromagnetism – which he called the two kingdoms, Maxwell's and Einstein's – in the use of light signals in empirical criterion of causal processes, even in relation to psychology, in psychophysical parallelism and the critique of Gestaltic holism in terms of the standard set by local (contiguous) character of electromagnetic fields – in contrast to the appeal some Gestalt psychologists made to the analogy with electromagnetic theory in order to defend a physical model of Gestalt and a psycho-physical isomorphism (Cat 2007).
7. About Schlick's conception of causality in the context of physical theories, see Richard Dawid's chapter in this volume.
8. I owe the information about the letter of September 24, 1928, to Friedrich Stadler.
9. Like Kleinpeter, Frank also mentioned Goethe's phenomenalism, with the caveat, Frank added, that Goethe was not the good physicist Maxwell was. Kleinpeter (1913, 5) discussed both at some length, even mentioning both on the same page.
10. Also, Reichenbach pointed to a Machian analysis of causality and determinism by Petzold (Howard 1996; Ryckman 1991).
11. The event was a pre-conference to the 8th International Congress of Philosophy (Stadler 2001, 356–357).
12. Frank's view on partial analogies is discussed further by Adam Tamas Tuboly in this volume.

References

Ariew, R. and Barker, P. (1986), 'Duhem on Maxwell: A Case-Study in the Interrelations of History of Science and Philosophy of Science', in *PSA: Proceedings of the Biennial Meeting of the Philosophy of Science Association*. Vol. 1. Chicago: University of Chicago Press, pp. 145–156.

Banks, E. C. (2003), *Ernst Mach's World Elements*. Dordrecht: Kluwer.

Barker, P. (1980), 'Hertz and Wittgenstein', *Studies in History and Philosophy of Science Part B*, 11 (3): 243–256.

Berg, A. (2016), *Phenomenalism, Phenomenology, and the Question of Time: A Comparative Study of the Theories of Mach, Husserl, and Boltzmann*. Lanham: Lexington Books.

Boltzmann, L. (1891/1893), *Vorlesungen über die Maxwells Theorie der Elektricität und des Lichtes*. 2 vols. Leipzig: Barth.

——— (1892/1974), 'On the Methods of Theoretical Physics', in *Theoretical Physics and Philosophical Problems*. Dordrecht: Reidel, pp. 5–12.

——— (1897/1974), 'On the Indispensability of Atomism in Natural Science', in *Theoretical Physics and Philosophical Problems*. Dordrecht: Reidel, pp. 41–53.

——— (1897), *Vorlesungen über die Principe der Mechanik*. Leipzig: Barth.

——— (1899/1974a), 'On the Development of the Methods of Theoretical Physics in Recent Times', in *Theoretical Physics and Philosophical Problems*. Dordrecht: Reidel, pp. 77–100.

——— (1899/1974b), 'On the Fundamental Principles and Equations of Mechanics, I, II', in *Theoretical Physics and Philosophical Problems*. Dordrecht: Reidel, pp. 101–128.

——— (1902/1974), 'Model', in *Theoretical Physics and Philosophical Problems*. Dordrecht: Reidel, pp. 213–222.

Carnap, R. (1922/2019), 'Space: A Contribution to the Theory of Science', in A. W. Carus et al. (eds.), *The Collected Works of Rudolf Carnap, Volume I: Early Writings*. Oxford: Oxford University Press, pp. 21–171.

——— (1923/2019), 'On the Task of Physics and the Application of the Principle of Maximal Simplicity', in A. W. Carus et al. (eds.), *The Collected Works of Rudolf Carnap, Volume I: Early Writings*. Oxford: Oxford University Press, pp. 209–241.

——— (1926/2019), 'Physical Concept Formation', in A. W. Carus et al. (eds.), *The Collected Works of Rudolf Carnap, Volume I: Early Writings*. Oxford: Oxford University Press, pp. 339–427.

——— (1928), *Der logische Aufbau der Welt*. Berlin: Weltkreis-Verlag.

——— (1931/1995), *The Unity of Science*. Translated by Max Black. London: Thoemmes Press.

——— (1934/1937), *Logical Syntax of Language*. London: K. Paul, Trench, Trubner & Co.

Carnap, R., Neurath, O., and Hahn, H. (1929/1973), 'The Scientific Conception of the World: The Vienna Circle', in M. Neurath and R. S. Cohen (eds.), *Empiricism and Sociology*. Dordrecht: D. Reidel, pp. 299–318.

Cat, J. (2001), 'On Understanding: Maxwell on the Methods of Illustration and Scientific Metaphor', *Studies in History and Philosophy of Science Part B: Studies in History and Philosophy of Modern Physics*, 32 (3): 395–441.

——— (2007), 'Switching Gestalts on Gestalt Psychology: On the Relation Between Science and Philosophy', *Perspectives on Science*, 15 (2): 131–177.

——— (2019), 'Neurath and the Legacy of Algebraic Logic', in Cat and Tuboly (2019), pp. 241–338.

Cat, J. and Tuboly, A. T. (eds.) (2019), *Neurath Reconsidered: New Sources and Perspectives*. Cham: Springer.

Corry, L. (2006), 'The Origin of Hilbert's Axiomatic Method', in J. Renn (ed.), *The Genesis of General Relativity*. Vol. 4: Theories of Gravitation in the Twilight of a Classical Physics: The Promise of Mathematics and the Dream of a Unified Theory. Dordrecht: Springer, pp. 139–236.

Darrigol, O. (1993), 'The Electrodynamic Revolution in Germany as Documented by Early German Expositions of "Maxwell's Theory"', *Archive for History of Exact Sciences*, 45: 189–280.

——— (2000), *Electrodynamics from Ampère to Einstein*. Oxford: Oxford University Press.

——— (2018), *Atoms, Mechanics and Probability: Ludwig Boltzmann's Statistico-Mechanical Writings: An Exegesis*. Oxford: Oxford University Press.

Duhem, P. (1902), *Les Théories Électriques de J. Clerk Maxwell: Étude Historique et Critique*. Paris: Hermann.

——— (1906), *La Théorie Physique. Son Object et Sa Structure*. Paris: Rivière.

Einstein, A. (1905a), 'Über einen die Erzeugung und Verwandlung des Lichtes betreffenden heuristischen Gesichtspunkt', *Annalen der Physik*, 17: 132–148.

——— (1905b), 'Zur Elektrodynamik bewegter Körper', *Annalen der Physik*, 17: 891–921.

Feigl, H. (1929), *Theorie und Erfahrung in der Physik*. Karlsruhe: Verlag G. Braun.

Frank, Ph. (1907/1941), 'The Law of Causality and Experience', in *Between Physics and Philosophy*. Cambridge, MA: Harvard University Press, pp. 17–27.

——— (1917/1941), 'The Importance of Ernst Mach's Philosophy of Science for Our Times', in *Between Physics and Philosophy*. Cambridge, MA: Harvard University Press, pp. 28–54.

——— (1930/1941), 'Physical Theories of the Twentieth Century and School Philosophy', in *Between Physics and Philosophy*. Cambridge, MA: Harvard University Press, pp. 55–103.

——— (1934/1941), 'Is There a Trend Today towards Idealism in Physics?', in *Between Physics and Philosophy*. Cambridge, MA: Harvard University Press, pp. 104–126.

——— (1935/1941), 'The Positivistic and the Metaphysical Conception of Physics', in *Between Physics and Philosophy*. Cambridge, MA: Harvard University Press, pp. 127–138.

——— (1937), 'The Mechanical versus the Mathematical Conception of Nature', *Philosophy of Science*, 4 (1): 41–74.

——— (1937/1949), 'Mechanical "Explanation" or Mathematical Description?', in *Modern Science and Its Philosophy*. Cambridge, MA: Harvard University Press, pp. 138–143.

——— (1941). 'Introduction: Historical Background', in *Between Physics and Philosophy*. Cambridge, MA: Harvard University Press, pp. 3–16.

Friedman, M. (2001), *Dynamics of Reason*. Stanford, CA: CSLI.

Heidelberger, M. (1998), 'From Helmholtz's Philosophy of Science to Hertz's Picture-Theory', in D. Baird, R. I. G. Hughes, and A. Nordmann (eds.), *Heinrich Hertz (1857–1894): Classical Physicist, Modern Philosopher*. Dordrecht: Kluwer, pp. 9–24.

Hertz, H. (1892/1893), *Electric Waves*. London: Macmillan.

——— (1894/1899), *The Principles of Mechanics*. London: Macmillan.

Howard, D. (1996), 'Relativity, Eindeutigkeit, and Monomorphism: Rudolf Carnap and the Development of the Categoricity Concept in Formal Semantics', in N. Giere and A. W. Richardson (eds.), *Origins of Logical Empiricism*. Minneapolis, MN: University of Minnesota Press, pp. 115–164.

Janik, A. and Toulmin, S. (1973), *Wittgenstein's Vienna*. New York: Simon and Schuster.

Kjaergaard, P. C. (2002), 'Hertz and Wittgenstein's Philosophy of Science', *Journal for General Philosophy of Science*, 33: 121–149.

Kleinpeter, H. (1913), *Der Phänomenalismus, eine naturwissenschaftliche Weltanschauung*. Leipzig: Barth.

Lorentz, H. A. (1895), *Versuch einer Theorie der elektrischen und optischen Erscheinungen in bewegten Körpern*. Leiden: E. J. Brill.

Mach, E. (1883/1907), *The Science of Mechanics*. Chicago: Open Court.

—— (1885/1897), *Contributions to the Analysis of Sensations*. Chicago: Open Court.

—— (1894/1895), *Popular Scientific Lectures*. Chicago: Open Court.

—— (1896/1986), *Principles of the Theory of Heat*. Dordrecht: Reidel.

—— (1905/1976), *Knowledge and Error*. Dordrecht: Reidel.

Maxwell, J. C. (1856/1890), 'On Faraday's Lines of Force', in W. D. Niven (ed.), *The Scientific Papers of James Clerk Maxwell*. Vol. 1. Cambridge: Cambridge University Press, pp. 155–229.

—— (1861/1890), 'On Physical Lines of Force: Parts I and II', in W. D. Niven (ed.), *The Scientific Papers of James Clerk Maxwell*. Vol. 1. Cambridge: Cambridge University Press, pp. 451–188.

—— (1862/1890), 'On Physical Lines of Force: Parts III and IV', in W. D. Niven (ed.), *The Scientific Papers of James Clerk Maxwell*. Vol. 1. Cambridge: Cambridge University Press, pp. 489–513.

—— (1865/1890), 'A Dynamical Theory of the Electromagnetic Field', in W. D. Niven (ed.), *The Scientific Papers of James Clerk Maxwell*. Vol. 1. Cambridge: Cambridge University Press, pp. 586–597.

—— (1876), *Matter and Motion*. London: Society for Promoting Christian Knowledge.

Miller, A. I. (1998), *Albert Einstein's Special Theory of Relativity*. New York: Springer.

Neurath, O. (1910/2004), 'War Economy', in Th. Uebel and R. S. Cohen (eds.), *Otto Neurath: Economic Writings, Selections 1904–1945*. Dordrecht: Kluwer, pp. 153–199.

—— (1915/1973), 'On the Foundations of the History of Optics', in M. Neurath and R. S. Cohen (eds.), *Empiricism and Sociology*. Dordrecht: D. Reidel, pp. 101–112.

—— (1916/1983), 'On the Classification of Systems of Hypotheses', in R. S. Cohen and M. Neurath (eds.), *Otto Neurath: Philosophical Papers 1913–1946*. Dordrecht: D. Reidel, pp. 13–31.

—— (1919/1973), 'Utopia as a Social Engineer's Construction', in M. Neurath and R. S. Cohen (eds.), *Empiricism and Sociology*. Dordrecht: D. Reidel, pp. 150–155.

—— (1925/2004), 'Economic Plan and Calculation in Kind', in Th. Uebel and R. S. Cohen (eds.), *Otto Neurath: Economic Writings: Selections 1904–1945*. Dordrecht: Kluwer, pp. 405–465.

—— (1931/1983a), 'Physicalism: The Philosophy of the Viennese Circle', in R. S. Cohen and M. Neurath (eds.), *Otto Neurath: Philosophical Papers 1913–1946*. Dordrecht: D. Reidel, pp. 48–51.

—— (1931/1983b), 'Sociology in the Framework of Physicalism', in R. S. Cohen and M. Neurath (eds.), *Otto Neurath: Philosophical Papers 1913–1946*. Dordrecht: D. Reidel, pp. 58–90.

—— (1936/1983), 'Individual Sciences, Unified Science, Pseudo-Rationalism', in R. S. Cohen and M. Neurath (eds.), *Otto Neurath: Philosophical Papers 1913–1946*. Dordrecht: D. Reidel, pp. 132–138.

—— (1973), 'Memories of Otto Neurath', in M. Neurath and R. S. Cohen (eds.), *Empiricism and Sociology*. Dordrecht: D. Reidel, pp. 1–83.

Planck, M. (1900), 'Zur Theorie des Gesetzes der Energieverteilung im Normalspektrum', *Verhandlungen der Deutschen Physikalischen Gesellschaft*, 2: 237–245.

Poincaré, H. (1890/1891), *Electricité et Optique*. 2 vols. Paris: Blondin and Bruhnes.

—— (1902/1905), *Science and Hypothesis*. New York: The Science Press.

Ruyer, R. (1933), 'La Psychologie, la "Desubjéctivation" et le Parallélisme', *Revue de Synthèse*, 6: 167.

Ryckman, Th. (1991), 'Condition Sine Qua Non? Zuordnung in the Early Epistemologies of Cassirer and Schlick', *Synthese*, 88 (1): 57–95.

—— (2005), *The Reign of Relativity: Philosophy in Physics 1915–1925*. New York: Oxford University Press.

—— (2007), 'Logical Empiricism and the Philosophy of Physics', in A. Richardson and Th. Uebel (eds.), *The Cambridge Companion to Logical Empiricism*. Cambridge: Cambridge University Press, pp. 193–277.

Schlick, M. (1917), *Raum und Zeit in der gegenwärtigen Physik*. Berlin: Springer.

—— (1918), *Allgemeine Erkenntnislehre*. Berlin: Springer.

—— (1919/1920), *Space and Time in Contemporary Physics*. Oxford: Oxford University Press.

—— (1920/1979), 'Philosophical Reflections on the Causal Principle', in H. Mulder and B. F. B. van de Velde-Schlick (eds.), *Moritz Schlick: Philosophical Papers, Volume I, 1909–1922*. Dordrecht: D. Reidel, pp. 295–321.

—— (1925/1974), *General Theory of Knowledge*. 2nd ed. Vienna: Springer.

—— (1925/1979), 'Outlines of the Philosophy of Nature', in H. Mulder and B. F. B. van de Velde-Schlick (eds.), *Moritz Schlick: Philosophical Papers, Volume II, 1925–1936*. Dordrecht: D. Reidel, pp. 1–90.

—— (1931/1979), 'Causality in Contemporary Physics', in H. Mulder and B. F. B. van de Velde-Schlick (eds.), *Moritz Schlick: Philosophical Papers, Volume II, 1925–1936*. Dordrecht: D. Reidel, pp. 176–209.

—— (1932/1979), 'Form and Content: An Introduction to Philosophical Thinking', in H. Mulder and B. F. B. van de Velde-Schlick (eds.), *Moritz Schlick: Philosophical Papers, Volume II, 1925–1936*. Dordrecht: D. Reidel, pp. 285–369.

—— (1934/1979), 'On the Foundations of Knowledge', in H. Mulder and B. F. B. van de Velde-Schlick (eds.), *Moritz Schlick: Philosophical Papers, Volume II, 1925–1936*. Dordrecht: D. Reidel, pp. 370–387.

—— (1935/1979a), 'Introduction and on "Affirmations' from Sur le Fondement de la Connaissance"', in H. Mulder and B. F. B. van de Velde-Schlick (eds.), *Moritz Schlick: Philosophical Papers, Volume II, 1925–1936*. Dordrecht: D. Reidel, pp. 405–413.

—— (1935/1979b), 'On the Relation between Psychological and Physical Concepts', in H. Mulder and B. F. B. van de Velde-Schlick (eds.), *Moritz*

Schlick: Philosophical Papers, Volume II, 1925–1936. Dordrecht: D. Reidel, pp. 420–436.

Siegel, D. M. (1991), *Innovation in Maxwell's Electromagnetic Theory*. Cambridge: Cambridge University Press.

Spielberger, H. (1982), *The Phenomenological Movement: A Historical Introduction*. The Hague: Nijhoff.

Stadler, F. (2001), *The Vienna Circle: Studies in the Origins, Development, and Influence of Logical Empiricism*. Vienna and New York: Springer.

Wien, W. (1900), 'Über die Möglichkeit einer elektromagnetischen Begründung der Mechanik', in *Recueil de Travaux Offerts par les Auteurs à H.A. Lorentz*. The Hague: Nijhoff, pp. 96–107.

Wilson, A. D. (1989), 'Hertz, Boltzmann and Wittgenstein Reconsidered', *Studies in History and Philosophy of Science Part A*, 20: 245–263.

3   Kurt Grelling and the Idiosyncrasy of the Berlin Logical Empiricism

Nikolay Milkov

> How outlandish he [Grelling] approaches everything, how traditional. Fries, Nelson, Oppenheim, etc. One can't assimilate it easily, a lot of thereof remains undigested.
>
> Otto Neurath's letter to Rudolf Carnap,
> March 16, 1935 (RC 029–09–70)

3.1. Introduction

In the last decades, philosophy of science often concentrated its attention on intricate epistemological problems: scientific explanations as different from scientific descriptions, scientific truth, the relation between phenomena and data of science, the relation between scientific theories and scientific models, and similar. Arguably, the result is "the decline of the philosophy of science in the academy and in the public intellectual sphere" (Howard 2003, 77). Here is a recent statement of the physicist and cosmologist of repute Lawrence Kraus:

> The only people, as far as I can tell, that read work by philosophers of science are other philosophers of science. It has no impact on physics whatsoever and I doubt that other philosophers read it because it's fairly technical.
>
> (quoted in Andersen 2012)[1]

We remember quite well, however, that at the beginning of the last century, the philosophical writings of David Hume, Ernst Mach, and Henri Poincaré helped Einstein to formulate his special theory of relativity. In historical context, this development was described by Don Howard in the following words:

> In the 1950s and 1960s the philosophy of science was one of the most exciting places to be in the academy. Not just in the philosophy departments but from many other places on the typical North

American campus, the philosophy of science was seen as one of the most important centers of intellectual activity. . . . The times, however, have changed. I have the impression that today, on more and more campuses, the philosophy of science is moving toward the periphery.

(2003, 74 f.)

The question raises itself: Why this change for the worse?

As we see it, this state of the art is a long-term outcome of the logical positivists' highly influential approach to philosophy of science. In a chapter titled "Against Philosophy" in his book *Consilience*, the Nobel Prize winner for physics for 1979, Steven Weinberg, wrote: "The positivist concentration on observables like particle positions and momenta has stood in the way of a 'realist' interpretation of quantum mechanics, in which the wave function is the representation of physical reality" (1992, 181).[2]

Unfortunately, the post-positivist philosophers of science didn't fare better. Above all, Quine's programmatic claim that philosophy is a part of natural science didn't help to really put science closer to philosophy. The same can be said about the philosophy of science as autonomous discipline that was established and also institutionalized in North America in the 1960s and then in the whole world. Furthermore, there are good reasons to maintain that the road to it was paved by the appearance of Hempel and Oppenheim's "Studies in the Logic of Explanation" (1948) (see Section 3.4). Today, after seven decades of research, the exploration of the explanation in science continues full steam ahead. Unfortunately, science itself learned less about its philosophy from this discussion. The new movement of "scientific metaphysics" that came to the scene in the 1970s, inspired by works of Saul Kripke and Hilary Putnam, proved even less helpful. In short, the conclusion can be drawn that "no new paradigm has come to the fore to define the field following the demise of neo-positivism" (Howard 2003, 76).

In this chapter, we shall try to track down an alternative venue that, we hope, can help to bring philosophy and science back together. To this purpose, we are going to rationally reconstruct the project of the Berlin Group, led by Hans Reichenbach. We are going to do this by a close analysis of the works of "the third man" of the group (Peckhaus 2013), Kurt Grelling (1886–1942). We shall see that he made several instructive gestures towards promising sides of the project of the Berlin Group.

By this endeavor, we follow a specific historical-philosophical approach. In short, we are not going to only logically analyze the philosophical ideas of Kurt Grelling and their relatedness to other philosophers of the time, including other members of the Berlin Group. We shall also explore the historical context of these ideas, including their psychological side. Following this approach, we hope to reveal aspects of the philosophical development of the Berlin Group that otherwise remain unnoticed.

3.2. Kurt Grelling

3.2.1. Kurt Grelling Between Leonard Nelson and Hans Reichenbach

Hans Reichenbach insisted that we are to investigate science not only from a logical point of view but also sociologically and psychologically (Milkov 2015, xl). Clearly, we can explore this way also the philosophical development of the members of the Berlin Group. In the case of Kurt Grelling, a psychological analysis is especially appropriate. In the sequel, we are going to see why was this so.

Kurt Grelling started his study in 1906 in Göttingen with Leonard Nelson, a Neo-Friesian philosopher who in 1919 received, with the decisive support of David Hilbert, an "associate professorship [*außerordentliche Professur*] for systematic philosophy of exact sciences". In 1908, Grelling discovered the famous today "Grelling paradox"; in 1910, he wrote a dissertation under, perhaps, the best pure mathematicians of the time, David Hilbert and Ernst Zermelo, on the Göttingen-typical topic of axiomatization of arithmetic (Grelling 1910a).[3] Also in 1910, at the age of 24, Grelling published an important paper on theory of probability (Grelling 1910b). In the same year, on August 3, 1910, he wrote a letter to Bertrand Russell in which he communicated problems that he supposedly discovered in Russell's ramified theory of types (Bertrand Russell Archive, McMaster University).

Instead of fighting for a university position in logic, however, following Leonard Nelson's advice, Grelling started studying political economy at the University of Munich for three years in 1910. Apparently enhancing his knowledge in this realm was more important to him than the dogged pursuit of an academic career. Among other things, in 1912/13, Grelling organized an informal philosophical seminar in Munich in which Reichenbach could also have taken part – in that year, Reichenbach was also studying in Munich. Be this as it may, it is well documented that while in Munich, Grelling and Reichenbach came in contact as members of the left-oriented "Free Students" Society.

After he returned to Göttingen, Grelling found Nelson occupied mainly with political issues – a development in 1913 caused by the upcoming Great War and the social changes of the time. Indeed, Grelling continued to work on philosophy of mathematics. In January 20, 1914, for example, he lectured in Hilbert's Mathematical Colloquium in Göttingen on "Recent Works in Philosophy and Mathematics", and on December 18 the same year, delivered "Additions and Exact Formulations" as part of Heinrich Behmann's lecture course on Russell's and Whitehead's *Principia mathematica* (Corry 2004, 319 n. 11). But he also engaged with political philosophy, producing, among other things, a paper on the "Philosophical Foundations of Politics" (1916a) and the brilliant pamphlet *Anti-j'accuse* (1916b),[4] in which he tried to explain the causes and reasons for the Great War.

Grelling's academic priority, to achieve ever more knowledge instead of fighting for academic posts, made him most informed in the field of scientific philosophy, in fact, better informed than either Nelson or Reichenbach. For example, while Reichenbach had practically no idea of the role of Frege as a logician,[5] Grelling knew much about Frege's logic already when he worked with Nelson. This knowledge was demonstrated, for example, in Grelling's (1932/1933) critical discussion of Dubislav's book *The Definition*, in which he, among other things, sharply criticized Dubislav's interpretation of Frege. Grelling also wrote papers on Russell, Gödel, Carnap, Tarski, and Leśniewski. Based on that knowledge, in the late 1920s and in the 1930s, Grelling successfully played the role of a very well-informed analytic "gadfly" (to use here this happy expression of Richard Rorty) who criticized the latest developments of logic, science, philosophy of science, and the philosophy of nature of his time.

Grelling was also an avid translator of scientifically informed philosophers of the time. In 1910, he had already translated a book of Federigo Enriques from Italian. Between 1927 and 1930, Grelling translated four books by Bertrand Russell from English and in 1930 a book by Émile Meyerson from French. We can judge the quality of Grelling's translations from the words of Meyerson in the "Preface" of the German translation of his *Identität und Wirklichkeit* (Meyerson studied in Germany and was fluent in German, so he could read Grelling's translation in the original): "Dr. Grelling fulfilled the obligations of the translator a in seldomly achieved perfect way. I can confirm that I read some pages [of the translation] with a real pleasure" (Meyerson 1930, xi).

Eventually, this side of Grelling's academic character made him, as already noted, "the third man" of the Berlin Group, behind Reichenbach and Dubislav. Typically, Grelling never delivered a lecture at the Society of Empirical/Scientific Philosophy which, after June 1929, was run by the Berlin Group of which Grelling himself was a core member. Being at the time only a high-school teacher in Berlin, he consigned this task to Reichenbach and Dubislav. A few years later, however, apparently hoping that this would help him to finally find an academic or research position away from Nazi Germany, he delivered talks at the International Congresses for the Unity of Science in Prague (1934) and Paris (1936 and 1937).

Grelling's career is sometimes seen as movement from being a collaborator of Nelson's to being a collaborator of Reichenbach's (Peckhaus 1994). This, however, is anything but the whole truth about him as a scientifically informed philosopher. Exactly because of the peculiarities of his academic character, he incorporated the spirit of the Berlin Group like nobody else. His encyclopedic knowledge in logic, mathematics, and politics made him "the great unknown" of the Berlin Group, the hidden figure behind its façade. In the film industry, there is an idiosyncratic concept of "best supporting role". The artist assisting the main actor gives her the opportunity to unfold her talent. Kurt Grelling played similar role to the developing philosophy of logical empiricism in Berlin.

3.2.2. Fries and Nelson

In order to become better acquainted with Grelling's position in philosophy of science, we shall first go back to his training in the context of the Jacob Fries Society led by Leonard Nelson (cf. Milkov 2013a, 12). Against the conventional view, which heavily underlines the role of Ernst Cassirer, we see logical empiricism as practiced by the Berlin Group as deeply rooted in that Society.

Traditionally, Jacob Friedrich Fries (1773–1843) is characterized as a Kantian who tried to reorient Kant towards psychology. Fries, however, was anything but a proponent of psychologism in the sense of Frege or Husserl. He simply defended the famous regressive method already formulated by Kant (1880/1968, §105), according to which philosophy is to start its analysis from complex and confused data. In contrast to Kant, however, Fries didn't start from the data of experience but from the data of the sciences and this in order to go back (to "regress") to their grounds – that is, to discover their principles (*sie herauszuschälen*).

At the beginning of the 20th century, Fries' philosophy of science was rediscovered by Leonard Nelson. Nelson (1905) fought the "obscurantism" of the Neo-Kantians that he saw as being not in the spirit of science (cf. Grelling 1907). Similarly, Fries, whose ideas were positively attested to by scientists of the time like Carl Friedrich Gauß and Alexander von Humboldt, criticized Kant in that the latter only pays lip service to science. To be more exact, he discusses a small area of mathematics and science. "For example, Kant never seriously undertook the philosophical analysis or justification of calculus, of formal algebra, of the theory of probability, or of analytical mechanics" (Pulte 2013, 45). Also beyond Kant's sphere of interest was chemistry, which he considered reducible to physics. One of Fries' objectives was to extend Kant's philosophy to all sciences.

We know today, thanks to Michael Friedman (2001), that in 1920, Hans Reichenbach adopted the conception of relative a priori (Reichenbach 1920/1965). To be more specific, he replaced Kant's a priori, which is valid for all kinds of science once and forever, by "relativized, and dynamic, constitutive principles, which change incoherently from one theory to another" (Friedman 2005, 125). The key point was the rejection of the Kantian thesis that knowledge as such, and scientific knowledge in particular, has eternal general principles. "There are no general presuppositions of knowledge, only presuppositions of particular hypothesis" – or of current theory (Reichenbach 1931/1978, 362), and exactly these individual presuppositions have to be brought out by the logical analysts of science.

Recent studies show, however, that Jacob Fries had already adopted a kind of relative a priori. Prima facie, Fries followed Kant's synthetic a priori. But he also specified that, according to him, every specific scientific theory, "the theory of electricity or magnetism, for example, may have its own maxims that can gain constitutive relevance" (Pulte 2013, 46). Furthermore, these maxims are relative – they can be revised by new scientific discoveries and theories.

3.2.3. *Grelling and Reichenbach*

We have already noted (in 3.2.1) that it is possible that Reichenbach visited Grelling's philosophical seminar in Munich (1912/13). However, there are independent reasons to believe that Kurt Grelling animated Reichenbach to deal with probability. Recall that as far back as in 1910, Grelling published the paper "The Philosophical Foundations of the Probability Calculus" (1910b), in which he defended the objective interpretation of this discipline against Carl Stumpf's subjectivism. Besides, Grelling linked his thoughts on probability to the problem of induction. These two approaches were the kernel of Reichenbach's theory of probability during his whole career.

In his dissertation (1915), Reichenbach discussed Grelling's paper on induction as well as Fries' *Essay on the Critique of the Principles of the Probability Calculus* (1842) and Ernst Friedrich Apelt's (one of Fries' students) *Theory of Induction* (1854). This cannot be surprising, since, during Reichenbach's work on his dissertation in 1914, Grelling was on hand with help and advice, a fact confirmed in an autobiographical note of Reichenbach dated in 1927: "Probability has to be introduced as a foundation – this objection was already made to me by Grelling in 1914" (HR 044–06–21).

Although Grelling's and Reichenbach's paths did not cross between 1914 and 1926, they kept in touch. Moreover, as we learn from Grelling's postcard to Reichenbach from October 10, 1921, he answered positively Reichenbach's request to use Grelling's "formulation" (HR 015–54–06). We don't know exactly which formulation Reichenbach meant here, though what is clear is that the formulation in question comes from Grelling's paper "Theory of Relativity and Critical Philosophy", which he delivered at the Jacob Fries Society on August 15, 1921, of which Grelling apparently sent a copy to Reichenbach.

Grelling's paper was substantially based on Reichenbach (1920/1965) and was, in fact, the most positive reaction to Reichenbach's book at the time, showing Grelling once again as a man who closely followed the new events in scientifically informed philosophy. His conclusion was that "the theory of relativity seriously undermines the conception of the a priori of the critical philosophy".[6] This position meant an actual break with Leonard Nelson and his group, who clearly failed to adopt "the conventionalist and fallibilist elements of Fries' philosophy of science [as well as] his theory of space". (Pulte 2013, 51) Nelson's philosophy of science remained "radically *conservative*" (original emphasis). It continued to pursue a "certistic" (ibidem, 53) conception of scientific knowledge.

Reichenbach's high regard for Grelling in these years was confirmed in the beginning of February 1923 when Reichenbach started, with Carnap, to prepare the famous Erlangen-workshop on exact philosophy that took place in March the same year. Reichenbach first sent invitation letters to four persons only: Schlick, Paul Hertz, Kurt Lewin, and Kurt Grelling.[7] Grelling's answer to Reichenbach is not preserved, but apparently he declined the invitation.[8] Our guess is that the negative answer came because exactly in the first months of 1923, Grelling had to move from Göttingen to Berlin.

Grelling and Reichenbach resumed their regular meetings only when Reichenbach moved from Stuttgart to Berlin in October 1926. In fact, these were the first steps in setting up the real Berlin Group. This suggestion is supported by the Preface of Reichenbach's *The Philosophy of Space and Time*, which informed the reader that Grelling read the book in manuscript and made "friendly criticism concerning some details" (1928, iii).

3.2.4. *Grelling's Ontological Turn*

After 1936, and until 1939, when he was interned in the south of France, Grelling concentrated his efforts on formal ontology. In this short period, he developed as a systematic scholar, partly abandon his old preoccupation with being a philosophical interpreter. In these few years, he wrote six papers on formal ontology:

1. "The Concept of Gestalt in the light of Modern Logic", with Paul Oppenheim (1937/38/1988);
2. "Logical Analysis of 'Gestalt' as 'Functional Whole'", with Paul Oppenheim (1939/1988);
3. "A Logical Theory of Dependence" (1939/1988);
4. "Melody as Gestalt" (1975);
5. "On Definitions by Equivalence and By Group Invariants" (1969);
6. "On the Logical Relations between Groups and Equivalence" (1970).

Grelling started (with Paul Oppenheim) his ontological investigation from Christian von Ehrenfels' programmatic paper "On Gestalt Qualities" (1890). A typical example of a Gestalt discussed by Ehrenfels is melody: it cannot be reduced to its parts. But Grelling knew quite well that the concept of Gestalt had also been explored by two Berlin psychologists and philosophers who worked in close association with the Berlin Group, Wolfgang Köhler and Kurt Lewin. Among other things, with Max Wertheimer and Kurt Koffka, the two developed the most influential school of Gestalt psychology of the time. More than this, Köhler was the supervisor of the Ph.D. dissertation of two core members of the Berlin Group: Walter Dubislav and Carl Hempel. Besides, as we will see in 3.3.2, between 1917 and 1925, Kurt Lewin worked very closely with Reichenbach and exercised considerable influence on him.

In their paper (1937/38/1988), Grelling and Oppenheim first try to clarify the concept of Gestalt. To this purpose, they adopt the concept of *classifier*, coined by Carl Hempel, which determines the classifications of Gestalt terms, for example, the "pitch" of a musical piece. In a first attempt to define the concept of Gestalt, Grelling and Oppenheim state that "in general, 'Gestalt' can be represented as a classifier whose arguments are complexes and whose values are Gestalt-individuals.[9] The names of sensory qualities, such as colour, smell, taste, and so on, can be seen as classifiers" (1937/38/1988, 196 f.). Second, Gestalts like musical pieces are complexes which have specific articulation. When a melody is played with "the same tempo and dynamics, but on different sorts of instrument and in a different key", we say that these complexes *correspond* to one another. In musical theory, the passage from one key to another is called *transposition*. Grelling's and Oppenheim's final definition of Gestalt is the "invariant of transpositions". According to it, a melody is the Gestalt of a *tone sequence*.

Significantly, the concept of Gestalt is richer in content than the concept of totality. Totality "is completely determined by its parts and is of the same type as these. In this it differs from complexes. . . . As a result, meaningful statements can be made about tone sequences which cannot be made about totalities"[10] (Grelling and Oppenheim 1937/38/1988, 198 f.). In short, Grelling and Oppenheim's "ontological ranking" is: aggregates, complexes, totalities, Gestalts. Still different is what Grelling calls a "determinational system" in which

> there is a division of the whole such that every part of this division stands in the relation *R* to every other, and every object of which stands in the relation *R* to at least one part is itself a part of the whole.[11]
>
> (1937/38/1988, 199)

Importantly, this concept was advanced by the Berlin Gestalt-psychologists Köhler and Koffka and was not used by Ehrenfels. But it had already been used by Grelling's old master in philosophy – Fries (1822, 597).

Grelling and Oppenheim underline that the concept of Gestalt is successfully used not only in psychology and in musicology but also in science, including in physics. "In physics our concept of Gestalt could probably be applied to any field, in structural chemistry to a molecule" (1937/38/1988, 197). For example, cases of determinational systems are "self-regulating wholes" like charged and isolated conductors, atoms, atomic nuclei, molecules, cells, organisms, economies – apparently, both objects of living and inorganic matter as well as artifacts.[12] Following David Hilbert, Grelling calls the sciences which explored them "sciences of the real" (*Realwissenschaften*) (we shall return to them in Section 3.5). Only the objects of "formal sciences", logic and mathematics, cannot be ordered in determinational systems.

This brings us back to the lead claim of the present chapter. In the last years of his life, Grelling advanced a project in ontology that can be successfully used in philosophy of nature (*Naturphilosophie*). Importantly, this project was based on elements that Reichenbach had already developed in the 1920s, to which, as we are going to see in the next section, he also returned in the last days of his life. In other words, Grelling's intensive work on formal ontology after 1936 reveals essential features of the authentic project of the Berlin Group. In what follows, we are going to substantiate this view.

3.3. Hans Reichenbach

3.3.1. Reichenbach's Program for Comparative Analysis of Science

Reichenbach's former students remember him as possessing

> a unique talent for going to the heart of any issue, clearing away peripheral matters and also cutting through the irrelevancies, which made his teaching a model of clarity. It also endowed his creative philosophical work with genuine profundity.
>
> (Salmon 1977, 8)

This characteristic of Reichenbach's approach, however, posed its problems. Pursuing "the heart of any issue" often made him overlook some detail, making him, in this way, a one-dimensional philosopher of nature. More often than not, he followed one line of exploration, shutting his eyes to all others. Among other things, this explains why he didn't recognize influences of Richard von Mises on himself and repeatedly failed to mention books that were close to his conception (for example, Robb 1914, 1924; Lewis 1929). "Strikingly absent is [also] any reference to Birkhoff and von Neumann's 1936 paper on the logical structure of quantum mechanics" (Glymour and Eberhardt 2016).

In 3.2.3, we have already spoken about the principle of relative a priori in science, introduced by Fries and later rediscovered by Reichenbach. One of its implications is that philosophers have to "logically analyze" every new important theory of science. Second, we cannot strictly discriminate one science from another, so that the principles that are valid for one science are void for others. This means that the new philosophers of nature (*Naturphilosophen*) have to examine the principles of all sciences. Ostensibly, Reichenbach hoped that the "logical analysis" of different sciences could bring to light connections between their ever-changing principles (Milkov 2011, 151). He presented this idea in several papers published around 1930: "New Approaches in Science" (1929/1978), "The Philosophical Significance of Modern Physics" (1930/1978), and "Aims and Methods of Modern Philosophy of Nature" (1931/1978) (the latter is sometimes considered the manifesto of the Berlin Group).

Among other things, this task led Reichenbach to transform the Berlin-based Society for Empirical Philosophy, which Reichenbach took over from Joseph Petzoldt in June 1929, into the Berlin Society of Scientific Philosophy. First, by the end of 1931, the name change was carried out, allegedly following a suggestion by David Hilbert. But not just the title of the Society was changed. After February 1932, and until Reichenbach left Berlin in July 1933, the term "empiricism" appeared no longer in any of the titles of the 32 presentations delivered. This tendency persisted in the ten lectures presented after Reichenbach's departure, when the Society was led by Dubislav. Of course, it was not the case that Reichenbach abandoned the method of empiricism in his philosophy of nature – it was his unwavering position. Rather, to Reichenbach, the joint philosophical effort of the best-informed, innovative scientists of the time that had presentations at his Society was of prime importance.[13] Apparently, the tentative task was no less than to discover the relative a priori not only of one science alone but of science in general. This conception perfectly harmonized with the pronounced interdisciplinary profile of the presentations at the Society.

3.3.2. Kurt Lewin's Conception of Genidentity

It is little known today, and even less discussed, that between 1917 and 1925, Reichenbach collaborated with the psychologist, gestalt theorist, and former student of Carl Stumpf, Kurt Lewin.[14] In fact, this was the first variant of the Berlin Group. Reichenbach and Lewin knew each other from 1911 when both had studied in Munich. In 1917, during the Great War, the two worked together for the Prussian Ministry of War and even published a joint "Draft for an Aptitude Test for Radio-Telegraphists" (co-authored by Otto Lipmann) (HR 024–16–02).

Lewin explicitly advanced a theory of science that was to replace the conventional theory of knowledge. The deficiency of the Neo-Kantians is, said Lewin, that their analyses "were still not sufficiently concrete. . . . The[ir] examples often carry the character of mere illustrations" (Lewin 1925/1981, 61). Significantly, as we have already seen (in 3.2.3), the same reproach against Kant was made, but from a different perspective, by Jacob Fries, and against Hermann Cohen and Ernst Cassirer by Nelson (1905) and Grelling (1907).

As Reichenbach wrote in his review (1921) of Lewin's (1920) paper, in Lewin's philosophy of science, the fundamental, "scientific-theoretical equivalent" concepts of different sciences are compared, or *analyzed*. A product of this analysis is the concept of *genidentity*. Lewin introduced it as constitutive both in biology, in physics, and also in chemistry. The point is that the objects and events of both biological and physical world develop in time but at the same time retain their identity. For example, the relation between the egg and the hen is that of "biological genidentity": it represents different stages of development of the same biological individual. An important point is that genidentity is not a logical identity but a relation of the *existing* objects. The problem here is ontological, and Grelling immediately understood this (see Grelling 1925).

But while Lewin insisted that different sciences have structural similarities, he also maintained that we cannot speak of identity between their terms. These refer to different angles, or modalities, of reality. That is why the differences between these structures are considerable. This also concerns the concept of genidentity. In particular, Lewin speaks about "individual genidentity", or about "simple genidentity" in biology, and about "complete genidentity" in physics. Whereas with complete genidentity, we have a monotonic sequence of the slices of objects and events one after another, with simple genidentity, this is not the case. Reichenbach (1924), who, as we are going to see in the next section, adopted Lewin's concept, maintained that whereas the series in physics are continuous, those in biology are discrete.[15]

3.3.3. Reichenbach on Genidentity

Reichenbach grasped the importance of Lewin's conception immediately after he read Lewin's paper (1920) in manuscript. He promptly adopted it in his first book, *The Theory of Relativity and A Priori Knowledge* (1920/1965). In it, the concept of genidentity is presented as an a priori constitutive principle of human knowledge which indicates "how physical concepts are to be connected in sequences in order to define 'the same thing remaining identical with itself in time'" (1920/1965, 53). In fact, this concept refers to a characteristic of physical objects and events that is more fundamental than the simple temporal order. But it is a principle since it is not a necessary condition: we simply assume that it is correct.[16]

Shortly afterwards, however, in *The Axiomatization of the Theory of Relativity* (1924/1969), Reichenbach dropped the concept of genidentity. Arguably, this was an implication of the replacement of the coordinative principles by coordinative definitions and of the conventionalism which Reichenbach adopted under Schlick's influence. Now Reichenbach eliminated the measuring rods and clocks he postulated in his philosophy of nature in (1920/1965, 20) and derived the topological properties of space and time from light-signals.[17] To be more explicit, genidentity was substituted by the "mark principle", which maintains that when an event "is marked at *P*, the mark can also be observed at *P'''*" (1924/1969, 27).

Immediately after finishing the *Axiomatization*, Reichenbach sent a copy of the book to his collaborator at the time, Kurt Lewin. Lewin, however, criticized some of its theses and, importantly, Reichenbach agreed with him. As a result, under Lewin's influence, Reichenbach introduced in his philosophy of nature the fork asymmetry account of time, which came to replace the simple causal chains. The causal chains are to be open. Significantly, this "account relies quite consistently on Lewin's analysis of the splitting and intersecting series" (Padovani 2013, 119).

These changes in Reichenbach's understanding, made under the influence of a "scientist of the real" (see on this concept Section 3.2.4), Kurt Lewin, are clearly discernible in "The Causal Structure of the World" (1925/1978a), where he used the notion of *direction of time* (of time-asymmetry) for the first time in his writings. In parallel, Reichenbach returned to philosophical realism, embracing

> a metaphysical axiom, a belief in the uniformity of the world that cannot be proven and that nonetheless makes a positive assertion about the world. We formulate, in this principle, the most universal feature of the real. It cannot be established upon the basis of the knowing mind; rather, it asserts something about things-in-themselves.
>
> (1925/1978b, 292)

As can be expected, this position brought on severe criticism from Schlick.

After *The Philosophy of Space and Time* was published, Reichenbach was silent on genidentity for years. He returned to this concept on a large scale only in what is often considered his "philosophical testament", *The Direction of Time* (1956).[18] However, whereas in *The Philosophy of Space and Time*, Reichenbach was interested in applying the concept of genidentity to the theory of relativity, in *The Direction of Time*, he applied it to quantum mechanics.

In more concrete terms, in *The Direction of Time*, Reichenbach changed his position from 1925 in the following way. According to the laws of classical mechanics, time can be both reversible and irreversible. Time is only irreversible in quasi-closed (isolated) physical systems which have finite lifetimes. This was his position from 1928. Now, however, Reichenbach held that it is possible that time has different sections, in some of which it has "directions which would be statistically expressed in the same way as for temporally open universe" (1956, 134). In support of this position, Reichenbach referred to a paper by the young Richard Feynman (1949), who argued for particles moving back in time.

This last point brings us to an important characteristic of Reichenbach as a philosopher of nature. This "evangelist of science" (van

Fraassen) managed to discuss, with philosophical methods, problems of the cutting-age science of his time. To start with, he was one of the first philosophers to immediately realize the paradigm-changing significance of Einstein's theory of relativity. Second, as Reichenbach himself maintained, the principle of indeterminacy in quantum mechanics "was initially suggested by [his] philosophical considerations concerning the principle of causality"[19] (1931, 377). His task as philosopher was simply "to point out this possibility" (ibid.). Third, as we just saw, Reichenbach immediately realized the importance of the revolutionary ideas about causality in quantum mechanics of Richard Feynman.

All these facts support Reichenbach's claim that his philosophy works in direct connection with the results achieved by science. However, philosophy is not identical with science. It has its own methods like that of logical and ontological analysis which help to better appreciate the results of science and even to predict them.

3.4. Carl Hempel, the Logic of Explanation, and the General Philosophy of Science

Another scientist who worked with Reichenbach in the early 1920s was the chemist and private scholar Paul Oppenheim (1885–1977). Reichenbach met him around 1921 in Stuttgart and actively encouraged him to develop his ideas in print. In his first book (1926), Oppenheim warmly thanked Reichenbach for his support. Above all, Oppenheim explored the order of sciences. In particular, he was adamant about demonstrating that there is no break between natural science and the humanities. Between 1934 and 1936, Oppenheim worked with Hempel and between 1936 and 1938 with Grelling. After the Second World War, he collaborated with a series of young, talented philosophers like Hilary Putnam and Nicholas Rescher.

In order to better understand the kind of collaboration Oppenheim was engaged in, it can be helpful to turn attention to the information about it provided by Nicholas Rescher, who worked with him on a joint paper (Rescher and Oppenheim 1955):

> What Oppenheim principally contributed to our collaboration were two things: (1) the topic of the investigation, and (2) a guiding concern for structural issues. . . . But beyond this guiding concern for a theory of ordering concepts, Oppenheim made very little substantive contribution to the investigation. He was like an Aristotelian first mover, having set a project in motion (in a rather generally indicated) direction, he stood back and let nature take its course, with minimal interference as long as things kept on track.
>
> (1997, 159)

By the considerable freedom Oppenheim's collaborators enjoyed in their work, it deserves notice that although Oppenheim's joint work with Hempel and Grelling took place in practically the same period of time, it brought about markedly different results. In short, while Hempel pushed a project close to logic, Grelling advanced the already-discussed (in Section 3.2.4) explorations in ontology. This brings us to a discrete split in the Berlin Group into two wings – philosophers of nature (*Naturphilosophen*) (Reichenbach and Grelling) and philosophers of science (Dubislav and Hempel).

To be more explicit, Hempel's object of investigation in (Hempel and Oppenheim 1936a, 1936b/2015) was the *logical* analysis of personal psychology – of the psychological types. He, with Oppenheim, observed a recent change in the concept construction of all sciences but in particular of contemporary typology, which moves from concepts of classification to concepts of topological order, and, further, to metrical concept construction in physics (1936/2015, 373). Hempel also maintained that sciences make use of different kinds of concept formation simply because they are at different stages of development. He, however, underlined that the concept formation of diverse sciences is not of logically different types. Hempel further held that the conventional explorations of concept formation with the help of "traditional logic", but also with the "theory of propositional function", cannot help. In any case, it is inappropriate to treat concepts with the help of classes, propositional functions, or other "rigid conceptual schemes" (Hempel and Oppenheim 1936/2015, 366). Hempel's project was to replace them with "elastic" notions (a term Hempel adopts from Bergson). By way of closing, it is important to notice that precisely the pursuit of logically analyzing the concepts of science later led Hempel in the mid-1940s, motivated by Oppenheim, to explore the logic of confirmation and explanation (Rescher 1997, 161 f.).

As a matter of historical fact, Hempel's slant to logic was developed under the influence of Walter Dubislav, with whom Hempel worked while in Berlin. Among other things, this connection is confirmed by the fact that Hempel read the proofs of Dubislav's *Die Philosophie der Mathematik in der Gegenwart* (1932, v). Another point supporting this claim is that Hempel (1933, 1934) wrote two reviews of Dubislav's *Philosophy of Nature* (1933), assessing this work as "extremely stimulating, concise and clearly written" (Hempel 1933, 56). He was explicit that what distinguished Dubislav's volume from studies in the philosophy of nature by the likes of the Viennese logical positivists (in fact, also of Reichenbach and Grelling) was that it didn't primarily concern itself with specific problems of science. Instead, "it systematically explores the logical and methodological problems of scientific knowledge" (Hempel 1934, 760). As it turned out, this was a program that Hempel himself would follow in his *Philosophy of Natural Science* (Hempel 1966), regarded today as a standard work in the general philosophy of science.

3.5. Epilogue

Some 50 years after the paper of Grelling and Oppenheim was published in *Erkenntnis*, Peter Simons translated it, along with two other papers, into English and commented on them (in Smith 1988, 191–225). In Simons' judgment, the paper of Grelling and Oppenheim "cut ontological ice" (Simons 1988, 161).

As a matter of fact, Simons was part of a new wave of formal ontologists (Barry Smith, Kevin Mulligan, and Dale Jacquette among them) inspired by the metaphysical turn in philosophy in the 1970s and 1980s, epitomized by David Lewis. The new ontologists tried to outline a new direction for philosophy. In particular, they advanced an alternative conception of logic. The problem is that the standard first-order logic as starndardly conceived is not helpful when it comes to treating objects of the real world such as universals, types, and processes. To be sure, its "universe of discourse consists of particular items" only (Smith 2008, 110). One can successfully use the conventional logic mainly in mathematics, the objects of which are not situated in space and time. For the objects of the real world, we will need an alternative logic, for example, a logic along the lines sketched in Smith (2008); but also formal ontology.

It is true that the new wave of ontologists didn't produce works in the realm of philosophy of nature or philosophy of science and had practically no influence on them. Be this as it may, their arguments indicate the direction in which Reichenbach's wing of the Berlin Group developed many years before – the realist ontology of nature. It is a matter of fact that the members of the Berlin Group in general worked with representatives of what Grelling has called "sciences of the real" (*Realwissenschaften*) (see Section 3.2.4), the objects of which *are* situated in space and time. This claim is also supported by the fact that, as we have seen in Section 3.3.2, between 1920 and 1925, Reichenbach collaborated with the psychologist and gestalt theorist Kurt Lewin (in fact, much more closely than with Moritz Schlick). Grelling, on his side, collaborated for years with the medical doctor Heinrich Poll (1933). Finally, the Berlin Society for Scientific Philosophy, as different from the Berlin Group, was led formaly for years by the Nobel Prize laureat in physiology and medicine Friedrich Kraus. The meetings of the Society took part in the famous Charité hospital whose director was Kraus.

We see the specific approach of the philosoph of nature wing of the Berlin Group as a promising venue that can also help to improve the relation between philosophy and science today. It is an alternative both to the philosophy of science developed in the wake of the logical positivist and to Hempel's general philosophy of science exploring, among other things, explanations in science. In this sense, our hope for a new start of the philosophy of nature is connected with a turn *zurück zu Reichenbach*. There could be no better guide in such a twist than Kurt Grelling.

Notes

1. A similar position is described and defended in Clark Glymour's chapter in this volume.
2. About this interpretation of quantum mechanics and what role positivism played in the story, see Jan Faye and Rasmus Jaksland's chapter in this volume.
3. After the discovery of the paradox of classes by Russell, David Hilbert and his group concentrated efforts in avoiding paradoxes by axiomatization of mathematical disciplines and making the definitions of their terms more precise.
4. The book was immediately translated into Swedish and published in Stockholm in the same year and into French in the next year in Zürich.
5. This is clear, for example, from Reichenbach's short review of the history of logic in Schilpp's Russell volume, in which he states that "we may date the modern period of logic" from George Boole who started the first phase of the modrn logic. Russell's *The Principles of Mathematics* (1903) signals "the second phase of modern logic" (Reichenbach 1944, 24 f.) No word about Frege.
6. Nachlass Nelson, 90 Ne 1, Nr. 388, pp. 243–46; here p. 243.
7. Cf. Reichenbach's letter to Schlick of 03.02.1923. (Schlick's Archive)
8. This is clear from the prospective program of the Workshop Carnap prepared two weeks later, on 19.02.1923, in which Grelling's name is not included. (HR-015–50–03)
9. "Gestalt-individual" designates, among other things, every particular artifact, like Mozart's *A Little Night Music* or Schubert's *Ave Maria*.
10. It deserves notice that Wittgenstein's famous criticism of Russell's *Theory of Knowledge* project in June 1913, based on ideas of Frege, was made from a related perspective of criticism of the complexes in defense of totalities; see Milkov (2013b). Grelling knew that the radical discrimination between complexes and totalities was also drawn by the Polish logicians Tarski and Leśniewski.
11. In another place, we called this type or relation *reciproca tantum*, see Milkov (2020, 98, 106, 234).
12. Moritz Schlick (1935/1979), in contrast, refused to accept wholes in his philosophy of science.
13. On the qualification of the presenters at the Berlin Society of Scientific Philosophy, see Milkov (2013a, 10).
14. In 1911/12, Reichenbach himself studied psychology with Carl Stumpf in Berlin.
15. To be precise, in (1921), Reichenbach underestimated the difference between these two types of identity but soon (in 1924) corrected his view with Lewin's support; see Padovani (2013, 111–112, n. 42)
16. Grelling held the problem of genidentity at the center of his attention all the time. In Grelling (1936), for example, he connected genidentity with Leibnitz's principle of identity of indiscernibles. Grelling defended the latter against the formalist critic of Wittgenstein and Friedrich Waismann. Importantly, Grelling's 1936 paper can be seen as a prelude to his ontological papers of 1937–39 we discussed in Section 3.2.4.
17. See Reichenbach's letter to Schlick of January 22, 1922 (Schlick collection). The only author of the time who noted this change in Reichenbach's view was Grelling, who mentioned it in his review of Lewin (1922) (1925, columns 688–689). Significantly, Grelling underlined in it the strong relation of Lewin's book to Reichenbach's studies.

18. Be this as it may, between 1928 and 1953, Reichenbach continued to be engaged with ontological analyses. In *Experience and Prediction*, for example, he insisted, in an important discussion with the logical positivists, that "the transition from external things to impressions [perception] cannot be interpreted as reduction; it is of another type of [onto]logical structure" (Reichenbach 1938, 105). As an illustration, as if starting from Grelling's and Oppenheim's (1937/38) arguments for Gestalt complexes we discussed in Section 3.2.4, Reichenbach suggested the example with the melody: it is not reducible to its elements.
19. It is worth mentioning that other scientists and scientific theorists denied this claim.

References

Andersen, R. (2012), 'Has Physics Made Philosophy and Religion Obsolete?', *The Atlantic*, April 23.

Apelt, E. F. (1854), *Theorie der Induktion*. Leipzig: Engelmann.

Corry, L. (2004), *David Hilbert and the Axiomatization of Physics (1898–1918)*. Dordrecht: Kluwer.

Dubislav, W. (1932), *Die Philosophie der Mathematik in der Gegenwart*. Berlin: Junker und Dünnhaupt.

—— (1933), *Naturphilosophie*. Berlin: Junker und Dünnhaupt.

Ehrenfels, Ch. von (1890), 'Über Gestaltqualitäten', *Vierteljahrschrift für wissenschaftlichen Philosophie*, 14: 249–292.

Feynman, R. (1949), 'The Theory of Positrons', *Physical Review*, 76: 749–759.

Friedman, M. (2001), *Dynamics of Reason*. Stanford, CA: CSLI Publications.

—— (2005), 'Ernst Cassirer and the Contemporary Philosophy of Science', *Angelaki*, 10: 119–128.

Fries, J. F. (1822), *Mathematische Naturphilosophie*. Heidelberg: Winter.

—— (1842), *Versuch einer Kritik der Prinzipien der Wahrscheinlichkeitsrechnung*. Leipzig: Braunschweig.

Glymour, C. and Eberhardt, F. (2016), 'Hans Reichenbach', *The Stanford Encyclopedia of Philosophy*, E. N. Zalta (ed.), https://plato.stanford.edu/entries/reichenbach/.

Grelling, K. (1907), 'Das gute, klare Recht der Freunde der anthropologischen Vernunftkritik verteidigt gegen Ernst Cassirer', *Abhandlungen der Fries'schen Schule*, 2 (2): 155–190.

—— (1910a), *Die Axiome der Arithmetik mit besonderer Berücksichtigung der Beziehungen zur Mengenlehre*. Ph.D. Thesis. Dieterichsche Universitäts-Buchdruckerei, Göttingen.

—— (1910b), 'Die philosophischen Grundlagen der Wahrscheinlichkeitsrechnung', *Abhandlungen der Fries'schen Schule*, n.s., 3 (3): 440–478.

—— (1916a), 'Philosophische Grundlagen der Politik', *Sozialistischen Monatshefte*, 22 (3): 1045–1065.

—— (1916b), *Anti-j'accuse. Eine deutsche Antwort*. Zürich: Orell.

—— (1925), 'Kurt Lewin, *Der Begriff der Genese in Physik, Biologie und Entwicklungsgeschichte*', *Deutsche Literaturzeitung*, 14: 685–690.

—— (1932/1933), 'Bemerkungen zu Dubislavs "Die Definition"', *Erkenntnis*, 3: 189–200.

—— (1936), 'Identitas Indiscernibilium', *Erkenntnis*, 6: 252–259.

—— (1939/1988), 'A Logical Theory of Dependence', in B. Smith (ed.), *Foundations of Gestalt Theory*. Munich and Vienna: Philosophia Verlag, pp. 217–226.

—— (1969), 'On Definitions by Equivalence Classes and by Group Invariants', *Methodology and Science*, 2: 116–122.

—— (1970), 'On the Logical Relations between Groups and Equivalence Relations', *Methodology and Science*, 3: 5–17.

—— (1975), 'Melody as Gestalt', Methodology and Science, 8: 13–23.

Grelling, K. and Oppenheim, P. (1937/38/1988), 'The Concept of Gestalt in the Light of Modern Logic', in B. Smith (ed.), *Foundations of Gestalt Theory*. Munich and Vienna: Philosophia Verlag, pp. 191–205.

—— (1939/1988), 'Logical Analysis of "Gestalt" as "Functional Whole"', in B. Smith (ed.), *Foundations of Gestalt Theory*. Munich and Vienna: Philosophia Verlag, pp. 210–216.

Hempel, C. G. (1933), 'Walter Dubislav, *Naturphilosophie*', *Jahrbuch über die Fortschritte der Mathematik*, 59 (1): 56–57.

—— (1934), 'Walter Dubislav, *Naturphilosophie*', *Deutsche Literaturzeitung*, 55: 759–762.

—— (1966), *Philosophy of Natural Science*. Englewood Cliffs, NJ: Prentice-Hall.

Hempel, C. G. and Oppenheim, P. (1936a), *Der Typusbegriff im Lichte der neuen Logik: wissenschaftstheoretische Untersuchungen zur Konstitutionsforschung und Psychologie*. Leiden: Sijthoff.

—— (1936b/2015), 'Die logische Bedeutung des Typusbegriffs', in N. Milkov (ed.), *Die berliner Gruppe*. Hamburg: Felix Meiner Verlag, pp. 365–375.

—— (1948), 'Studies in the Logic of Explanation', *Philosophy of Science*, 15 (2): 135–175.

Howard, D. (2003), 'Two Left Turns Make a Right: On the Curious Political Career of North American Philosophy of Science at Midcentury', in G. Hardcastle and A. W. Richardson (eds.), *Logical Empiricism in North America*. Minnesota, MN: University of Minnesota Press, pp. 25–93.

Kant, I. (1880/1968). *Logik: Ein Handbook zu Vorlesungen, Herausgeben von G. B. Jäsche: Gesammelte Schriften, Akademie-Ausgabe, Bd. IX*. Berlin: Walter de Gruyter.

Lewin, K. (1920), 'Die Verwandtschaftsbegriffe in Biologie und Physik und die Darstellung vollständiger Stammbäume', *Abhandlungen zur theoretischen Biologie*, 5: 38–73.

—— (1922), *Der Begriff der Genese in Physik, Biologie und Entwicklungsgeschichte*. Berlin: Springer.

—— (1925/1981), 'Über Idee und Aufgabe der vergleichenden Wissenschaftslehre', *Symposion*, 1 (1): 61–93.

Lewis, C. I. (1929), *Mind and the World Order: Outline of a Theory of Knowledge*. New York: Charles Scribners.

Meyerson, É. (1930), *Identität und Wirklichkeit*. Translated by Kurt Grelling. Leipzig: Akademische Verlagsgesellschaft.

Milkov, N. (ed.) (2011), *Hans Reichenbach: Ziele und Wege der heutigen Naturphilosophie*. Hamburg: Felix Meiner Verlag.

—— (2013a), 'The Berlin Group and the Vienna Circle: Affinities and Divergences', in N. Milkov and V. Peckhaus (eds.), *The Berlin Group and the Philosophy of Logical Empiricism*. Dordrecht: Springer, pp. 3–32.

—— (2013b), 'The Joint Philosophical Program of Russell and Wittgenstein and Its Demise', *Nordic Wittgenstein Review*, 2: 81–105.

——— (2015), 'Einleitung: Die Berliner Gruppe des logischen Empirismus', in N. Milkov (ed.), *Die Berliner Gruppe: Texte zum Logischen Empirismus. Eine Anthologie.* Hamburg: Felix Meiner Verlag, pp. ix–lxi.

——— (2020), *Early Analytic Philosophy and the German Philosophical Tradition.* London: Bloomsbury.

Nelson, L. (1905), '*Logik der reinen Erkenntnis*, von Hermann Cohen', *Göttingische gelehrte Anzeigen*, 167: 610–630.

Oppenheim, P. (1926), *Die natürliche Anordnung der Wissenschaft.* Jena: Gustav Fischer.

Padovani, F. (2013), 'Genidentity and Topology of Time: Kurt Lewin and Hans Reichenbach', in N. Milkov and V. Peckhaus (eds.), *The Berlin Group and the Philosophy of Logical Empiricism.* Dordrecht: Springer, pp. 97–122.

Peckhaus, V. (1994), 'Von Nelson zu Reichenbach: Kurt Grelling in Göttingen und Berlin', in L. Danneberg et al. (eds.), *Hans Reichenbach und die Berliner Gruppe.* Braunschweig: Vieweg, pp. 53–86.

——— (2013), 'The Third Man: Kurt Grelling and the Berlin Group', in N. Milkov and V. Peckhaus (eds.), *The Berlin Group and the Philosophy of Logical Empiricism.* Dordrecht: Springer, pp. 231–243.

Poll, H. (1933), 'Genetik und Melistik als Grundlage des ärztlichen Denkens', in *Einheitsbestrebungen in der Medizin.* Dresden and Leipzig: Steinkopff, pp. 219–228.

Pulte, H. (2013), 'J.F. Fries' Philosophy of Science, the New Friesian School and the Berlin Group: On Divergent Scientific Philosophies, Difficult Relations and Missed Opportunities', in N. Milkov and V. Peckhaus (eds.), *The Berlin Group and the Philosophy of Logical Empiricism.* Dordrecht: Springer, pp. 43–66.

Reichenbach, H. (1920/1965), *The Theory of Relativity and A Priori Knowledge.* Berkeley and Los Angeles: University of California Press.

——— (1921), 'Kurt Lewin, *Die Verwandtschaftsbegriffe in Biologie und Physik und die Darstellung vollständiger Stammbäume*', *Die Naturwissenschaften*, 9: 51.

——— (1924/1969), *The Axiomatization of the Theory of Relativity.* Berkeley and Los Angeles: University of California Press.

——— (1924), 'Kurt Lewin, *Der Begriff der Genese in Physik, Biologie und Entwicklungsgeschichte*', *Psychologische Forschung*, 5: 188–190.

——— (1925/1978a), 'The Causal Structure of the World and the Difference between Past and Future', in M. Reichenbach and R. S. Cohen (eds.), *Hans Reichenbach: Selected Writings 1909–1953, Volume Two.* Dordrecht: D. Reidel, pp. 81–119.

——— (1925/1978b), 'Metaphysics and Natural Science', in M. Reichenbach and R. S. Cohen (eds.), *Hans Reichenbach: Selected Writings 1909–1953, Volume One.* Dordrecht: D. Reidel, pp. 283–297.

——— (1928), *Philosophie der Raum-Zeit-Lehre.* Berlin: De Gruyter.

——— (1929/1978), 'New Approaches in Science', in M. Reichenbach and R. S. Cohen (eds.), *Hans Reichenbach: Selected Writings 1909–1953, Volume One.* Dordrecht: D. Reidel, pp. 245–257.

——— (1930/1978), 'The Philosophical Significance of Modern Physics', in M. Reichenbach and R. S. Cohen (eds.), *Hans Reichenbach: Selected Writings 1909–1953, Volume One.* Dordrecht: D. Reidel, pp. 304–323.

—— (1931/1978), 'Aims and Methods of Modern Philosophy of Nature', in M. Reichenbach and R. S. Cohen (eds.), *Hans Reichenbach: Selected Writings 1909–1953, Volume One*. Dordrecht: D. Reidel, pp. 359–388.

—— (1938), *Experience and Prediction*. Chicago: University of Chicago Press.

—— (1944), 'Bertrand Russell's Logic', in P. Schilpp (ed.), *The Philosophy of Bertrand Russell*. Evanston, IL: Northwestern University Press, pp. 23–54.

—— (1956), *The Direction of Time*. Berkeley: University of California Press.

Rescher, N. (1997), 'H2O: Hempel-Helmer-Oppenheim, an Episode in the History of Scientific Philosophy in the 20th Century', *Philosophy of Science*, 64 (2): 334–360.

Rescher, N. and Oppenheim, P. (1955), 'Logical Analysis of Gestalt Concepts', *British Journal for the Philosophy of Science*, 6: 89–106.

Robb, A. A. (1914), *A Theory of Time and Space*. Cambridge: Cambridge University Press.

—— (1924), *The Absolute Relations of Time and Space*. Cambridge: Cambridge University Press.

Russell, B. (1903), *The Principles of Mathematics*. London: Allen & Unwin.

Salmon, W. (1977), 'The Philosophy of Hans Reichenbach', in W. C. Salmon (ed.), *Hans Reichenbach: Logical Empiricist*. Dordrecht: Reidel, pp. 1–84.

Schlick, M. (1935/1979), 'On the Concept of Wholeness', in H. Mulder and B. F. B. van de Velde-Schlick (eds.), *Moritz Schlick: Philosophical Papers, Volume II, 1925–1936*. Dordrecht: D. Reidel, pp. 388–399.

Simons, P. M. (1988), 'Gestalt and Functional Dependence', in B. Smith (ed.), *Foundations of Gestalt Theory*. Munich and Vienna: Philosophia Verlag, pp. 158–190.

Smith, B. (ed.) (1988), *Foundations of Gestalt Theory*. Munich and Vienna: Philosophia Verlag.

—— (2008), 'The Benefits of Realism: A Realist Logic with Applications', in K. Munn and B. Smith (eds.), *Applied Ontology*. Frankfurt: Ontos, pp. 109–124.

Weinberg, S. (1992), *Dreams of a Final Theory*. New York: Vintage Books.

4 The Selection of Facts in Poincaré and Neurath

Katherine Dunlop

4.1. Introduction

The influence of Henri Poincaré's conventionalist view of geometry on the logical empiricists has been seen in their writings on physical theory, which draw on Chapters III and V of *Science and Hypothesis* (*SH*). Important studies of the reception of Poincaré's ideas by Carnap, Schlick, and Reichenbach show that these thinkers tended to conflate Poincaré's conventionalism with Duhem's holism. In Section 4.2 of this chapter, I show how close Poincaré's view appears to Duhem's from the narrowly focused perspective of Reichenbach and Schlick, which elides the broader epistemological context of Poincaré's arguments.

The fuller understanding of Poincaré's view that has emerged in recent decades recognizes the limited scope of his conventionalism, in contrast to Duhemian holism, and the importance of group-theoretic considerations for his argument.[1] My aim in this chapter is to show how Poincaré's view of the conventionality of geometry, thus understood, exemplifies a broader view of the relationship of theory to data. In Section 4.3, I argue that for Poincaré, the perceptual experiences which are partitioned into equivalence classes in the various metric geometries are themselves the products of certain conventional assumptions. This renders futile any hope of basing geometry directly on experience (without any appeal to convention), and more generally tells against the existence of unprejudiced data on which to base choices between theories.[2] I do not take Poincaré's point to be that data are theory laden (which would obscure the distinction between his view and Duhem's), but rather that before experience can confront theory, it must undergo regimentation, which involves a selective neglect of some perceptual input.

One would not necessarily expect Otto Neurath to have had more regard than Schlick or Reichenbach for the distinctive features of Poincaré's view. Neurath is known above all for the simile of sailors rebuilding their boat at sea, as used by Quine to illustrate an antifoundationalist holism more radical than Duhem's.[3] Moreover, in his mature writings, Neurath frames the question of foundationalism in terms of the relationship between "protocol statements" and "content statements", whereas Poincaré has little regard for such formalizations of reasoning. The burden of Section 4.4 of this chapter is to show that, nonetheless, Neurath's denial of an empirical foundation for science can be explicated in terms very similar to Poincaré's. Neurath's early writings express the impossibility of an empirical basis in terms of the "selection" to which phenomena or "facts" are subject, terminology which is also used by Poincaré and, I show, derives from Ernst Mach.

This important point of agreement between Neurath and Poincaré – that experience must undergo a selective sort of regimentation before it can confront theory – raises the question of influence. I think the evidence permits us to conclude only that both Poincaré and Neurath took Mach's view in the same novel direction, not that Poincaré's group-theoretic treatment of geometry was a significant influence on Neurath. But I think it is still worth observing that Neurath gives expression (even if inadvertently) to Poincaré's distinctive version of conventionalism. Looking beyond Poincaré's reception by Reichenbach and Schlick gives us a fuller understanding of conventionalism's legacy in logical empiricism.

4.2. Chapter V of *Science and Hypothesis* as a Source for the Logical Empiricists

In the logical empiricists' best-known accounts of Poincaré's conventionalist view, his main argument for denying that the geometry of (physical) space could be empirically determined is that empirical measurement of spatial dimensions is possible only through conventions specifying the effects of physical "factors" on measured magnitudes. Thus, nothing precludes the adoption of an alternative coordinative scheme, within which the same empirical findings (readings of measuring instruments) would be evidence that a different geometry holds.

Reichenbach's *Philosophy of Space and Time* (1928/1958) contains a particularly clear presentation of these ideas. Reichenbach emphasizes that a statement to the effect that a particular geometry is true of physical space is "meaningless" taken in isolation or, more specifically, apart from a definition of congruence. The application of geometry to physical space requires, first, the designation of a "physical length" as unit of length and the stipulation that certain physical objects shall be considered equal (to one another) in length when they are at different places (1928/1958, 15–16). Second, the claim that a geometry G applies is parsed as the claim that a geometry G' is found by measurement to apply and a force F is defined as the cause of the difference between G and G'. Now F can be set equal to 0 in the equation $G = G' + F$ by defining a rigid body as a solid body "not affected by differential forces, or concerning which the influence of differential forces has been eliminated by corrections" (where a "differential" force has different effects on different materials).

Reichenbach indeed claims that "the whole system of physics" is based on this definition (which is nonetheless "not explicitly given in the literature on physics", 1928/1958, 23), since it permits generalizations to be formulated with the universality demanded of physical laws.[4] But the demand for universality can just as well be met by positing "universal forces" which act indifferently on all materials to deform them in the same proportions. As Reichenbach writes in another context, the assumption of universal forces "means merely a change in the coordinative definition of congruence" (1951, 133). In particular, universal forces can be supposed to deform bodies in whatever proportion is required to make the sum $G' + F$ equal to a certain geometry G: "there is nothing wrong with a coordinative definition established on the requirement that a certain kind of geometry is to result from the measurements".

Although Moritz Schlick does not speak in terms of "coordinative definitions", he likewise maintains, in *Space and Time in Contemporary Physics*, that "there is no meaning in talking of an absolute geometry of 'space', omitting all reference to physics and the behavior of physical bodies" (1917/1920, 47). All measurement of spatial distance, in particular, "is performed by placing one body against another", and for such a comparison "to become a measurement, it must be *interpreted* by taking due account of certain [physical] principles" (1917/1920, 50). Two years earlier, in what Michael Friedman identifies as "the first article on relativity theory within the tradition of logical positivism", Schlick makes the point that the physical principles in question can be chosen so that a certain geometry is to be made to agree with experience:

> We are always measuring, as it were, the mere product of two factors, namely the spatial properties of bodies and their physical properties in the narrower sense, and we can assume one of these two factors as we please, so long as we merely take care that the product agrees with experience, which can then be attained by a suitable choice of the other factor.
>
> (1915/1979; quoted and translated in
> Friedman 1995–1996/1999, 72)

It is easy to see the roots of this conception in Chapter V of *SH*, "Experience and Geometry", which is dedicated to showing "that the principles of geometry are not experimental facts and that in particular Euclid's postulate cannot be proven experimentally" (1902/1929, 81). In this chapter, Poincaré explicitly asserts that "[e]xperiments only teach us the relations of bodies to one another; none of them bears or can bear on the relations of bodies with space, or on the mutual relations of different parts of space" (1902/1929, 86). Since empirical measurement is always only of "the relations of bodies", any putative empirical determination of geometrical properties necessarily involves presuppositions about the behavior of bodies. Poincaré raises the possibility that "physical properties in the narrower sense" can be suitably chosen (to use Schlick's phrasing) with respect to the question of whether "astronomical observations might enable us to decide between the three geometries":

> If Lobachevski's geometry is true, the parallax of a very distant star will be finite; if Riemann's is true, it will be negative. . . . But in astronomy "straight line" means simply "path of a ray of light". If therefore negative parallaxes were found, or if it were demonstrated that all parallaxes are superior to a certain limit, two courses would be open to us; we might either renounce Euclidean geometry, or else modify the laws of optics and suppose that light does not travel rigorously in a straight line.

According to Poincaré, it is "needless to add that everyone would regard the latter solution as the more advantageous", and it follows that the "Euclidean geometry has nothing to fear from fresh experiments" (1902/1929, 81).

This discussion of putative empirical determination of geometry is brief (roughly ten lines of text), but Poincaré also develops a detailed illustration of how the same empirical data can be taken to establish different geometries, under appropriate assumptions about the effects of forces on bodies.

Poincaré's argument in Chapter V relies on the explanation of geometry's possibility given in the preceding chapter, "Space and Geometry", according to which geometry is not "a frame imposed on *each* of our representations, considered individually" but rather "the résumé of the laws according to which these images succeed each other" (1902/1929, 75). To show that we can imagine "a series of representations, similar in all points to our ordinary representations, but succeeding each other according to laws different from those to which we are accustomed", Poincaré introduces the famous example of a "world enclosed in a giant sphere", in which temperature varies as a certain function of distance from the sphere's center and makes all bodies expand and contract, immediately upon being moved, in the same proportion. Passing over details for now, the gist of Poincaré's argument is that the possibility of geometry requires that at least some objects must be regarded as undergoing changes of position without changing their form, that is, as rigid solids. Poincaré maintains that the inhabitants of the world enclosed in the sphere would treat certain changes in the external environment as merely changes of position and thus develop a geometry; however, their geometry "will not be, as ours is, the study of the movements of our rigid solids" but rather of that of solids "subjected to unequal dilatation in exact conformity to [this] law of temperature" (1902/1929, 77–78).

This scenario seems to have inspired, for instance, Reichenbach's discussion of the relevance of physical factors to geometry,[5] for Reichenbach begins by considering a deformation of measuring rods "caused by a physical factor, for instance by a source of heat under the plane, the effects of which are concentrated in the middle area" (1928/1958, 12). Reichenbach then argues that such a deformation would be noticeable in virtue of the differing expansion of measuring instruments made of different materials. Thus, the forces that are supposed to deform bodies in whatever proportion is required to make measurements bear out a particular geometry must instead be universal forces (1928/1958, 13).

Although Chapter V of *SH* is thus a source for the logical empiricists' accounts of geometric conventionalism, it is by now widely recognized that they conflate Poincaré's conventionalism with Duhemian holism. Michael Friedman, who has particularly emphasized how these accounts overlook the limited scope of Poincaré's argument and the special status it confers on geometry, puts this point in terms of the "observational equivalence" of Euclidean and non-Euclidean geometries (see also Stump 1989; Ben-Menahem 2006):

> [T]he logical positivists' argument from observational equivalence is in no way a good argument for the conventionality of geometry, at least as this was understood by Poincaré himself. For the argument from observational equivalence has no particular relevance to physical geometry and can be applied equally well to *any* part of our physical theory. The argument shows only that geometry considered in isolation has no empirical consequences: such consequences are only possible if we also add further hypotheses about the behavior of bodies. But this point is completely general and is today well-known as the Duhem-Quine thesis: *all* individual physical hypotheses require further auxiliary hypotheses in order to generate empirical consequences. Poincaré's own conception, in contrast, involves a very special status for physical geometry.
>
> (1995–1996/1999, 73)

4.3. Poincaré's Conventionalist Foundations for Geometry

4.3.1. *Duhem's Criticism of Poincaré: Mistaking Theory for Convention*

A different sort of argument for geometric conventionalism comes into view when we consider the distinction that both Duhem and Poincaré draw between "theoretical" or "scientific" facts, on the one hand, and "practical" or "crude" ones, on the other. That Neurath was in a position to appreciate this kind of argument is suggested by the importance he places on the distinction, as motivating conventionalist views. Neurath writes that the "very fact that the relatively great stability of common language statements is not linked [*ne s'accompagne pas*] with a precision equal to that of scientific formulas, will doubtlessly continue to provoke research like that originated by Mach, Poincaré, Duhem" and others "before one had the present resources of logico-scientific analysis at one's disposal" (1936b/1983, 151). That Neurath presents this distinction as a common inspiration for Poincaré and Duhem is striking, because as I show in this section, Duhem himself highlights the difference between his and Poincaré's views of the relationship between practical and theoretical facts. I will argue in sections 4.3.2 and 4.3.3 that Duhem correctly identifies the point of disagreement.

In *Aim and Structure of Physical Theory*, Duhem stresses the "extremely great difference" between the "concrete facts" observed by the physicist and the theorist's "numerical symbols". According to Duhem, the theorist "introduces the circumstances of an experiment" into calculations by replacing the concrete facts with the numerical symbols. Conversely, "in order to verify the result that a theory predicts for that experiment, a translation exercise must transform a numerical value into a reading formulated in experimental language". The "rendering of these two translations in either direction" is made possible by the "method of measurement", which can thus be considered as a "dictionary". But, Duhem maintains, "translation is treacherous: *traduttore, traditore*. There is never a complete equivalence between two texts when one is a translated version of the other" (1914/1991, 133).

In Duhem's account, "complete equivalence" is impossible because the imprecision of "practical facts" makes it possible to translate any practical fact by "an infinity of different theoretical facts". Duhem takes as his example the practical fact whose translation is "The temperature is to be distributed in a certain manner over a certain body". In the practical fact, the body is not a "geometrical solid" but a "concrete block": "However sharp its edges, none is a geometrical intersection of two surfaces; instead, these edges are more or less rounded and dented spines". Hence, the thermometer does not give us "the temperature at each point" (which the theoretical fact expresses) but "a sort of mean temperature relative to a certain volume whose extent cannot be too exactly fixed". Even more importantly, the practical fact does not "assert that this temperature is a certain number to the exclusion of any other number": rather than asserting "that this temperature is strictly equal to ten degrees", we "can only assert that the difference between this temperature and ten degrees does not exceed a certain fraction of a degree depending on the precision of our thermometric methods" (1914/1991, 134). Thus, different theoretical facts concerning the numerical value of a quality such as temperature or quantity such as length all translate the same practical fact if the measuring instrument is not precise enough to discriminate between them.

Duhem takes Poincaré to hold, in contrast, that physical theory "should be simply a vocabulary permitting one to translate concrete facts into a simple and convenient conventional language". (The language of physics would then be like any "technical language employed in the diverse arts and trades" not only in that "the initiated can translate it into [concrete] facts" but also in that "a given sentence of a technical language expresses a specific operation performed on very specific objects"; 1914/1991, 149).

Duhem's main objection to this view is that it does not acknowledge the paramount importance of physical theory for the measurement procedure that translates between theoretical and practical facts. In the view Duhem attributes to Poincaré, to say that a current is passing through a wire is just "a conventional manner of expressing the fact that the magnetized little bar of the galvanometer has deviated". Duhem first objects that the (theoretical) fact of the current's occurrence cannot be equivalent to the (practical) fact of the magnet's deviation, because the experimenter may assert that "The current is on, and yet the magnet has not deviated; the galvanometer shows some defect". This could happen if, for instance,

> he has observed that in a voltameter, placed in the same circuit as the galvanometer, bubbles of gas were being released; or else, that an incandescent lamp inserted on the same wire was glowing; or else, that a coil around which this wire was wrapped was becoming warm; or else, that a break in the conductor was accompanied by sparks; and because, in virtue of accepted theories, each of these facts as well as the deviation of the galvanometer may be translated by the words 'the current is on'.
>
> (Duhem 1914/1991, 150)

According to Duhem, Poincaré recognizes that the "technical" formulation "a certain wire carries a current of so many amperes" expresses not a "single" concrete fact but "an infinity of" possible facts like these. For in his 1902 article, "Sur la valeur objective des théories physiques" (reprinted under the title "Is Science Artificial?" as Ch. X of 1905/1929), Poincaré claims that "[t]here is in this circuit a current of so many amperes" is "a statement which agrees with a very large number of absolutely different brute facts" (about the behavior of various measuring instruments). Poincaré holds that what makes it possible to express this diversity of facts by the same proposition is our acceptance of "a law according to which every time a certain mechanical effect is produced a certain chemical effect is produced on its side" (quoted in Duhem 1914/1991, 151). Duhem objects that the relations among the "diverse experimental laws" are "precisely what everybody calls 'the theory of the electric current'", and "it is precisely because this theory is assumed constructed that the words 'there is a current of so many amperes in the wire' may condense so many different significations". Thus, the existence of the "clear and precise" language in which scientific facts are expressed "presupposes the creation of a physical theory", and therefore the scientist's role is "not limited to creating" such a language (*ibid.*).

4.3.2. How Geometrical Conventions Are Subject to Empirical Test for Poincaré

Duhem charges, in summary, that Poincaré "paid insufficient attention to the all-pervasiveness of what we might call the 'irreducible theoreticity' of scientific reasoning" (as Thomas Uebel puts it; 1998, 80). I will argue, first of all, that the disagreement between Duhem and Poincaré appears less sharp when we take a broader view of Poincaré's philosophy. But I hold that ultimately Duhem is right to claim that he finds a role for theory where Poincaré finds only convention.

Poincaré does indeed claim in the article cited by Duhem that "*[t]he scientific fact is only the crude fact translated into a convenient language*" and that "*all the scientist creates in a fact is the language in which he enunciates it*" (1905/1929, 330–332, original emphasis). But his objective is to distinguish his view from the "nominalism" of Édouard LeRoy, according to which "science is only a rule of action" rather than a "means of knowledge" (1905/1929, 323–324).[6] Against LeRoy, Poincaré insists that the "rules" that constitute science are not arbitrary conventions. Thus, when Poincaré claims that the scientist only creates the language in which scientific facts are expressed, he does not mean to deny that the scientist also creates a theory, but only to deny that the scientist "creates without restraint" the scientific fact itself (1905/1929, 325).

Poincaré holds, more specifically, that the scientist does not create the scientific fact because "it is the crude fact which imposes it upon him" (*ibid.*). The scientist's freedom to arbitrarily create a language, or rules, is constrained by the requirement that the rules must "succeed", "generally at least". To succeed is to yield fulfilled predictions. For example,

> [i]f I say, to make hydrogen cause an acid to act on zinc, I formulate a rule which succeeds; I could have said, make distilled water act on gold; that also would have been a rule, only it would not have succeeded.

It is because science thus "foresees" that "it can be useful and serve as rule of action" (1905/1929, 324). In the case of the "law according to which every time a certain mechanical effect is produced a certain chemical effect is produced on its side", its assumption is licensed by "very numerous previous experiments" whose outcome it correctly predicted. If "the law should one day be found false", that is, if "it was perceived that the concordance of the two effects, mechanical and chemical, is not constant", then "it would be necessary to change the scientific language to free it from a grave ambiguity" (329).

Every law or principle governing the use of scientific terminology, however general, is held to this same standard. But the utility of principles that are more general or abstract manifests itself (or fails to) in more subtle ways. As Uebel points out, "all along" Poincaré "recognized that [conventionally adopted principles] might conceivably cease to be useful in extending the application of theories and aiding their unification". Thus even definitions are controlled by "crude facts" or experience, but only as they are incorporated into theories; they can be regarded as "holistically refutable hypotheses", as on Duhem's view, rather than as "irrefutable singular principles" (Uebel 1998, 82).

Although Poincaré's view thus agrees with Duhem's at this general level, for our purposes the differences between them are much more important. Poincaré himself makes very clear that geometrical conventions are subject to tests of their utility, but in a less direct manner than physical principles. He concludes Part III of *SH*, "Force", by remarking that insofar as the "postulates of mechanics possess a generality and a certainty lacked by the experimental verities whence they are drawn, this is because they reduce in the last analysis to a mere convention". While we have "the right to make" this convention, "because we are certain beforehand that no experiment can ever contradict it", the convention nonetheless "does not spring from our caprice; we adopt it because certain experiments have shown us that it would be convenient". This is "how experiment could make the principles of mechanics, and yet why it cannot overturn them". Poincaré then draws an analogy with geometry, which he says appears complete "at first blush", so that one

> will be tempted to say: Either mechanics must be regarded as an experimental science, and then the same must hold for geometry; or else, on the contrary, geometry is a deductive science, and then one may say as much of mechanics.

It is to block this conclusion that Poincaré argues:

> The experiments which have led us to adopt as more convenient the fundamental conventions of geometry bear on objects which have nothing in common with those geometry studies; they bear on the properties of solid bodies, on the rectilinear propagation of light. They are experiments of mechanics, experiments of optics; they cannot in any way be regarded as experiments of geometry. . . . On the contrary, the fundamental conventions of mechanics, and the experiments which prove to us that they are convenient, bear on exactly the same objects or on analogous objects. The conventional and general principles are the natural and direct generalization of the experimental and particular principles.
>
> (1902/1929, 124–125)

In describing the "general principles" of mechanics as "natural and direct generalizations" of experimental and particular ones, Poincaré strongly implies that these principles are more directly related to experience, and hence more liable to be overturned on the basis of experience, than are the principles of geometry. So here we have clear evidence of Friedman's point that Poincaré's conventionalism does not generalize in the same way as Duhem's holism.

4.3.3. The Conventionality of the Classification of "External Changes" (as "Displacements")

The narrower scope of Poincaré's conventionalism, as compared to Duhem's holism, is explained by the difference noted by Duhem: that Poincaré does not link the conventional character (which both thinkers understand in terms of the multiplicity of possible translations) of scientific or "theoretical" facts with their theoreticity. (For principles of mechanics and optics, just like those of geometry, confront experience only as they are incorporated in theories together with assumptions about measurement.) This difference in approach is particularly clear in Poincaré's 1898 article "On the Foundations of Geometry" (*FG*), which contains the fullest presentation of his argument for the group-theoretic basis of geometry and in which numerous interpreters find needed supplementation for *SH*'s arguments for geometric conventionalism (in addition to Dunlop 2016, §8, see Ben-Menahem 2006, 51; Heinzmann 2009).

FG elaborates upon the account given in Chapter IV of *SH* of how we arrive at the idea of geometrical space. Its genesis requires special explanation because the conceivability of non-Euclidean geometries shows that this idea is not imposed on the mind *a priori* (1902/1929, 64); meanwhile, the boundedness, non-homogeneity, and anisotropy of visual, tactual, and motor "spaces" (i.e. the systems of relations between, respectively, visual, tactual, and motor sensations) show that the idea is not already given with particular sensations, bound as they are to these modalities (1902/1929, 67–70). As briefly noted previously, Poincaré holds that geometry is not "a frame imposed on *each* of our representations, considered individually" but "the résumé of the laws according to which these images succeed each other" (1902/1929, 75). Geometry (and the idea of geometrical space) arises, specifically, by distinguishing among changes in our qualitative states, that is, in the aggregates of impressions (visual, tactual, and/or motor) that we experience at each given time.

Poincaré notes that we first distinguish between "internal" changes, which are voluntary and accompanied by muscular sensation, and "external" changes, which lack these traits. Key to Poincaré's account is a further distinction between two kinds of external changes: those "which are susceptible of being corrected by an internal change", which are termed "displacements" (changes of position), and those "which are not so susceptible", termed "alterations" (changes of state) (1898, 7; 1902/1929, 74). Poincaré illustrates the distinction with two examples:

> A sphere of which one hemisphere is blue and the other red, is rotating before our eyes and shows first a blue hemisphere and then a red hemisphere. Again, a blue liquid contained in a vase suffers a chemical reaction which causes it to turn red.

While "the impression of blue has given way to the impression of red" in both cases, the first change of impressions is classified as a displacement. For, in the first case, "it is sufficient for me merely to go around the globe to bring myself face to face again with the other hemisphere, and so to receive a second time the impression of blue" (1898, 6); obviously, the second change cannot be reversed in such a way. External changes that are capable of being corrected by the same internal change (that is, a change marked by the same muscular sensation) are considered the same displacement (1898, 8).

Poincaré claims that it is "the laws of" displacements, that is, changes of position or motions, which "constitute the object of geometry" (1902/1929, 74). On this account, the genesis of the idea of (geometrical) space requires both the experience of our own voluntary movement and the existence of rigid solids (or bodies that can be treated as such). An "*immobile* being could never have acquired" the idea because, "not being able to *correct* by his movements the effects of the changes of position of exterior objects, he would have had no reason whatever to distinguish them from changes of state"; nor could the being have acquired it "if his motions had not been voluntary or accompanied by any sensations" (1902/1929, 72). And the external body subject to displacement, as well as our own limbs in the compensatory internal change, must "be displaced as are rigid solids" (*ibid.*), so that after the internal change, the various parts of the external object can stand in the same relations to each other and to our sense organs as they did before the external change.

In Poincaré's account, the classification of external changes that are correctible (by internal changes) as displacements is "not a crude datum of experience, because the aforementioned compensation of the [internal and external changes] is never exactly realized". It thus stands as a theoretical or scientific fact. Poincaré claims it is "an active operation of the mind",

> which endeavors to insert the crude results of experience into a pre-existing form, a category. This operation consists in identifying two changes because they possess a common character, and in spite of their not possessing it exactly. Nevertheless, the very fact of the

mind's having occasion to perform this operation is due to experience, for experience alone can teach it that the compensation has approximately been effected.

(1898, 9)

Our present concern is not with Poincaré's Kantian-sounding view that this organizational scheme "pre-exists" in our mind as a "form" or "category" but with his account of the "active operation" by which we identify changes that we nonetheless experience as (qualitatively) distinct.

Poincaré next observes that the possibility of geometry requires that displacements form a group. (Thus the "pre-existing form" or "category" is "the general group concept", as Poincaré makes explicit in *SH*, 1902/1929, 79.) For the collection of displacements to form a group, it must be closed under the operation of combining displacements so that, in Poincaré's example,

> if the external change *A* is corrected by the internal change *A'*, and the external change *B* by the internal change *B'*, the resulting external change *A* + *B* will be corrected by the resulting internal change *B'* + *A'*;

in other words, the "resulting" external change will also be a displacement (1898, 9–10). Moreover, the operation of combining displacements should respect the identity of displacements, so that

> If two external changes α and α' are regarded as identical on the basis of the convention adopted above, or in other words, are susceptible of being corrected by the same internal change *A*; if, on the other hand, two other external changes β and β' can be corrected by the same internal change *B*, and consequently may also be regarded as identical, . . . the two changes $\alpha + \beta$ and $\alpha' + \beta'$ are susceptible of being corrected by the same internal change, and are consequently identical.

(Poincaré 1898, 10)

Poincaré argues that these properties cannot be ascertained *a priori*. Nor are they empirical, for in that case, geometry would be subject to empirical refutation, and "we may rest assured" that geometry will never be "overthrown" by experience. In his view, these properties are instead "laws" that we "impose upon nature", specifically by adopting a convention of reinterpreting experiences that do not conform to them.

> When experience teaches us that a certain phenomenon does not correspond *at all* to these laws, we strike it from the list of displacements. When it teaches us that a certain change obeys them *only*

approximately, we consider the change, *by an artificial convention*, as the resultant of two other component changes. The first component is regarded as a displacement *rigorously* satisfying the laws of which I have just spoken, while the second component, which is small, is regarded as a qualitative alteration. Thus we say that natural solids undergo not only great changes of position but also small flexions and small thermal dilatations.

(Poincaré 1898, 11, emphasis added)[7]

The first point I want to make about this passage is that while the second, conventionally assumed, component change may involve or presuppose the assumption of a physical *cause* (as in the example of "small thermal dilatations"), it is not assumed in the context of a physical *theory* and does not obviously involve or presuppose one. For the physical factors causing such assumed changes need not be subsumable under broader laws or systematically related in any other way. Recall that in Poincaré's account, the "*immobile* being could never have acquired" the idea of geometrical space because "he would have had no reason whatever to distinguish" changes of position from changes of state; what gives us reason to make this distinction, and to classify certain changes as displacements and as identical to certain other displacements, is the practical necessity of controlling our bodily movements in response to what we see, hear, smell, and so on in the heterogeneous sensory "spaces". Poincaré emphasizes that this process is informed by experience, in light of which one organizational scheme has proved most convenient. Thus, "if the education of our senses had been accomplished in a different environment, where we should have been subjected to different impressions, contrary habits [of associating ideas] would have arisen", and we would have adopted different laws relating our "muscular sensations" (1902/1929, 69). We may suppose that the experience in question has equipped us with an intuitive or "folk" physics, in terms of which we could conceive the compensating physical changes. ("Folk physics" may be considered a theory, in the loose sense that it serves as an aid to understanding and prediction, but I think it is clearly not a theory in the stricter sense relevant to Duhem's dispute with Poincaré.[8])

The second thing to observe about this passage is that while the "artificial convention" is introduced to guarantee that the resultants of identical changes α, α' and β, β' are also identical (we assume physical factors as needed to offset any differences between the state resulting from the internal change that corrects α combined with β, and the state resulting from the internal change that corrects α'+ β'), this same tactic is also what guarantees that external changes qualify as displacements in the first place. For we can just as well assume physical factors as needed to offset the difference between the experiential state immediately preceding external change A and the experiential state that immediately follows the compensatory internal change *A'*. I take Poincaré to hold that adopting such a convention is just what allows us to identify "two changes because they possess a common character, and in spite of their not possessing it exactly" (1898, 9).

On this point, *FG* is clearer than Chapter IV of *SH*, which claims that "we cannot foresee *a priori* that compensation is possible", and it is only "from the experimental fact that it sometimes happens that we start to distinguish changes of state from changes of position" (1902/1929, 72). Without the qualification that experience can show only that an external change is *approximately* compensated, experience would seem to be the sole basis for classifying external changes as displacements.[9] If this classification were made on the basis of experience, then the "variations of form" that solid bodies undergo "due to warming and cooling" would be important for it, and it would seem cavalier of Poincaré to claim these variations can be "neglect[ed] in laying the foundations of geometry, because, besides their being very slight, they are irregular and consequently seem to us accidental" (1902/1929, 76). But if these variations are posited in virtue of a convention, as the 1898 paper suggests, it is much easier to understand why we do not need to revert to theory to account for them (in sharp contrast to Duhem's view).

4.3.4. *Conventionality at Two Levels: The Classification of External Changes and the Choice of Metric Geometry*

In Section 4.4.3, I explain in more detail how, for Poincaré, the application of group notions is representative of the "translation" which renders crude facts into scientific ones. For now, it is important to see that because Poincaré argues similarly for the conventionality of the partitioning of external changes into classes of displacements and for geometric conventionalism, his theory of space has implications for the general issue of the relationship of theory to data, as well as for the epistemological status of geometry. We shall then be in position to draw a parallel between Poincaré's conventionalism and Neurath's antifoundationalism.

By "argument for geometric conventionalism", I mean specifically Poincaré's argument that the adoption of Euclidean geometry is conventional. Although we have already seen that Poincaré argues that an offsetting physical factor – namely the variation of temperature in the "world enclosed in a giant sphere" – can be assumed in order to correct measured deviations from Euclidean geometry, more consideration is needed to understand how the classification of external changes into displacements that accord with group-theoretic laws can depend on the same sort of convention as the adoption of a metric geometry.

For one thing, Poincaré emphasizes that we are at liberty to regard displacements as comprising either a Euclidean or a non-Euclidean group (which is what the thesis of geometric conventionalism states when expressed in terms of Poincaré's theory of space), yet the need to control our bodily movements obligates us to regard displacements as comprising some group or other. Poincaré's claim that "the general group concept pre-exists, at least potentially, in our minds", as a "form of our understanding", conveys this necessity (1902/1929, 79).[10] I have argued elsewhere, however, that for Poincaré, whether an organizing scheme or doctrine is conventional does not depend on whether there are alternatives to it (Dunlop 2017).

A more serious objection is that *FG*, which is Poincaré's most detailed account of how the group-theoretic scheme is conventionally adopted, seems to give a different picture of the preferred status of Euclidean geometry. While in *SH*, Poincaré claims "the general group concept" pre-exists in our minds, in *FG* he also says that notions of several particular continuous groups exist "in our mind prior to all experience". These include both the continuous group whose formal properties are studied in our (Euclidean) geometry and "that which corresponds to the geometry of Lobatchévski" (1898, 41).

> There are, accordingly, several geometries possible, and it remains to be seen how a choice is made between them. Among the continuous mathematical groups which our mind can construct, we choose that which deviates least from that rough group, analogous to the physical continuum, which experience has brought to our knowledge as the group of displacements. Our choice is therefore . . . guided by experience. But it remains free; we choose this geometry . . . because it is the more *convenient*.
>
> (Poincaré 1898, 42)[11]

Just as Poincaré explains most fully in *FG* how the general "laws" governing changes in experiential states (i.e. the axioms of group theory) are "imposed by us upon nature", here he also explains most fully why the Euclidean geometry is more convenient. It is chosen in virtue of its simplicity not "because it conforms best to some pre-existing ideal which already has a geometrical character" but "because certain of its displacements are interchangeable with one another, which is not true of the corresponding displacements of the group of Lobatchévski" (1898, 42–43). More technically, Euclidean geometry asserts the existence of "an *invariant* sub-group, of which all the displacements are interchangeable and which is formed of all translations", while "the group that corresponds to the geometry of Lobatchévski does not contain such an invariant sub-group" (1898, 21). Now, if the choice between these geometries is to be conventional in the same way as the adoption of the group-theoretic scheme, then we must have the option of instead assuming the other geometry, together with offsetting physical factors. In this account of why the Euclidean geometry is preferred, it seems that the factors would have to render certain displacements interchangeable (so as to form an invariant subgroup) by compensating for differences between the experiential states in which they terminate. But no such tactic can be used here, because there is no invariant subgroup whatsoever in Lobatchevskian geometry. This makes it hard to see how Poincaré can maintain, as he does in Chapter V in *SH*, that the adoption of Euclidean geometry is conventional in this way.

To reconcile the accounts given in *SH* and *FG*, we must appreciate the group-theoretic roots of Poincaré's models of non-Euclidean geometry. Beginning with the work of Jeremy Gray in the 1980s, scholars have shown how Poincaré's work on differential equations in the complex domain led him to a class of periodic functions, called "Fuchsian", thence to the "Fuchsian" groups that preserve these functions and tessellations of the complex plane which illustrate their periodicity. Poincaré first arrived at a model for non-Euclidean, specifically Lobachevskian or hyperbolic, geometry when he observed that functions mapping cells in the unit disk (in the complex plane) onto one another give rise to an appropriate expression for distance (see Stillwell 1996a, the introduction to 1996b or Heis ms.). In his first written presentation of these results, Poincaré articulates a conception of geometry now closely associated with Felix Klein's *Erlanger Programm*,[12] in which geometry studies those properties of objects which are invariant under transformations, and each geometry is characterized by a particular transformation group.[13] He states this view again in *FG* (1898, 41), *SH* (1902/1929, 79) and elsewhere (e.g. 1908/1929, 443).

Elie Zahar has shown how this background is important for understanding the models of non-Euclidean geometry presented in *SH*, namely the "dictionary" for translating expressions of Lobachevskian geometry into those of Euclidean geometry in Chapter III, and the "world enclosed in the giant sphere" described in Chapter IV. On a first reading, the dictionary may seem to change "self-evidently false" statements into true ones through "a bizarre redefinition of their primitive concepts" (Zahar 1997, 190). The translation of "distance between two points", in terms of the logarithm of the cross-ratio of these points and two others defined by the intersection of a certain circle with the "fundamental plane" in which all the defined objects and relations are located, clues us in that "we are dealing, not with a translation of [Lobachevskian] into [Euclidean] geometry, but with the construction of a model of [Lobachevskian] geometry within the complex plane" (Zahar 1997, 187–188). But the heuristic that unifies the disparate elements of this model becomes apparent only when we consider it as the image, under a certain mapping, of the unit disk model at which Poincaré first arrived.

Zahar observes that in the analytic manner typical of Riemann, Poincaré derives all the properties of the unit disk model from the formula for distance, in terms of the Euclidean distance modified by a certain factor; the "world enclosed in the giant sphere" (which is the three-dimensional counterpart of the unit disk model) provides this modifying factor with physical meaning, namely the temperature gradient. But while this analytic derivation makes sense of the model's features, it itself involves group-theoretic considerations characteristic of Klein's *Erlanger Programm*, for which "nothing prepares the reader" (Zahar 1997, 199). Poincaré's derivation involves, specifically, consideration of the group of Möbius transformations, which is central to the Kleinian treatment of the three geometries of constant curvature, for it includes the transformation groups by which these geometries are characterized (see, for instance, Poincaré 1882/1996 or Henle 2001).

What is crucial to understand is that the Kleinian approach unifies the treatment of the three metric geometries by developing them within the framework of projective geometry. Thus Klein claims, in one of his first papers on non-Euclidean geometry, that "projective geometry" supplies the "route" to "a clear general understanding" of the work of Bolyai, Lobachevsky, and Riemann (among others), "insofar as they relate to the theory of parallels". Klein proposes specifically to "construct a projective measure on ordinary space" in terms of "an arbitrary second degree surface" (1871/1996, 69).[14] By "ordinary space", Klein means a complex plane (or space of other dimension) as specified by projective coordinates; a "projective measure" is defined in terms of projective notions and is invariant under projective transformations that preserve the fundamental surface, and this surface is, in the case of the plane, a conic section. Depending on whether this conic is real, imaginary, or degenerate (in particular, two coinciding lines), the metrical properties are those of hyperbolic, elliptic, or Euclidean geometry (for more detail, see Torretti 1978 or A'Campo and Papadopoulos 2014).

The aspect of greater simplicity mentioned by Poincaré in *FG*, namely that of containing an invariant subgroup (comprised of all translations), already belongs at the projective or "pre-metrical" level to the group that is ultimately to be characterized as Euclidean (specifically, a group of transformations that preserves the fundamental surface and distance relations). Since metrical considerations are not yet in play here, there really is no possibility of arriving directly at non-Euclidean geometry by assuming some physical factor to offset measured values of spatial quantities. Yet even at this level, the adoption of the Euclidean group is conventional in an important and more general way: the simplifying assumption (the existence of an invariant subgroup) would be dropped if experience resisted being organized in accordance with it. As Poincaré says in the antepenultimate paragraph of *FG*, we "choose the geometry of Euclid because it is the simplest", but if "our experiences should be considerably different, the geometry of Euclid would no longer suffice to represent them conveniently, and we should choose a different geometry" (1898, 42).

Now it is not clear, to say the least, how we could find our experience to be at odds with Euclidean geometry before progressing to the level of metric geometry. The Kleinian approach arrives at metric geometry by limiting consideration to those projective transformations which leave unchanged both a certain "infinitely distant plane" and, most importantly, the fundamental surface (which is now taken as fixed) and then separating out, from this group of motions in space, the "subgroup of proper movements (rotations and translations) which, unlike the similarity transformations, leave the distance between two points wholly unchanged" (1925/1939, 161). Having progressed to this level, if our empirical measurements of geometrical quantities differed significantly from expected values, we would be forced to either abandon Euclidean geometry or keep it and assume a suitable offsetting factor. If we chose to avoid the complication of the offsetting factor at the metrical level, we would have to go back, as it were, and broaden the class of transformations considered at the projective level.[15] But Poincaré appears confident that we will never choose simplicity at the metrical level over the simplicity conferred by the existence of a subgroup. I take this to be Poincaré's point when he claims, at the end of Chapter III of *SH*, that Euclidean geometry "is, and will remain, the most convenient" because "it is the simplest; and it is so not only in consequence of our mental habits" but "in itself, just as a polynomial of the first degree is simpler than one of the second" (1902/1929, 65).

I think the comparison with Klein makes it easier to see how the adoption of a particular metric geometry can be conventional both in the sense of involving an assumed offsetting factor and in the sense of taking for granted the existence of a subgroup. But we must remember that for Poincaré, in contrast to Klein, geometry is not just a study of transformations and the properties they leave invariant, but specifically the study of the laws of *displacements* (laws "which mathematicians sum up in a word by saying that displacements form 'a group'", [1902/1929, 75]). Thus, the elements operated on by transformations in the different geometries (understood as groups) are already products of convention – in particular, they qualify as displacements, or as identical to one another, because we choose to regard discrepancies as resulting from physical factors. This undercuts any claim to empirically determine which geometry most accurately describes physical reality, even though the partitioning of external changes into equivalence classes of displacements cannot directly determine the choice between Euclidean and Lobachevskian geometry (as it would if we could arbitrarily make certain displacements all interchangeable with one another, so as to compose an invariant subgroup).

To express Poincaré's view of geometry in terms of Neurath's boat simile, not only are the "sailors" building a scientific edifice that is already in use (to organize experience and control bodily movements), but the "planks" they use to build it owe their "shapes" – what they contribute to the whole – to certain conventional assumptions. We will now see how the boat simile is used to make exactly this point in Neurath's own writings.

4.4. Selection of Facts and Economization of Thought in Poincaré, Neurath, and Mach

4.4.1. *Neurath on the Malleability of Ordinary-Language Expressions*

One would expect Poincaré's account of the arising of the concept of geo-metrical space (out of heterogeneous sensible "spaces") to have been of great interest to the Vienna Circle, given its members' avowed interest in the refinement of scientific concepts. But Neurath was not wholly com-mitted to that endeavor. In fact, he explicitly distances himself from the use of the word "philosophy", by "some representatives of the 'Vienna Circle'", to designate "the activity of clarifying concepts". In a 1931 piece, he rejects this usage both for the substantive reason that "it is impossible to separate the 'clarifying of concepts' from the 'pursuit of science to which it belongs'", leaving no room for philosophy as a separate disci-pline, and because it suggests that clarification of concepts is to take the place of a world-view, when "rather now 'science *without a world-view'* *confronts* all world-views" (1931b/1983, 58–59, original emphasis). Cartwright, Cat, Fleck, and Uebel identify 1931 as "the turning point in the evolution of Neurath's thinking within the Vienna Circle", in which Neurath turned away, in particular, from Carnap's project of creating "a system of concepts which provides the possibility of deriving concepts from one another" (1996, 184, n.53, quoting Neurath 1931a).

A case can be made, however, that Poincaré's account would have been relevant for Neurath's own purposes. Even if there is not sufficient evi-dence (which I think there is not) to conclude that Poincaré's view of the role of group concepts in geometry actually had a significant influence on Neurath, this intriguing possibility is still worth exploring, for the com-monalities it reveals between Poincaré's and Neurath's projects.

Instead of the system(s) proposed by Carnap, Neurath envisages a framework (1936b/1983) or "scientific cleaning machine" (1932/1983, 98) in which statements expressed in scientific language are checked against observation statements in ordinary language, leaving open which kind of statement to reject in the event of a contradiction.[16] In this view, the scientist starts "from masses of statements whose connection is only partly systematic, which we also discern only in part"; because these statements "use many vague terms", "'systems' can be separated only as abstractions" (1935b/1983, 122). Neurath later claims, more forcefully, that "everyday statements cannot really find a place" in a "closed" or "complete and definite" system (thus "the relationship linking scientific formulas with the formulas of everyday language" can be studied "much better" in the framework of an encyclopedia, which is neither hierarchical nor unrevisable [1936b/1983, 149]).

Against the background of early 20th-century movements toward formalization and systematization, Neurath stands out for his conviction that the "vagueness" and imprecision of ordinary language are not to be extirpated or overcome; he counsels us, rather, to understand their implications for the practice of science.[17] Neurath emphasizes the "stability" of "the not very precise [statements] of ordinary language", explaining how this stability permits theories to be checked against observations made before the theories' terms were formulated and defined and how it is increased by including the circumstances of an observation in its statement (1936b/1983, 149–151; also 1935b/1983, 129). (Including descriptions of observers' behavior in the observation statements cements the importance of the social sciences for unified science.) The fact that ordinary language "formulates the age-old results of the so often renewed experiences of everyday life" speaks to its "universal applicability" as well as its stability; thus "the lack of precision characteristic of ordinary language contributes . . . to the generalisation of its use" (1936b/1983, 155). Neurath also stresses the ambiguity of ordinary-language expressions, which he terms "*Ballungen*" ("clusters"), suggesting that different significations are "balled" or bundled up in them.

Cartwright, Cat, Fleck, and Uebel's account highlights one particularly momentous consequence of Neurath's complacence about the imprecision of ordinary language: the denial of "logically determinate connections between data and theory" (1996, 190). In particular, there are no determinate relations of logical deduction, inductive confirmation, or falsification between ordinary-language observation statements and scientific formulae.

They relate this indeterminacy to the ambiguity of ordinary-language expressions within the context of Neurath's "boat" simile. The point, in terms of the simile, is that the ordinary-language statements that make up the sailors' construction material[18] can be molded so as to fill arbitrarily shaped gaps. They argue for the later emergence of this point by contrasting an earlier version of the "boat" simile from 1921 with this one from 1932:

> There is no way to establish fully secured, neat protocol statements as starting points of the sciences. There is no *tabula rasa*. We are like sailors who have to rebuild their ship on the open sea, without ever being able to dismantle it in dry-dock and reconstruct it from the best components. . . . Imprecise 'verbal clusters' ['*Ballungen*'] are somehow always part of the ship. If imprecision is diminished at one place, it may well re-appear at another place to a stronger degree.
>
> (Neurath 1932/1983, 92)

In the earlier version of the simile, the sailors on board "had solid boards, cleanly sawn", so that "there was no question whether [or not] a given board would fit" into a gap needing to be filled. But in this version, gaps "are not so well defined"; the edges of any given gap "can be bent around a bit to make it accommodate a given piece". Just as two pieces can be used to fill differently shaped gaps, so differently shaped pieces "can be forced into the same gap" (Cartwright, Cat, Fleck and Uebel 1996, 191).

The malleability, as it were, of the ordinary-language expressions that constitute the basis for scientific endeavors can be understood in terms of the different precisifications to which "cluster" expressions are subject. Neurath makes explicit that "[w]hat we demand of a cluster-concept is that we can somehow make a 'formula' correspond to it, in connection with a theory" (1936b/1983, 148). Here Neurath says both that a formula can become a cluster and that there can be a cluster with "the same name" as a formula, leaving it unclear whether there is one concept that transitions from formula to cluster or a cluster concept that comes to be alongside a distinct formula concept, and in this passage, he discusses only the formula-to-cluster direction (using the example of hydrogen peroxide). But Neurath soon makes clear, with the example of water, that there is also development in the direction of greater precision:

> what corresponds in science to the common term 'water' has today a definition that is different from that of some centuries ago and even of a very short time ago when one did not know the difference between 'heavy' and 'light' water.

Such divergent definitions arise through the development of scientific theories, as Neurath indicates when he writes that "the terms of science must adapt themselves much more to the new theories than a cluster" (1936b/1983, 149). And for Neurath, the development of theories is to be understood in terms of sociological factors and even individual choice (as is clear from his attacks on "pseudorationalism", as exemplified, for instance, by Popper). So it can be said that the formatting that experiential input must undergo so that theories can be tested by it – in Neurath's terms, the articulation of "formulas" that correspond to "clusters" – is ineluctably conventional for Neurath, just as it is for Poincaré. Thus, Neurath's 1932 description of the sailors' predicament, as elaborated upon by Cartwright, Cat, Fleck, and Uebel, is strikingly apt to illustrate Poincaré's view of how the "crude results of experience" are smoothed into conformity with a theoretical scheme.

However, there are significant obstacles facing any attempt to draw a line of influence from Poincaré's account of the genesis of the concept of geometrical space to Neurath's rejection of determinate theory-data relationships. For one, there is no evidence (that I know of) that Neurath was familiar with the details of Poincaré's view. Second, Neurath takes himself to go beyond Poincaré and Duhem precisely by showing that *"multiplicity* and *uncertainty"* pertain not just to scientific theories but to the observation statements against which such theories are to be checked:

> Poincaré, Duhem and others have adequately shown that even if we have agreed on the protocol statements, there is not a limited number of equally applicable, possible systems of hypotheses. We have extended this tenet of the uncertainty of systems of hypotheses to all statements, including protocol statements that are alterable in principle.
> (1934/1983, 105)[19]

A further difficulty is that Cartwright, Cat, Fleck, and Uebel take Neurath to have given up on determinate data-theory relationships only in 1931, but, as we will shortly see, it is in Neurath's earliest publications that he seems to follow Poincaré most closely. Now if their dating is accepted, it could still be maintained that Neurath absorbed Poincaré's views in the first decade of the 20th century (when the members of the "first" Vienna Circle read Poincaré's books as their German translations appeared;[20] Haller 1982a/1991, 97), despite failing to cite them when he himself drew similar conclusions 20 years later. But the notion of *Ballungen* and the correspondingly open-ended bearing of data on theory is already illustrated by a suggestive example, though not yet explicitly articulated, in Neurath's 1916 paper on optics. So, *pace* Cartwright, Cat, Fleck, and Uebel, with respect to this example, we may side with the scholars who hold that the "essential traits of Neurath's thought can be found *in nuce"* in his early writings.[21] I want now to consider Neurath's 1916 paper on optics in order to show how the views common to Neurath and Poincaré are related to those of Ernst Mach. For my purposes, it is less important to settle the question of whether Poincaré influenced Neurath than to see how both developed Mach's view in the same novel direction.

4.4.2. Neurath and Poincaré on "Selecting Facts"

In his 1916 paper on optics, Neurath offers an early statement of holism: "Precisely in the field of optics, hypothesis and experience are entirely mixed up". He then turns to Goethe's criticism of Newton's theory of color to illustrate how one's attitude towards a hypothesis prejudices the experience that purportedly tests the hypothesis. To test the proposition that differently colored light has different degrees of refraction, Newton "looks through a prism at a strip of cardboard that is partly blue, partly red". According to Newton, the blue strip "appears more raised than the red one" if the refracting edge of the prism is on top.

> Goethe imputes a blue margin along the blue as well as along the red strip when looked at through the prism; whereas this [the blue

margin] is added to the width of the blue strip, it diminishes the red one because it is there in counteraction and hardly visible. . . .

Goethe points out that in Newton's illustration the red strip has fringes of which Newton makes no mention at all. "Why does he not mention this phenomenon in his text of which he has a careful, though not quite correct, engraving in copper? A Newtonian will probably answer: this is just a residue of the decomposed light which we can never get rid of entirely, and that still plays its tricks here".

(Neurath 1916/1983, 23–24)

In Neurath's view, Goethe is himself subject to the criticism he levels at Newton, namely that "[i]f something does not suit his view, it is neglected to begin with". So the example illustrates "how differently the optical phenomena can be grouped" (1916/1983, 23), rather than a flaw specific to Newton's way of proceeding.

Jordi Cat characterizes the "blurred edges" of the strip of refracted light as an illustration of Neurath's "metalinguistic doctrine of the inevitably limited precision of our scientific language in (linguistic) contact with empirical contents and ordinary context", which is just the doctrine of *Ballungen* (Cat 2019, 299 and *n*.). Gábor Zemplén, likewise, identifies the "'blurred margins' on every level of the investigation" with the *Ballungen* of Neurath's later work (2019, 230). Whether or not the phenomenon of the blurred edges or margins corresponds exactly to that of *Ballungen*, Neurath continued to use it to illustrate how the complexity of empirical phenomena outruns the precision of both scientific-language observation statements and scientific concepts. In a 1935 paper, he returns to the example to make the point that "to make good predictions, we could set out from different observation statements that we select from the large number at our disposal":

What one person neglects as unimportant – and then he shapes his concepts accordingly – may seem essential to another for the predictions. For example, Goethe strongly criticized Newton for omitting certain blurred margins of the spectrums as unimportant, whereas he himself started from this very point.

(Neurath 1935a/1983, 117)

Certainly, the phenomenon of the blurred margins is apt to illustrate how the same empirical phenomenon can bear out the predictions of different theories, depending on which aspects of its complexity are neglected. This is just how Poincaré, in my interpretation, understands the bearing of changes in experiential states (displacements) on the geometries that systematize them. (For the identification of two changes that do not "exactly" correspond (1898, 9) is a matter of neglecting their discrepancies.) This point of agreement between Neurath and Poincaré is recognized all the more easily because it is not yet a "metalinguistic" doctrine for Neurath, as it becomes in the context of the Vienna Circle. That scientific concepts are "shaped" by selectively neglecting aspects of what is experienced is a further point of agreement which is arguably even more central to Poincaré's view.

Neurath takes this example to show how, in general, "a hypothetical element is already introduced into a theory" by merely "neglecting or stressing some facts". For Neurath, the introduction of such a hypothetical element is inevitable because "[t]he whole fullness of a phenomenon can never be completely reflected" in a theory (1916/1983, 24). This language is very similar to Poincaré's, as we will shortly see. A respect in which Neurath's account follows Poincaré's yet more closely is that both understand the incompleteness and tentativeness of scientific theories in terms of a "choice of facts". Thus, Poincaré claims it is "incontestable" that "a choice [among facts] must be made", because "whatever be our activity, facts go quicker than we, and we can not catch them; while the scientist discovers one fact, there happen milliards of others in a cubic millimeter of his body" (1908/1929, 362–363). Neurath holds that on the basis of "more or less distinctly expressed hypotheses", one "sifts out certain facts, combines them into a whole and hopes to be able to incorporate the remaining facts". Thus, in classifying systems of hypotheses (Neurath's project in this paper),

> One should also always indicate which facts have been neglected, which favoured. The systems of hypotheses of physics, like all other systems of hypotheses, are *an instruction directing not only the connectedness, but also the selection of facts.*
> (Neurath 1916/1983, 24, original emphasis)

Like the example of the blurred margins, the notion of "selection of facts" also recurs in Neurath's later work. To illustrate "how much concrete research is hampered by metaphysical formulations", Neurath contends that Max Weber's *"metaphysical starting point* ... had an unfavourable influence on his selection of observation statements" in Weber's "powerful attempt to deduce the rise of capitalism from Calvinism" (1931b/1983, 84).

Neurath's and Poincaré's remarks about selecting facts have a clear antecedent in Mach's *Science of Mechanics.* Mach draws this general conclusion from his study of the development of statics:

> A rule, reached by the observation of facts, cannot possibly embrace the *entire* fact, in all its infinite wealth, in all its inexhaustible

manifoldness; on the contrary, it can furnish only a rough *outline* of the fact, one-sidedly emphasizing the feature that is of importance for the given technical (or scientific) aim in view.

(1933/1960, 90)

Similarly, in the chapter "The Economy of Science" Mach first claims that "the object of science" is "to replace . . . experiences by the reproduction and anticipation of facts in thought" (1933/1960, 577), and then that "[i]n the reproduction of facts in thought, we never reproduce the facts in full, but only that side of them which is important to us" (1933/1960, 578). So all attempts on the part of science to "embrace" the facts in full necessarily involves a selection, guided by whatever aim is view. The relevance of Mach's thought for Neurath's early papers on optics, and for these passages from Poincaré (1908/1929), is already noted in the literature.[22] I would add only that Poincaré's well-known classification of hypotheses, in Chapter IX of *SH*, appears indebted to Mach's distinction, later in the "Economy of Science" chapter, between "natural" ideas and "mental artifices".[23]

4.4.3. Language as the Instrument for Economizing Thought

The main fault lines between Mach, on the one hand, and Neurath and Poincaré, on the other, are also well known. As Philipp Frank later reported, Poincaré's conventionalism appealed to the logical empiricists precisely as a way to account for the "gap", which loomed large in Mach's view of principles as merely "abbreviated descriptions of sense observations", between the vague concepts employed in observation statements and the precision of mathematically formulated principles (1949, 7).[24] (Frank has it that for Poincaré, this gap is bridged by means of the "mental activity" needed to "formulate general statements about sense observations", in contrast to Mach's view of observations as passively recorded [*ibid.*]; this recalls Poincaré's assertion that external changes are classified as displacements by "an active operation of the mind", 1898, 9.) For Mach, the economization of thought is an absolute (ultimately biological) imperative, and theories tend to converge uniformly, as it were, on this goal.[25] Frank illustrates this view with the image of theories as "nets" connecting phenomena:

The known connections among phenomena form a network; the theory seeks to pass a continuous surface through the knots and threads of the net. Naturally, the smaller the meshes, the more closely is the surface fixed by the net. Hence, as our experience progresses the surface is permitted less and less play, without ever being unequivocally determined by the net.

(1949, 65)

While this account of science's goal is not disputed by Poincaré or Neurath, unlike Mach, they allow a multiplicity of approaches to the end of economizing thought. (Indeed, Neurath explicitly rejects, as "pseudorationalism", the "doctrine of a perfection, perhaps 'infinitely far away' to which science gets closer and closer", 1936a/1983, 137.)

It is striking that both Neurath and Poincaré allow that different theories can serve this end while conflicting with each other, even to the point of logical contradiction. Mach makes clear that it is a "logical necessity" that the concepts we form "purposely and consciously", through the endeavor to "adapt our ideas to our sensuous environment", must "agree with each other" (in context, his point is that there is no further necessity that these concepts must conform to relations obtaining between objects; 1933/1960, 318). On the contrary, both Neurath, in his early papers on optics, and Poincaré hold that contradiction in scientific theory can have a stimulating effect.

Neurath claims that it was precisely the inconsistency of Newton's theory "that was highly stimulating and gave posterity an opportunity to form hypotheses of many kinds, many of which have proved fertile" (1916/1983, 20). Poincaré endorses "the physicist's" response to a contradiction between two equally preferred theories, which is to "not bother about that, but hold firmly the two ends of the chain, though the intermediate links are hidden from us"; we need not assume that one theory must be false, because they may "both express true relations and the contradiction is only in the images wherewith we have clothed the reality" (1902/1929, 142). Poincaré uses the example of Maxwell to illustrate how "two contradictory theories, provided one does not mingle them, and if one does not seek in them the basis of things, may both be useful instruments of research", speculating that "the reading of Maxwell would be less suggestive if he had not opened up so many new and divergent paths" (1902/1929, 177).

While in this respect both Poincaré and Neurath are more pluralistic than Mach with regard to economizing thought, each of them also accepts certain of Mach's restrictions on means to this end. I will conclude by showing that the restrictions they accept are different in each case, for this difference deserves more emphasis than their points of agreement. The difference is apparent in the ways they develop Mach's view of language as, in his words, "the instrument" of the communicative process by which individuals share experience, thus sparing others "the trouble of accumulating it for [themselves]" (1933/1960, 577–578).

The filiations between Mach's and Neurath's views of language's role are readily apparent. Mach names language as the instrument of the communicative process at the beginning of the "Economy of Science" chapter and immediately proceeds to discuss the "universal character" and easy comprehensibility of the ideal language, which is approached by Chinese ideography. This discussion can be plausibly supposed to have inspired

Neurath's work on pictorial language. Most of all, Neurath's commitment to making language conducive to (unified) science by eliminating metaphysical expressions[26] speaks to the influence of Mach.[27] But there is also an important point on which Neurath disagrees with Mach: while Mach conceives the "sense observations" or phenomena that principles describe as psychological states of individuals, Neurath (along with Carnap) denies that scientific statements must be formulated in phenomenalistic language. This turns out to be an important difference with Poincaré, as well.

While Mach and Neurath thematize the role of language in economizing thought, Poincaré's account is not as developed. So I will be concerned to reconstruct his view as well as contrasting it with Neurath's.

Poincaré's most extensive discussion of the issue is in his plenary lecture to the 1908 International Congress of Mathematicians, reprinted in (Poincaré 1908/1929) as "The Future of Mathematics". Here he considers the word "group" as an example, and we see how the application of group notions models the "translation" which renders crude facts into scientific ones. Recall that for Poincaré, the use of the generic concept of group is not dispensable in the same way as the use of a particular group-concept. While in *SH* Poincaré characterizes the notion of group as an innate or "pre-existing" form, in the 1908 address he takes our use of it to illustrate the essential economizing tendency of science. Poincaré contends, first, that mathematicians, like other scientists, must "make a choice among facts": "the head of the scientist, which is only a corner of the universe, could never contain the universe entire; so that among the innumerable facts nature offers, some will be passed by, others retained" (1908/1929, 369). He then asserts that this choice is based on a "hidden analogy" that certain facts bear to others:

> [T]he facts which interest [physicists] are those capable of leading to the discovery of a law, and so they are analogous to many other facts, which do not seem to us isolated, but closely grouped with others. . . . [T]he genuine physicist alone knows how to see . . . the bond which unites many facts whose analogy is profound but hidden. . . . Facts would be sterile were there not minds capable of choosing among them, discerning those behind which something was hidden, and of recognizing what is hiding, minds which under the crude fact perceive the soul of the fact.
>
> We find just the same thing in mathematics. From the varied elements at our disposal we can get millions of different combinations; but one of these combinations, insofar as it is isolated, is absolutely void of value. . . . Quite otherwise will it be when this combination shall find place in a class of analogous combinations and we shall have noticed this analogy.
>
> (Poincaré 1908/1929, 370–371)

In this lecture, Poincaré explains the "value" of a fact or combination that "finds place in a class of analogous" ones[28] in terms of economy of thought:

> The celebrated Vienna philosopher Mach has said that the role of science is to produce economy of thought. . . . And that is very true. . . .
>
> The importance of a fact then is measured by its yield, that is to say, by the amount of thought it permits us to spare.
>
> In physics the facts of great yield are those entering into a very general law, since from it they enable us to foresee a great number of others, and just so it is in mathematics.
>
> (1908/1929, 371)

In Chapter IX of *SH*, Poincaré claims, similarly, that among facts the scientist chooses those that are apt for generalization. Here Poincaré makes clear that generalization is required for "foreknowledge" and thus for the "success" (in his terms) of science:

> the circumstances under which one has worked will never reproduce themselves all at once. The observed action then will never recur; [we can affirm only] that under analogous circumstances an analogous action will be produced. In order to foresee, then, it is necessary to invoke at least analogy.
>
> (1902/1929, 128)

Poincaré contends that this regard for analogy is distinctive to scientific thought. While "the historian" might say, with Carlyle, that only "facts" such as "John Lackland passed by here" are of importance, "[t]he physicist would say rather: 'John Lackland passed by here; that makes no difference to me, for he never will pass this way again'".

> A fact is a fact. A pupil has read a certain number on his thermometer; he has taken no precaution; no matter, he has read it, and if it is only the fact that counts, here is a reality of the same rank as the peregrinations of King John Lackland. Why is the fact that this pupil has made this reading of no interest, while the fact that a skilled physicist had made another reading might on the contrary be very important? It is because from the first reading we could not infer anything. [A] good experiment . . . is that which informs us of something besides an isolated fact; it is that which enables us to foresee, that is, that which enables us to generalize.
>
> (Poincaré 1902/1929, 128)

Neurath would surely have been unimpressed with Poincaré's distinction between the methods of the historian and the physicist. But he would have agreed with Poincaré that the important distinction between the physicist's instrument reading and the pupil's is that the former affords predictions, even though Neurath would have understood this difference in terms of the language (ordinary or scientific) used to express the reading.

In this chapter of *SH*, Poincaré explicitly claims not only that the scientist selects, for his or her attention, the facts or observation statements which are most apt to enter into analogies but that generalization of these facts also involves an element of choice or convention: "It is clear that any fact can be generalized in an infinity of ways, and it is a question of choice" (1902/1929, 131). Although he does not make this point in the 1908 lecture, it evidently holds for the example of the group concept. For the groups corresponding to the various geometries (Euclidean and non-Euclidean) each "generalize" particular displacements in different ways, in terms of how they combine with other displacements under geometrical operations.

In the 1908 lecture, Poincaré claims that to "fix", or more specifically "bring to light", the "kinship" obtaining between such "combinations", it often "suffices to make a new word" (1908/1929, 371). This claim gives content to Poincaré's view that what the scientist creates in a fact is the language in which it is expressed, since we now see that what the scientist contributes is not just terminology but a linkage among disparate facts. Poincaré credits to Mach the point that "a well-chosen word can economize thought" to a great extent, and claims that "group" belongs among those "words that have had the most fortunate influence" by making us "see the essence of many mathematical reasonings".

> [Such words] have shown us in how many cases the old mathematicians considered groups without knowing it, and how . . . they suddenly found themselves near [to one another] without knowing why. Today we should say that they had dealt with isomorphic groups . . . thanks to these words "group" and "isomorphism", which condense in a few syllables this subtle rule and quickly make it familiar to all minds, the transition [from one group to an isomorphic one] is immediate and can be done with all economy of thought effort.
>
> (Poincaré 1908/1919, 375–376)

In this case, it is clear that knowledge is not advanced by merely introducing the terms "group" or "isomorphic" but rather by thereby linking disparate phenomena.

Poincaré's reflections on the utility of language do not extend to ordinary language. Unlike Neurath, Poincaré shows no appreciation of ordinary language's stability, perhaps because he is not concerned with checking predictions or generalizations against empirical evidence. There is also a deeper reason that ordinary-language statements could not be a starting point for Poincaré, as they are for Neurath. For Neurath, the language in which protocol or observation statements are formulated must allow for "intersubjectivity", so that statements accepted at one time can be checked against those adopted earlier or later (or against aggregates in which those statements are incorporated, 1932/1983, 95). Statements reporting data from only a single sense modality do not satisfy this requirement[29] and thus should be understood to belong to a specialized phenomenal language. But Poincaré shares Mach's phenomenalist starting point. The experiential states that are ultimately systematized within geometrical space are first conceived as belonging to heterogeneous sensory spaces, and modality-specific phenomenal languages would be needed to express this initial data.

4.5. Conclusion

Neurath once pithily sums up his "insight" as that "a logically tenable multiplicity is reduced by life" (1935a/1983, 117). As he explains,

> Various possibilities offer themselves, and even more are only vaguely conjectured. . . . This restriction by life corresponds to the behaviour of the active man who chooses one of several possibilities – the act called planning. But such unambiguity of decision and action is *not* the logical result of some premises that lead to *one* single prediction about the success of the action, but rather the result of life taken as a whole.
> (Neurath 1935a/1983, 118, original emphases)

In context, Neurath appears to credit to Poincaré and Duhem the point that there can be a "multiplicity" of "systems of hypotheses" to account for given phenomena and to put it forth as his own contribution that "life" (i.e. practical exigency) narrows down these alternatives.

I have argued that for Poincaré, the idea of geometrical space arises through processes of classification (of experiential changes as displacements and of displacements into equivalence classes) which are undertaken for the purpose of directing bodily movements. The exigency of responding appropriately to stimuli both makes it imperative to adopt some classificatory scheme and steers the choice among schemes. Poincaré's view, on the reading offered here, is thus closer to Neurath's than Neurath acknowledges.

Notes

1. Duhem's holism is treated in detail, in the context of Albert Einstein and Philipp Frank's philosophy of science, in Don Howard's chapter in the present volume.
2. This point has also been made by Gerhard Heinzmann and Rudolf Haller. According to Heinzmann, "Poincaré knows even conventions that do not

only concern the choice between theories; rather, they come into play as classifications with respect to the interpretation of the basic concepts of each theory". Heinzmann aptly describes these classifications as "mediating elements between sensation and observation" (2010, 5). Haller (1982a/1991, 98) takes Poincaré, Mach, and Duhem to affirm "the intervention of rational operations [as ordering principles] in sensual experience" and thus to deny "the priority of observation to theory".

3. Quine's holism is based on a theory of meaning that applies to scientific theories in general, while Duhem's is based on considerations about measurement that apply primarily to physics. Moreover, Quine famously claims that all of science is subject to being confirmed or disconfirmed by a given empirical result, whereas Duhem allows that some hypotheses may remain insulated from empirical test (see Vuillemin 1979).

4. Reichenbach uses an example to illustrate how the assertion of physical laws (as true) depends on the definition of congruence by means of rigid bodies: "if a rubber band were used as the definition of congruence without any indication of its state of tension, the energy of closed systems would in general not be constant, since the measure of the energy would vary as a function of the rubber band. . . . The law of conservation of energy would be replaced by a law stating the dependence of the energy of closed systems on the state of the rubber band. But this law would be just as true as the law of the conservation of energy. The disadvantage would consist only in the fact that the biography of the rubber band would have to be included in all physical laws. It is one of the most important facts of natural science that it is possible to establish physical laws free from such complications; the significance of the rigid body is based on it" (1928/1958, 24).

5. As Friedman observes, Rudolf Carnap also adverts to Poincaré's example in (1966/1974). Carnap takes Poincaré to hold that if physicists "should discover that the structure of actual space deviated from Euclidean geometry", they could "either accept non-Euclidean geometry as a description of physical space, or they could preserve Euclidean geometry by adopting new laws stating that all solid bodies undergo certain contractions and expansions", just as in the ordinary practice of measurement, "we must make corrections that account for thermal expansions or contractions of the [measuring] rod" (1966/1974, 144–145).

6. In the "Introduction" to *Science and Hypothesis*, Poincaré cites a 1901 article by LeRoy as a source for "nominalism", which Poincaré characterizes as the suspicion that the *savant* is "the dupe of his own definitions", and "the world he thinks he discovers is . . . simply created by his own caprice" (1902/1929, 28).

7. In a concluding summary, Poincaré claims, similarly, that "we decompose the observed phenomenon conventionally into two others: a purely geometrical phenomenon which exactly obeys these laws, and a very minute disturbing phenomenon" (1898, 38).

8. It may be useful to note that "folk psychology", after which "folk physics" is named, has been variously described as a "model" (Malle 2001; *cf.* Elga 2007) or a "craft" (Dennett 1991).

9. In *FG*, Poincaré makes clear that experience cannot be a foundation for geometry precisely because it can only approximately verify geometrical propositions: "the experiences which serve to verify [formal properties of isomorphic groups] can never be more than approximate; that is to say, the experiences can never be the true foundation of these propositions" (1898, 13).

10. Here Poincaré opposes himself to the Kantian view that space is a "form of sensibility".

11. Similarly, in *SH* Poincaré claims experience "guides" the choice, "from among all the possible groups," of a "*standard* to which we shall refer natural phenomena" and thus informs us "not which is the truest geometry but which is the most convenient" (1902/1929, 79–80).

12. Hawkins (1984) and Gray (1992) argue that Poincaré's work was not influenced by Klein. Gray emphasizes (as does Heis [ms.]) that projective geometry does not have the importance for Poincaré that it does for Klein. I think that invoking the view now associated with Klein is helpful for fixing ideas, even if this view was not actually distinctive to Klein.

13. Poincaré writes that a geometry is "the study of the group of operations formed by the displacements to which one can subject a body without deforming it", and in Euclidean geometry "the group reduces to the rotations and translations", while "in the pseudogeometry of Lobachevskii it is more complicated" (quoted and translated in Gray and Walter 1997, 9).

14. As Klein makes clear, the construction of such a "measure" is originally due to Arthur Cayley. Cayley makes the oft-quoted claim that "the metrical properties of a figure are not the properties of a figure considered *per se*, apart from everything else, but its properties when considered in connection with another figure, *viz.* the conic termed the Absolute" (1859/1889, 592). But Cayley's main aim was to develop the metrical properties of Euclidean geometry using purely projective notions, and while he also supplied a metric for spherical geometry, he omitted to explain how the method generalizes to hyperbolic geometry.

15. The need for such a retrospective broadening is apparent from the development of metric geometry, "for purposes of general instruction", that Klein gives in (1925/1939). Having specified the group of motions, "whose invariant theory we shall look upon as the geometry of the plane" (162), Klein immediately narrows consideration to parallel translations, constructs "path curves" of points subject to these translations, and then "introduce[es] coordinates, in order thereafter to make our inferences entirely within the field of arithmetic", that is, analytic geometry (174). Klein later observes that since no non-Euclidean geometry contains a subgroup like that composed of the parallel translations, this development "*excluded non-Euclidean geometry once and for all and retained only Euclidean geometry*", although each non-Euclidean geometry in fact contains a group of transformations that "leave the relations of measure unchanged" and therefore could just as well be developed from the group of motions, if the existence of a subgroup were not assumed (184, original emphasis).

16. As a pluralist, Neurath admits multiple ways of configuring such a "machine" or "framework". Regarding the "machine" conception, he claims one can "not agree on a 'machine' that unambiguously produces 'inductions'. . . . The progress of science consists, as it were, in constantly changing the machine and in advancing on the basis of new decisions" (1935a/1983, 116). The "framework", for its part, is the "always variable" one of an encyclopedia, and "[t]he encyclopedia that is the model of science is in no way unique and select" (1936b/1983, 146).

17. See, for example, 1936b/1983, where Neurath claims to "follow quite a different course" by "accommodat[ing] the empirically given protocol statements" – which share the "medium complexity and uncertainty" of ordinary-language statements – "without, however, going so far as 'formalising' the common language".

18. Without using the "boat" simile, Neurath explicitly claims that "traditional statements", marked by "historically given stability", are the "basis" from

which we must depart; it is with respect to these statements that "[w]e can never make a *tabula rasa* from which to begin . . . a new life" (1936b/1983, 150).

19. Cf. "This is how things stand in *every 'layer'* of scientific work, not only in the narrower sphere of systems of hypotheses, as Poincaré and Duhem have pointed out with such intensity" (1935a/1983, 117, original emphasis).

20. *Science and Hypothesis* was published in translation in 1904, *The Value of Science* (1905/1929) in 1906, and *Science and Method* (cited here as 1908/1929), in 1914, all by B. G. Teubner (Leipzig, later Leipzig and Berlin). However, the essay "The Future of Mathematics" (*"L'avenir des mathématiques"*), chapter II of (1908/1929), was summarized in German journals shortly after it was delivered as a plenary address to the 4th International Congress of Mathematicians in 1908.

21. See Soulez (1997, 15), quoted (approvingly) by Ferrari (2001); see also Haller (1982b/1991), Zemplén (2019).

22. See Brenner (1998) on Poincaré; and see Stöltzner (2001) and Zemplén (2019) on Neurath.

23. Strictly, Mach's distinction is between kinds of scientific theories. But Poincaré here equates "hypotheses" with "generalizations", and since generalizations enable foreknowledge and involve a "choice" among the "infinity of ways" in which any given fact can be generalized (1902/1929, 131), they correspond to what Mach calls "theories". Poincaré classifies hypotheses into "natural", "neutral", and "fruitful" in Chapter IX of *SH*. (On the relationship of this taxonomy to the one given in the Introduction to *SH*, see de Paz 2015). His examples of "perfectly natural" hypotheses, "from which one can scarcely escape", include "that the influence of bodies very remote is quite negligible" and "that the effect is a continuous function of its cause" (1902/1929, 135). Mach claims that the formation of certain ideas, by "the principle of continuity", is "so natural that every child conceives [them]"; his examples are that "we imagine a moving body which has just disappeared behind a pillar, or a comet at the moment invisible, as continuing its motion and retaining its previously observed properties" (1933/1960, 588). Poincaré contrasts natural hypotheses with "neutral" ones which "may be useful, either as devices for computation, or to aid our understanding by concrete images" but are never necessary for obtaining results. One example is that "the analyst assumes at the beginning of his calculations either that matter is continuous or, on the contrary, that it is formed of atoms" (*loc. cit.*). Mach puts forth the atomic theory as an exemplary "mathematical *model* for facilitating the mental reproduction of facts", which did not arise "naturally and artlessly" but "is a product especially devised for the purpose in view" (1933/1960, 588–589).

24. Recall that Neurath claims that the research "originated by Mach, Poincaré, [and] Duhem" was stimulated by the "very fact that the relatively great stability of common language statements is not linked [*ne s'accompagne pas*] with a precision equal to that of scientific formulas" (1936b/1983, 151).

25. Thus the discussion of the one-sidedness of rules quoted in the preceding paragraph continues, "[h]ence there is always opportunity for the discovery of new aspects of the fact, which will lead to the establishment of new rules of equal validity with, or superior to, the old" (1933/1960, 90).

26. Neurath writes that the "great task of anti-metaphysical empiricism" is "to repulse traditional metaphysics", thus "sharpen[ing] the logical instrument in such a way" as to "create a unity of science" (1935a/1983, 119). See also Neurath (1946/1983).

27. Frank (1937/1938) shows how, from the logical empiricists' perspective, the elimination of metaphysics appears as Mach's main aim.
28. Later in the lecture, Poincaré argues more concisely that certain facts have value because they give cognitive access (by providing an analogy) to the facts that are otherwise "passed by": "This new fact is not merely precious by itself, but it alone gives value to all the old facts it combines. Our mind is weak as are the senses; it would lose itself in the world's complexity were this complexity not harmonious; like a near-sighted person, it would see only the details [and could not retain them]. The only facts worthy of our attention are those which introduce order into this complexity and so make it accessible" (1908/1929, 372).
29. According to Neurath, "'intersensual' and 'intersubjective'" language "expresses everything . . . that is common to the blind and the sighted, the deaf and those who hear"; it connects the statements someone makes "with his ears closed, with those he makes with his ear open" (1931b/1983, 62). See also Neurath (1941/1983, 221).

References

A'Campo, N. and Papadopoulos, A. (2014), 'On Klein's So-Called Non-Euclidean Geometry', in L. Ji and A. Papadopoulos (eds.), *Sophus Lie and Felix Klein: The Erlangen Program and Its Impact in Mathematics and Physics*. Zürich: European Mathematical Society, pp. 91–136.

Ben-Menahem, Y. (2006), *Conventionalism*. Cambridge: Cambridge University Press.

Brenner, A. (1998), 'Les voies du positivisme en France et en Autriche: Poincaré, Duhem et Mach', *Philosophia Scientiae*, 3: 31–42.

Carnap, R. (1966/1974), *Philosophical Foundations of Physics: An Introduction to the Philosophy of Science*. New York: Basic Books, Inc.

Cat, J. (2019), 'Neurath and the Legacy of Algebraic Logic', in J. Cat and A. T. Tuboly (eds.), *Neurath Reconsidered: New Sources and Perspectives*. Cham: Springer, pp. 241–338.

Cat, J., Cartwright, N., Fleck, L., and Uebel, Th. (1996), *Otto Neurath: Philosophy between Science and Politics*. Cambridge: Cambridge University Press.

Cayley, A. (1859/1889), 'A Sixth Memoir upon Quantics', *Philosophical Transactions of the Royal Society*. Vol. 149. Reprinted in *The Collected Mathematical Papers of Arthur Cayley*. Vol. 2. Cambridge: Cambridge University Press, pp. 568–606.

Dennett, D. C. (1991), 'Two Contrasts: Folk Craft versus Folk Science, and Belief versus Opinion', in J. D. Greenwood (ed.), *The Future of Folk Psychology*. Cambridge: Cambridge University Press, pp. 135–148.

de Paz, M. (2015), 'Poincaré's Classification of Hypotheses and Their Role in Natural Science', *International Studies in the Philosophy of Science*, 29: 369–382.

Duhem, P. (1914/1991), *The Aim and Structure of Physical Theory*. Princeton: Princeton University Press. Translation of *La théorie physique. Son object, sa structure*.

Dunlop, K. (2016). 'Poincaré on the Foundations of Arithmetic and Geometry. Part 1: Against "Dependence-Hoerarchy" Interpretations', *HOPOS: The Journal of the International Society for the History of Philosophy of Science*, 6 (2): 274–308.

————. (2017). 'Poincaré on the Foundations of Arithmetic and Geometry. Part 2: Intuition and Unity in Mathematics', *HOPOS: The Journal of the International Society for the History of Philosophy of Science*, 7 (1): 88–107.

Elga, A. (2007), 'Isolation and Folk Physics', in H. Price and R. Corry (eds.), *Causation, Physics, and the Constitution of Reality*. Oxford: Oxford University Press, pp. 106–119.

Ferrari, M. (2001), 'Recent Works on Otto Neurath', in M. Rédei and M. Stöltzner (eds.), *John von Neumann and the Foundations of Quantum Physics*. Dordrecht: Springer, pp. 319–327.

Frank, Ph. (1937/1938), 'Ernst Mach: The Centenary of His Birth', *Erkenntnis*, 7: 247–256.

————. (1949), *Modern Science and Its Philosophy*. Cambridge, MA: Harvard University Press.

Friedman, M. (1995–1996/1999), 'Poincaré's Conventionalism and the Logical Positivists', in *Reconsidering Logical Positivism*. Cambridge: Cambridge University Press. pp. 71–86.

Gray, J. (1992), 'Poincaré and Klein: Groups and Geometries', in L. Boi, D. Flament, and J.-M. Salanskis (eds.), *1830–1930: A Century of Geometry*. Berlin: Springer, pp. 35–44.

Gray, J., and Walter, S. (eds.) (1997), *Trois suppléments sur la découverte des fonctions fuchsiennes*. Paris/Berlin: Blanchard/Akademie Verlag.

Haller, R. (1982a/1991), 'The First Vienna Circle', in Th. Uebel (ed.), *Rediscovering the Forgotten Vienna Circle*. Dordrecht: Kluwer, pp. 95–108.

————. (1982b/1991), 'The Neurath Principle: Its Grounds and Consequences', in Th. Uebel (ed.), *Rediscovering the Forgotten Vienna Circle*. Dordrecht: Kluwer, pp. 117–130.

Hawkins, Th. (1984), 'The Erlanger Programm of Felix Klein: Reflections on Its Place in the History of Mathematics', *Historia Mathematica*, 11: 442–470.

Heinzmann, G. (2009), 'Hypotheses and Conventions in Poincaré', in M. Heidelberger and G. Schiemann (eds.), *The Significance of the Hypothetical in the Natural Sciences*. Berlin: De Gruyter, pp. 169–192.

————. (2010), 'Conventions in Geometry and Pragmatic Reconstruction in Poincaré', *Les Preprints de la MSH Lorraine*, 13: 1–16.

Heis, J. (ms.), *The Geometry behind Poincaré's Conventionalism*. Unpublished manuscript.

Henle, M. (2001), *Modern Geometries: Non-Euclidean, Projective, and Discrete*. 2nd ed. Upper Saddle River: Prentice Hall.

Klein, F. (1871/1996), 'On the So-Called Non-Euclidean Geometry', *Mathematische Annalen*, 4. Translated in Stillwell (1996b), pp. 69–112.

————. (1925/1939), *Elementary Mathematics from an Advanced Standpoint: Geometry*. New York: Dover. Translation of *Elementare Mathematik vom höhere Standpunkt*. Vol. 1. 3rd ed.

Mach, E. (1933/1960), *The Science of Mechanics: A Critical and Historical Accont of its Development*. Translated by T.J. McCormack. La Salle: Open Court.

Malle, B. (2001), 'Folk Explanations of Intentional Action', in B. Malle, L. J. Moses, and D. A. Baldwin (eds.), *Intentions of Intentionality*. Cambridge, MA: MIT Press, pp. 265–286.

Neurath, O. (1916/1983), 'On the Classification of Systems of Hypotheses (with Special Reference to Optics)', in R. S. Cohen and M. Neurath (eds.), *Otto Neurath: Philosophical Papers 1913–1946*. Dordrecht: D. Reidel, pp. 13–31.

—— (1931a/1981), *Empirische Soziologie*. Reprinted in R. Haller and H. Rutte (eds.), *Gesammelte philosophische und methodologische Schriften*. Vienna: Hölder-Pichler-Tempsky, pp. 423–527.

—— (1931b/1983), 'Sociology in the Framework of Physicalism', in R. S. Cohen and M. Neurath (eds.), *Otto Neurath: Philosophical Papers 1913–1946*. Dordrecht: D. Reidel, pp. 58–90.

—— (1932/1983), 'Protocol Statements', in R. S. Cohen and M. Neurath (eds.), *Otto Neurath: Philosophical Papers 1913–1946*. Dordrecht: D. Reidel, pp. 91–99.

—— (1934/1983), 'Radical Physicalism and the "Real World"', in R. S. Cohen and M. Neurath (eds.), *Otto Neurath: Philosophical Papers 1913–1946*. Dordrecht: D. Reidel, pp. 100–114.

—— (1935a/1983), 'The Unity of Science as a Task', in R. S. Cohen and M. Neurath (eds.), *Otto Neurath: Philosophical Papers 1913–1946*. Dordrecht: D. Reidel, pp. 115–120.

—— (1935b/1983), 'Pseudorationalism of Falsification', in R. S. Cohen and M. Neurath (eds.), *Otto Neurath: Philosophical Papers 1913–1946*. Dordrecht: D. Reidel, pp. 121–131.

—— (1936a/1983), 'Individual Sciences, Unified Science, Pseudo-Rationalism', in R. S. Cohen and M. Neurath (eds.), *Otto Neurath: Philosophical Papers 1913–1946*. Dordrecht: D. Reidel, pp. 132–138.

—— (1936b/1983), 'Encyclopedia as "Model"', in R. S. Cohen and M. Neurath (eds.), *Otto Neurath: Philosophical Papers 1913–1946*. Dordrecht: D. Reidel, pp. 145–158.

—— (1941/1983), 'Universal Jargon and Terminology', in R. S. Cohen and M. Neurath (eds.), *Otto Neurath: Philosophical Papers 1913–1946*. Dordrecht: D. Reidel, pp. 213–229.

—— (1946/1983), 'The Orchestration of the Sciences by the Encyclopedism of Logical Empiricism', in R. S. Cohen and M. Neurath (eds.), *Otto Neurath: Philosophical Papers 1913–1946*. Dordrecht: D. Reidel, pp. 230–242.

Poincaré, H. (1882/1996), 'Theory of Fuchsian Groups', *Acta Mathematica*, 1. Reprinted in Stillwell (1996b), pp. 123–130.

—— (1898), 'On the Foundations of Geometry', *The Monist*, 9 (1): 1–43.

—— (1902/1929), 'Science and Hypothesis', in *Foundations of Science*. New York: Science Press. Translation of *La science et le hypothèse*.

—— (1905/1929), 'The Value of Science', in *Foundations of Science*. New York: Science Press. Translation of *La valeur de la science*.

—— (1908/1929), 'Science and Method', in *Foundations of Science*. New York: Science Press. Translation of *Science et méthode*.

Reichenbach, H. (1928/1958), *The Philosophy of Space and Time*. New York: Dover.

—— (1951), *The Rise of Scientific Philosophy*. Berkeley and Los Angeles: University of California Press.

Schlick, M. (1915/1979), 'The Philosophical Significance of the Principle of Relativity', in H. L. Mulder and B. van de Velde-Schlick (eds.), *Moritz Schlick: Philosophical Papers*. Dordrecht: Reidel, pp. 153–189.

—— (1917/1920), *Space and Time in Contemporary Physics*. Oxford: Oxford University Press.

Soulez, A. (1997), 'Qui était Otto Neurath?', in A. Soulez, F. Schmitz, and J. Sebestik (eds.), *Otto Neurath, un philosophe entre guerre et science*. Paris: L'Harmattan, pp. 11–17.

Stillwell, J. (1996a), 'Introduction to Poincaré's Theory of Fuchsian Groups, Memoir on Kleinian Groups, on the Applications of Noneuclidean Geometry to the Theory of Quadratic Forms', in Stillwell (1996b), pp. 113–122.

—— (1996b), *Sources of Hyperbolic Geometry*. Providence, RI: American Mathematical Society.

Stöltzner, M. (2001), 'Otto Neurath 1913–1915', in J. T. Blackmore, R. Itagaki, and S. Tanaka (eds.), *Ernst Mach's Vienna 1895–1930*. Dordrecht: Springer, pp. 105–122.

Stump, D. (1989), 'Henri Poincaré's Philosophy of Science', *Studies in History and Philosophy of Science*, 20: 335–363.

Torretti, R. (1978), *Philosophy of Geometry from Riemann to Poincaré*. Dordrecht: Kluwer.

Uebel, Th. (1998), 'Fact, Hypothesis, and Convention in Poincaré and Duhem', *Philosophia Scientiae*, 3: 75–94.

Vuillemin, J. (1979), 'On Duhem's and Quine's Theses', *Grazer Philosophische Studien*, 9: 69–96.

Zahar, E. (1997), 'Poincaré's Philosophy of Geometry, or Does Geometric Conventionalism Deserve Its Name?', *Studies in History and Philosophy of Science Part B*, 28: 183–218.

Zemplén, G. (2019), 'Neurath's Theory of Theory Classification: History, Optics, and Epistemology', in J. Cat and A. T. Tuboly (eds.), *Neurath Reconsidered: New Sources and Perspectives*. Cham: Springer, pp. 217–240.

5   The Philosopher Physicists
Albert Einstein and Philipp Frank

Don Howard

5.1. Introduction

Much has been written about Albert Einstein, logical empiricism, and the Vienna Circle, including such classic essays as Gerald Holton's "Mach, Einstein, and the Search for Reality" (Holton 1968) and my several essays on the subject (for example, Howard 1984, 1994, 2014).[1] Surprisingly little attention has been paid, however, to Einstein's longest and, in many ways, most important relationship with one of the central figures in the Vienna Circle, the physicist and philosopher of science Philipp Frank. There are good reasons for this neglect, foremost among them being that Einstein's frequent interactions with Frank had less impact on the development of Einstein's philosophy of science than did Einstein's relationships with Moritz Schlick and Hans Reichenbach and less impact on the development of core logical empiricist doctrine. But Einstein had a profound impact on the development of Frank's philosophy of science, and Frank's many writings about Einstein provide us with a unique window through which to appreciate aspects of Einstein's philosophy of science that are still not widely enough understood.[2]

Most important in this respect is Einstein's profoundly important endorsement of key themes in Pierre Duhem's philosophy of science, including his holism and his underdeterminationist version of conventionalism (Duhem 1906; Howard 1990, 1993). Frank was far more alert to the Duhemian moment in Einstein's thinking because he shared that sympathy for Duhem as one of the most prominent members of what is known as the "left wing" of the Vienna Circle, along with that other famous fan of Duhem, Otto Neurath (Howard 2019), and, thanks to that shared sympathy, Frank ventured what was, at the time, a novel and insightful reading of Einstein's philosophical views. Thus, in order for us better to understand both Einstein and Frank, along with the philosophical fine structure of the logical empiricist movement, we must take a closer look at this 48-year professional friendship.

5.2. In the Beginning: "Simplicity of Nature"

Philipp Frank was born in Vienna in 1884, five years after Einstein's birth in Ulm. He studied physics and mathematics in Göttingen and Vienna under Ludwig Boltzmann, Felix Klein, and David Hilbert, completing his doctorate in physics in Vienna in 1906 (see Stadler 2001, 631–636). The first direct encounter between Einstein and Frank took place in 1907 after Einstein read an essay of Frank's on the law of causality in which Frank presented an analysis of the law based on Henri Poincaré's conventionalist philosophy of science (Poincaré 1902; Frank 1907/1949). Einstein wrote to Frank with complimentary remarks about the paper and one important, friendly suggestion. Here is how Frank recalled the key point many years later:

> When I first published this paper (1907) it aroused a certain amazement among scientists. Among the comments were those of two men of world-wide fame, although in different fields: Einstein and Lenin. Einstein's letter was my first personal contact with him. He approved the logic of my argument, but he objected that it demonstrates only that there is a conventional element in the law of causality and not that it is merely a convention or definition. He agreed with me that, whatever may happen in nature, one can never prove that a violation of the law of causality has taken place. One can always introduce by convention a terminology by which this law is saved. But it could happen that in this way our language and terminology might become highly complicated and cumbersome. What is *not* conventional in the law of causality is the fact that we can save this law by using a relatively *simple* terminology: we are sure that a state *A* has recurred when a small number of state variables have the same values that they had at the start. This "simplicity of nature" is the observable fact which cannot be reduced to a convention on how to use some words. These remarks had a great influence on my thought on the future course of the philosophy of science. I realized that Poincaré's conventionalism needs qualifications. One has to distinguish between what is logically possible and what is helpful in empirical science. In other words, logic needs a drop of pragmatic oil.
> (1949d, 10–11, original emphases)

One is not surprised to find Einstein focusing already at this early date, in his own work, on the role of simplicity in science, for that was an abiding feature of Einstein's thinking, memorably expressed in his 1933 Herbert Spencer Lecture at Oxford, where he said, "[o]ur experience hitherto justifies us in believing that nature is the realisation of the simplest conceivable mathematical ideas" (Einstein 1934, 167; see also Norton 2000).[3]

It is difficult to document the development of Einstein's relationship with Frank in these early years because none of their correspondence from this period has survived. But it is clear that Einstein was reading Frank's work carefully, including his 1909 Vienna habilitation thesis on the place of the relativity principle in mechanics and electrodynamics (Frank 1909) and that Einstein was deeply impressed by Frank as both a physicist and a philosopher, for, when Einstein left the Charles University in Prague in 1912 to return to Zurich and a new appointment at the Swiss Federal Polytechnic, he strongly recommended Frank as a top candidate to be his successor. In their "Report to the Philosophical Faculty of the German University on a Successor to the Chair of Theoretical Physics" on May, 24 1912, Einstein and his Prague colleagues, Anton Lampa and Georg Pick, wrote:

> It turns out that, among the people under consideration, two are to be mentioned in the first rank because of their distinguished scientific accomplishments, namely, Paul Ehrenfest and Philipp Frank, both from Vienna. . . .
>
> Philipp Frank. He was born in Vienna in 1884 and studied also in Vienna and Göttingen. He was promoted in Vienna in 1906 and has lectured, with great success, as a Privatdozent at the University of Vienna for two years. The abundance of capable scientific work that this only 28-year-old person has achieved is to be admired. He unites a rare mastery of the mathematical tools with a good understanding of the problems in physics.
>
> Frank's most important papers in physics concern themselves with the theory of relativity. In a paper written in 1908, he showed how, in a simple way, one can get to Minkowski's equations for the electrodynamics of moving bodies starting from Lorentz's electron theory. In two papers appearing in 1908 and 1909 the space-time transformations of classical mechanics, on the one hand, and those of the theory of relativity, on the other hand, are placed in the foreground, and it is shown how the former lead to classical mechanics and to Hertz's electrodynamics of moving bodies and the latter lead to the corresponding equations of the theory of relativity. Another investigation of comparable, essential, systematic significance is that published in 1911 together with H. Rothe, entitled "On the Transformation of Space-Time-Coordinates from Resting to Moving Systems." In this paper it is shown that there are only three types of transformations that together form a one-parameter, linear, homogenous group. One of these groups of transformations is that of classical mechanics, another is that of the Lorentz transformations.
>
> We mention briefly that Frank has written several original essays of an epistemological nature ("The Law of Causality and Experience," "Mechanism or Vitalism?"), that give evidence of the multifaceted

nature of their author, as well as of his effort to engage general problems of knowledge. A few papers of a purely mathematical nature are also to be mentioned, as well as a few papers from the domain of analytical mechanics, which are probably of more mathematical than physical interest.

Herr Frank is an excellent lecturer. From his papers it is clear that he is a capable mathematician and, at the same time, an excellent student of theoretical physics, and that his is an unusually multifaceted aptitude.

(CPAE 5, Doc. 400, 470–473)

Frank was duly hired, and he occupied the physics professorship at the Charles University until the German takeover of Czechoslovakia in 1938 forced his departure for the United States.

5.3. Frank's Interpretation of Einstein's Philosophy of Science: Mach, Poincaré, and Duhem

Einstein and Frank shared more than just an interest in physics. They were both, in nearly equal measure, philosopher-physicists, scientists for whom the pursuit of physics required deep engagement with questions of method in physics and the philosophical foundations of physical theory. Einstein gave a compelling expression of this attitude in his 1916 memorial essay on Ernst Mach, writing:

How does it happen that a properly endowed natural scientist comes to concern himself with epistemology? Is there no more valuable work in his specialty? I hear many of my colleagues saying, and I sense it from many more, that they feel this way. I cannot share this sentiment. When I think about the ablest students whom I have encountered in my teaching, that is, those who distinguish themselves by their independence of judgment and not merely their quick-wittedness, I can affirm that they had a vigorous interest in epistemology. They happily began discussions about the goals and methods of science, and they showed unequivocally, through their tenacity in defending their views, that the subject seemed important to them. Indeed, one should not be surprised at this. . . .

Concepts that have proven useful in ordering things easily achieve such an authority over us that we forget their earthly origins and accept them as unalterable givens. Thus they come to be stamped as "necessities of thought," "a priori givens," etc. The path of scientific advance is often made impassable for a long time through such errors. For that reason, it is by no means an idle game if we become practiced in analyzing the long commonplace concepts and exhibiting those circumstances upon which their justification and usefulness depend, how they have grown up, individually, out of the givens of experience. By this means, their all-too-great authority will be broken. They will be removed if they cannot be properly legitimated, corrected if their correlation with given things be far too superfluous, replaced by others if a new system can be established that we prefer for whatever reason.

(1916, 101–102)

But it was not just Mach from whom Einstein and Frank drew inspiration and ideas. They were both equally indebted to Poincaré and Duhem.

About the decade after his first encounter with Einstein, Frank later wrote:

During the interval (1907–1917) . . . my interest was directed mainly toward any possible advance in the logic of science. I was convinced that the solution must be sought by starting from the ideas of such men as Mach and Poincaré.

At first glance these two authors seemed to contradict each other flagrantly. I soon realized that any advance in the philosophy of science would consist in setting up a theory in which the views of Mach and of Poincaré would be two special aspects of one more general view. To summarize these two theories in one single sentence, one might say: According to Mach the general principles of science are abbreviated economical descriptions of observed facts; according to Poincaré they are free creations of the human mind which do not tell anything about observed facts. The attempt to integrate the two concepts into one coherent system was the origin of what was later called logical empiricism.

(1949d, 11–12)

This might not be a complete and correct origin story about logical empiricism, but Frank's comment highlights what, for him, was the key challenge facing the philosophy of science at that time, namely, how to reconcile Mach's positivism with Poincaré's conventionalism. By the way, the dates that Frank gives are noteworthy, spanning exactly the period from that early paper on the law of causality, where he was defending a conventionalist view, and his important 1917 paper on the significance of Mach's epistemology of science for the present (Frank 1917/1949).

The first steps toward that hoped-for reconciliation require our disentangling what Frank terms the "logical viewpoint" on physical theory from questions of its physical or empirical interpretation:

The axiomatic or structural system, including its conclusions, is merely an arbitrary convention if only the purely logical viewpoint is maintained without going into the physical interpretation. It was clear to Poincaré that the structural system is logically arbitrary because it cannot be demonstrated by logical means. It is not psychologically arbitrary, however, because in practice we construct only those systems that can be interpreted in terms of physical facts and that are therefore helpful for the formulation of natural laws.

If this line of reasoning is followed, we can see, in a perfunctory way at least, how Mach's and Poincaré's ideas about the general principles of science can be integrated. The axiomatic system, the set of relations between symbols, is a product of our free imagination; it is arbitrary. But if the concepts occurring in it are interpreted or identified with some observational conceptions, our axiomatic system, if well chosen, becomes an economical description of observational facts.

Now the presentation of the law of causality as an arbitrary convention [cf. Frank 1907/1949] can be freed of its paradoxical appearance. The law of causality, as a part of an axiomatic system, is an arbitrary convention about the use of terms like "the recurrence of a state of a system," but if interpreted physically it becomes a statement about observable facts. In this way, the philosophy of Mach could be integrated into the "new positivism" of men like Henri Poincaré, Abel Rey, and Pierre Duhem.

(1949d, 13–14)

Indeed, for Frank, it was especially Duhem who pointed the way to reconciling the philosophies of Mach and Poincaré:

The property of the structural system of not telling us anything about the world of observable physical facts was particularly emphasized by the French scientist, philosopher, and historian Pierre Duhem. His writings exerted a strong influence upon our group and, particularly, upon my own thinking. . . . Duhem says, much as Mach has done,

"A theory of physics is not an explanation; it is a system of mathematical propositions deduced from a small number of principles the aim of which is to represent as simply, as completely, and as exactly as possible, a group of experimental laws."

This formulation is a great step on the way toward an integration of Mach and Poincaré. Duhem understood very well that no single proposition of a physical theory can be said to be verified by a specific experiment. The theory as a whole is verified by the whole body of experimental facts.

(Frank 1949d, 15, quoting from Duhem 1906, 24)

It is Duhem's holism that connects Mach's empiricism with Poincaré's conventionalism. Yes, theoretical science is grounded in experience but

not in the rigid manner suggested by those who caricature Mach's position as a version of reductionalist phenomenalism,[4] which would imply
that every individual proposition is vested with its own, determinate,
empirical content and, thus, can be tested in isolation from the rest of the
theory in which it is embedded. Instead, only whole theories are subjected
to empirical testing, which leaves room to make changes in various places
in the theory, as emphasized by Poincaré's conventionalism, should the
theory as a whole be found to be in conflict with the totality of the available, empirical evidence.

The group to which Frank is referring in that last-quoted remark, the
group now often dubbed "the first Vienna Circle" (Haller 1991), had at
its core, in addition to Frank, the economist Neurath[5] and the mathematician Hans Hahn. Frank fondly recalls their regular Thursday-evening
meetings in a Viennese coffee house, where they discussed late into the
night the most urgent questions in the philosophy of science. The one
question that most dominated those discussions was a perceived loss of
public faith in science brought about, Frank thought, by the decline of the
long influential mechanistic point of view, under challenge on every front
by the new physics of relativity and the quantum. Frank characterized the
problem as a "crisis" in the "scientific conception of the world":

> The decline of the belief in mechanistic science seemed to favor [the]
> organismic view, which has been attractive to many because of its
> religious and social implications. In this way there had arisen at the
> turn of the century what some called a crisis in science or, more accu-
> rately, in the scientific conception of the world. For more than two
> centuries the idea of progress in science and human life had been con-
> nected with the advance of the mechanistic explanation of natural
> phenomena. Now science itself seemed to abandon this mechanistic
> conception, and the paradoxical situation arose that one could fight
> the scientific conception of the world in the name of the advance of
> science.
>
> (1949d, 4)

Thus, Frank's wrangling with questions about the nature of science and
its empirical warrant concerned far more than just arcane issues in technical philosophy of science. As with the prominent, progressivist commitments of the more famous Vienna Circle that grew up around Schlick in
the 1920s and 1930s, so, too, this first Vienna Circle struggled to preserve
the cultural authority of science in the service of progressive ends.

Those of later generations who first encountered Duhem through
W.V.O. Quine's appropriation of some of his ideas in his influential essay,
"Two Dogmas of Empiricism" (1951), might be surprised by Frank's
suggestion that Mach's empiricism and Duhem's holism could be com-
fortably wed, because Quine was one of those who wrongly read Mach

as a reductionist phenomenalist and argued explicitly that Duhem-style holism was antithetical to what Quine termed "reductionism," the second of the two "dogmas" that he attacked. The fact of the matter is that Mach himself highlighted and celebrated the community of purpose between himself and Duhem and the harmony of their views.

Not well enough known is the fact that it was Mach who championed the German translation of Duhem's *La Théorie physique, son objet et sa structure* [*The Aim and Structure of Physical Theory*] (1906) and contributed to it a foreword in which he explained the several deep points of agreement, including, notably, Duhem's holism and his anti-metaphysical approach to science (Duhem 1908; Mach 1908; note that Frank translated Duhem's *L'évolution de la mécanique* in 1912 into German). But he had already recorded his sympathy with Duhem immediately after the publication of *La Théorie physique* when, in 1906, he brought out a second edition of the 1905 book *Erkenntnis und Irrtum* [*Knowledge and Error*] (Mach 1905) that differs from the first mainly only in the addition of several footnotes and one glowing remark in the foreword where he talks about the many points of agreement between himself and Duhem:

> I was very pleased by Duhem's work, "La théorie physique, son objet et sa structure" (1906). I had not yet hoped to find such thoroughgoing agreement on the part of physicists. Duhem repudiates any metaphysical conception of questions in physics; he views the conceptually-economical determination of the factual as the aim of physics; he regards the historical-genetic representation of theories as the uniquely correct and didactically most effective one. Those are views that I have amply defended with reference to physics for three decades. The agreement between us is all the more precious to me, since Duhem arrived at the same results wholly independently.
>
> (Mach 1906, x)

> Claude Bernard advises us to disregard all theory in experimental investigations, to leave theory at the door. Duhem rightly objects that this is impossible in physics, where experiment without theory is incomprehensible. . . . In fact, one can only recommend that attention be given to whether or not the experimental result is on the whole compatible with the assumed theory. Cf. Duhem (La Théorie physique, pp. 297f).
>
> (Mach 1906, 202)

> Duhem (La Théorie physique, pp. 364f) explains that hypotheses are not so much *chosen* by the researcher, arbitrarily and at will, but rather *force* themselves *upon* the researcher in the course of historical

development, under the impress of facts that are gradually becoming known. Such a hypothesis usually consists of a whole complex of ideas. If a result then arises, e.g., through an "experimentum crucis" that is incompatible with a hypothesis, then for the time being one can only regard it as contradicting the *entire complex of ideas*. On this latter point cf. Duhem, l.c., pp. 311f.

(Mach 1906, 244)

Mach is, of course, not here anticipating Frank's specific point about Duhem's having shown us how to marry the views of Mach and Poincaré. But Mach's strong endorsement of Duhem's philosophy of science is conformable with Frank's reconciliation thesis and shows that commonly deployed later twentieth-century maps of this philosophical landscape are seriously in error, requiring correction through Frank's thoughtful reading of the early twentieth-century philosophy of science dialectic.

The issue of reconciling Mach's empiricism with Poincaré's conventionalism is a kind of leitmotif running through many of Frank's writings over the years. But it might appear to some a curious fact that, whereas earlier he assigned the role of the peacemaker to Duhem, in an important late essay on Einstein's philosophy of science published in 1949 in *Reviews of Modern Physics*, he assigned that very same role to Einstein:

At the first glance, Mach's and Poincaré's views on the principles of science seem to be antagonistic and even contradictory to each other. In Einstein's philosophy of science these two views appear as two aspects of one integrated view.

According to Mach the principles of science offer an "economic" (practical) description of a great diversity of sense observations. According to Poincaré, however, these principles are free creations of the human mind which are neither true nor false but may be convenient or inconvenient.

(Frank 1949a, 350)

After pointing out that there were aspects of special relativity that reflected Mach's empiricist way of thinking, such as Einstein's stressing the empirical meaninglessness of the notion of absolute, distant simultaneity, but also aspects that reflected Poincaré's emphasis on the role of conventions in physical theory, as with Einstein's conventional definition of what count as equal time intervals, he makes a similar remark about general relativity's having both empiricist and conventionalist moments. He then remarks that "the validity of both characteristics within one and the same physical theory was a clear demonstration of the compatibility of Mach's and Poincaré's views if they were correctly understood" (Frank

1949a, 350). And he writes that it was Einstein who showed how the reconciliation was to be done:

> We can summarize Einstein's contribution by stressing that he gave the last and decisive touch to the program of physics that was outlined by men like Mach and Poincaré. What these scientists had in mind as a mere program and what they were not able to formulate in full precision Einstein achieved and presented as a full-fledged scientific theory. It assigned its correct and satisfactory place to non-Euclidean geometry and defined precisely the mutual relation between convention and facts in science.
>
> (Frank 1949a, 352)

It is more than a little puzzling that Frank makes no mention of Duhem in this paper, all the more so because the historical memoir that was quoted previously where, thinking back to his early years, Frank assigned the conciliatory role to Duhem was published in the same year! But in thus giving joint credit to both Einstein and Duhem, Frank evinced an understanding of Einstein's philosophy of science unlike anyone else at that time or for decades thereafter.

In that same year of 1949, Frank published another essay on Einstein's philosophy of science, his contribution to the Library of Living Philosophers volume, *Albert Einstein: Philosopher-Scientist* (Schilpp 1949). Entitled "Einstein, Mach, and Logical Positivism," the essay rehearses many of the same themes as the paper in *Reviews of Modern Physics* but now with the focus more narrowly on Mach and logical empiricism, and it carefully catalogues Einstein's own, changing appreciations of the influence that Mach had upon him. Perhaps most interesting, however, is the way in which this second essay offers an interpretation of Mach's philosophical project strongly at odds with what was by then the common misconstrual of his views as a version of reductionist phenomenalism. Thus, after discussing an 1894 lecture of Mach's on the principle of comparison in physics (Mach 1894) and selected passages from Auguste Comte's *Cours de philosophie positive* (Comte 1830–1842), Frank writes:

> From these quotations it seems to become clear that even the "classical positivism" of Comte or Mach did not hold the opinion that the laws of nature could be simply "derived" from experience. These men knew very well that there must be a theoretical starting-point, a system of principles constructed by the human imagination in order to compare its consequences with observations. This feeling was so strong that Comte accepted even the theological principles as a starting-point to "get science going."
>
> (1949b, 278)

Frank's point is that the positivism of Comte and Mach does not tie theory to experience so tightly as to disallow the "free creation" of scientific concepts that both Einstein and Poincaré regarded as essential. Thus, in this essay, too, Frank is promoting his message of consilience.

What was Einstein's own view of the matter? As I have argued elsewhere (Howard 1990, 1993, 1994), Einstein's philosophy of science owed more of a debt to Duhem than to any other single thinker, including Mach and Poincaré. Einstein's appreciation and appropriation of Mach's empiricism and Poincaré's conventionalism predated his first acquaintance with Duhem's work by as much as a decade. It was during his student years at the Swiss Federal Polytechnic that Einstein first read Mach, and by the time of his first paper on special relativity (Einstein 1905), Einstein had read and assimilated not only Mach's classic book on the analysis of sensations (Mach 1900) but also his magisterial books on the critical history of mechanics (Mach 1897) and the theory of heat (Mach 1896). We can date his reading of Poincaré's *La science et l'hypothèse* [*Science and Hypothesis*] (Poincaré 1902), probably in the form of the German translation (Poincaré 1904) to his early years working in the Swiss Federal Patent Office in Bern between 1902 and 1905.[6] Traces of his reading of both Mach and Poincaré are plainly evident in his first paper on special relativity (Einstein 1905).

Einstein's first encounter with Duhem seems to have come in the fall of 1909 or shortly thereafter. When Einstein moved from Bern to Zurich to assume his first regular academic appointment at the University of Zurich, his family chanced to rent an apartment in the very same building where his old student friend, the physicist Friedrich Adler, also lived. By coincidence, Adler had been the other finalist for Einstein's position at the University of Zurich but withdrew his candidacy when he heard of Einstein's being considered for the job. We know from Adler's letters to his father, Victor Adler, founder of the Austrian Social Democratic Workers' Party, that Adler and Einstein set up a shared study in the attic where they spent many long hours discussing physics and the philosophy of science. As it happens, Adler was the translator for the recently published German edition of Duhem's *La Théorie Physique* (Duhem 1908; see also Howard 1990).

That Einstein immediately grasped the philosophical significance of Duhem's holism and underdeterminationism is obvious from a remark in his electricity and magnetism lectures at the University of Zurich in the winter semester of 1910–1911, where he explained how the concept of electrical charge in the interior of a solid charged body, such as an electron, made empirical sense even though we cannot introduce a test charge inside the solid body to measure directly the strength of the interior charge:

> We have seen how experience led to the introduction of the concept of electrical charge. It was defined with the help of forces that electrified

bodies exert on each other. But now we extend the application of the concept to cases in which the definition finds no direct application as soon as we conceive electrical forces as forces that are exerted not on material particles but *on electricity*. We establish a conceptual system whose individual parts do not correspond immediately to experiential facts. Only a certain totality of theoretical materials corresponds again to a certain totality of experimental facts.

We find that such an el[ectrical] continuum is always applicable only for representing relations inside ponderable bodies. Here again we define the vector o[f] el[ectrical] field strength as the vector of the mech[anical] force that is exerted on a unit of pos[itive] electr[ical] charge inside a ponderable body. But the force thus defined is no longer immediately accessible to exp[eriment]. It is a part of a theoretical construction that is true or false, i.e., corresponding or not corresponding to experience, only *as a whole*.

(1910–1911, 325)

One could not ask for a more concise statement of the main points in Duhem's philosophy of science (see Howard 1993).

Holism and an underdeterminationist version of conventionalism is a theme running through virtually all of Einstein's writings on the philosophy of science from this early point to the very end of his life. One salient example comes from Einstein's important 1921 paper "Geometrie und Erfahrung" ["Geometry and Experience"], where he analyzes the respective roles of purely formal geometry and physically or empirically interpreted geometry in the general theory of relativity and argues that we cannot assess the status of geometry alone but only in conjunction with the dynamics of general relativity:

We feel impelled toward the following more general view, which characterizes Poincaré's standpoint. Geometry (G) predicates nothing about the behavior of real things, but only geometry together with the totality (P) of physical laws can do so. Using symbols, we may say that only the sum of (G) + (P) is subject to experimental verification. Thus (G) may be chosen arbitrarily, and also parts of (P); all these laws are conventions. All that is necessary to avoid contradictions is to choose the remainder of (P) so that (G) and the whole of (P) are together in accord with experience. Envisaged in this way, axiomatic geometry and the part of natural law which has been given a conventional status appear as epistemologically equivalent.

Sub specie aeterni Poincaré, in my opinion, is right. The idea of the measuring rod and the idea of the clock coordinated with it in the theory of relativity do not find their exact correspondence in the real world. It is also clear that the solid body and the clock do not in the conceptual edifice of physics play the part of irreducible elements, but that of composite structures, which must not play any independent part in theoretical physics. But it is my conviction that in the present stage of development of theoretical physics these concepts must still be employed as independent concepts; for we are still far from possessing such certain knowledge of the theoretical principles of atomic structure as to be able to construct solid bodies and clocks theoretically from elementary concepts.

(1921, 8)

Two pages later, Einstein adds:

> According to the view advocated here, the question whether the continuum has Euclidean, Riemannian, or any other structure is a question of physics proper which must be answered by experience, and not a question of a convention to be chosen on the grounds of mere expediency.

(1921, 10)

These remarks have occasioned some puzzlement, much of it owing to the curious and, to me, still inexplicable fact that the view ascribed to Poincaré at the first mention of his name is really that of Duhem, namely that geometrical and physical propositions cannot be tested in isolation from one another but only conjointly, and that the view referenced at the second mention of Poincaré's name is, again, Duhem's, whereas the view being repudiated in that second paragraph is, in fact, Poincaré's, specifically the idea that geometrical primitive concepts are linked to experience, one by one, via conventional, physical coordinating definitions, in this case defining the infinitesimal metrical interval by means of ideal measuring rods and clocks.

Those exegetical challenges notwithstanding, the point of view that Einstein here defends is exactly the one that Frank held to be central to the new logical empiricism, with Mach's empiricism and Poincaré's conventionalism being reconciled by recognizing, *à la* either Duhem or Einstein, that it is only whole theories that possess empirical significance and are subject to empirical test. No doubt that is why Frank devoted considerable attention to this essay in his previously quoted 1949 paper on Einstein's philosophy of science:

> Einstein formulated this underlying philosophy of space and time explicitly in his lecture "Experience and Geometry" which he gave in 1921 to the Prussian Academy of Science in Berlin.
> This lecture is a historic landmark in the long and torturous approach to philosophical clarity. The relation between experience and logic in geometry, and in all science altogether, was presented,

the first time, in a satisfactory way. Einstein did it with great simplicity and directness leaving no dark angle where remnants of obscurity could take a hiding. . . .

Einstein's General Theory of Relativity was the first physical theory in which what one called formerly "geometry" was completely integrated into the frame of physics in general. Only within such a frame the relation between reasoning and experience in geometry can be clearly stated. The axioms of geometry by themselves are, as Hilbert and Poincaré pointed out, purely formal systems from which we can draw results without knowing the significance of the words "straight line" etc. But by interpreting a straight line as a light ray or as a rigid rod, the axioms of geometry become physical hypotheses; they become statements within mechanics or optics. All conclusions can be checked with the same precision and with the same uncertainty as all statements of physics. Einstein characterized this situation by the famous dictum that, as far as geometry is precise, it does not tell us anything about the world (it is a formal system); if, however, it tells us something about the world, it is not precise (it has the same status as a statement of physics).

(1949a, 351)

Einstein confuses Poincaré and Duhem; Frank transposes the title of Einstein's 1921 lecture, turning it into the title of the specific essay by Poincaré to which Einstein was responding, "L'expérience et la géométrie" (1902). Both would have been helped by competent copy editors and proofreaders. But the point remains that Einstein's 1921 essay is, by Frank's own lights, the purest expression of the synthesis of Mach and Poincaré mediated by Duhem and Einstein that Frank held to be the most important achievement of early twentieth-century philosophy of science.

That Einstein-Duhem version of logical empiricism promoted by Frank and other members of the left wing of the Vienna Circle must be clearly distinguished from the importantly different version of logical empiricism offered up by the right wing, especially in the early work of Reichenbach (after he had abandoned his original, neo-Kantian position) and Rudolf Carnap (Reichenbach 1924/1969, 1928/1958; Carnap 1928). The right-wing view derived directly from Poincaré, holding that each individual concept acquires its own determinate, empirical content by way of a conventional, coordinating definition. This is the origin of the notion of verificationism in right-wing, logical empiricist philosophy of science. It follows that each individual proposition likewise acquires its own empirical content and can, therefore, be tested independently of the rest of the theory to which it belongs (see Howard 1994). By contrast, the Einstein-Duhem position denies that individual concepts and propositions possess determinate empirical content.

Helpful additional light on the most confusing parts of Einstein's "Geometrie und Erfahrung" essay, and a crucial further demonstration of his commitment to Duhemian holism, came three years later in an obscure but very interesting book review. The book in question, *Kant und Einstein. Untersuchungen über das Verhältnis der modernen Erkenntnistheorie zur Relativitätstheorie* [*Kant and Einstein: Investigations into the Relationship of Modern Epistemology to Relativity Theory*] was by a minor neo-Kantian, Alfred Elsbach, one of many such seeking at the time to respond to general relativity's challenge to the Kantian doctrine of the a priori, especially regarding the allegedly necessary a priori, Euclidean structure of space as a form of outer intuition (Elsbach 1924; see also Howard 1994). In his review, Einstein explained his dissent from Kantian orthodoxy in a perhaps surprising way:

> This does not, at first, preclude one's holding at least to the Kantian *problematic*, as, e.g., Cassirer has done. I am even of the opinion that this standpoint can be rigorously refuted by no development of natural science. For one will always be able to say that critical philosophers have until now erred in the establishment of the a priori elements, and one will always be able to establish a system of a priori elements that does not contradict a given physical system. Let me briefly indicate why I do not find this standpoint natural. A physical theory consists of the parts (elements) A, B, C, D, that together constitute a logical whole that correctly connects the pertinent experiments (sense experiences). Then it tends to be the case that the aggregate of fewer than all four elements, e.g., A, B, D, *without* C, no longer says anything about these experiences, and just as well A, B, C without D. One is then free to regard the aggregate of three of these elements, e.g., A, B, C as a priori, and only D as empirically conditioned. But what remains unsatisfactory in this is always the *arbitrariness in the choice* of those elements that one designates as a priori, entirely apart from the fact that the theory could one day be replaced by another that replaces certain of these elements (or all four) by others.
>
> (1924, 1688–1689)

Einstein's subtle point against Kant is that, when a whole theory meets experience, there are multiple possible choices for what to hold fixed, hence a priori, and what to allow to be exposed to the test of experience, hence contingent, and that the choice among these alternatives is arbitrary. Einstein's view should not be confused with the notion of a contingent or relativized a priori such as had been promoted by Reichenbach (1920/1965) four years previously in his book on relativity theory and the a priori, for Reichenbach's a priori is fixed, for the time being, but can be revised in light of new experience, whereas Einstein's point is that the very notion of the a priori is rendered suspect because of the arbitrariness in one's designation of what to regard as the a priori elements of scientific cognition. In fact, Einstein is here, yet again, asserting the point of view that Frank took to be the essence of his version of logical empiricism, namely that whole bodies of theory are tested empirically, not individual propositions, and that the particular manner in which one arrays the formal elements of a theory is a matter of convention. He is essentially reiterating what he wrote in "Geometrie und Erfahrung" about how it is never geometry alone that possesses empirical content but only geometry together with physics, "(G) + (P)."

The other puzzling comment in "Geometrie und Erfahrung," found in the "*Sub specie aeterni*" paragraph, spoke of how, in principle, rods and clocks, as physical objects, must be represented by solutions to the field equations, but that, in the current, underdeveloped state of the theory, that was impossible, implying that, until we find a theory of everything, we are compelled to employ stipulated rods and clocks as affording, in effect, a coordinating definition for the infinitesimal metrical interval. Does this mean that Einstein was abandoning his principled Duhemian view and, on pragmatic grounds, endorsing the Poincaré-Reichenbach view? Einstein's review of Elsbach's book sheds additional light on this as well.

Einstein takes up an argument that Elsbach had borrowed from the neo-Kantian Paul Natorp to the effect that we can always save Euclidean geometry from seeming empirical refutation by making changes in our physics and that we can never directly determine the metrical structure of space because space is not real but ideal (Natorp 1910). Einstein responds as follows:

> The position that one takes with respect to these theses depends on whether one grants reality to the practically-rigid body. If yes, then the concept of the interval [the infinitesimal metrical interval] corresponds to something experiential. Geometry then contains assertions about possible experiments; it is a physical science that is directly underpinned by experimental testing (standpoint A). If the practically-rigid measuring body is accorded no reality, then geometry alone contains no assertions about experiences (experiments), but instead only geometry with physical sciences taken together (standpoint B). Until now physics has always availed itself of the simpler standpoint A and, for the most part, is indebted to it for its fruitfulness; physics employs the latter in all of its measurements. . . . But if one adopts standpoint B, which seems excessively cautious [übertrieben vorsichtig] at the present stage of the development of physics, then geometry alone is not experimentally testable. There are then no geometrical measurements whatsoever. . . . Only a *complete* scientific conceptual system comes to be univocally coordinated with sensory experience. . . .

> Viewed from standpoint B, the choice of geometrical concepts and relations is, indeed, determined only on the grounds of simplicity and instrumental utility. . . . Concerning the metrical determination of space, nothing can then be made out empirically, but not "because it is not real," but because, on this choice of a standpoint, geometry is not a *complete* physical conceptual system, but only a part of one such.
>
> (1924, 1690–1691)

Einstein's point is that the "practically-rigid measuring body" is not real, because no real physical object can maintain its shape and size perfectly, which points us toward "standpoint B," Duhemian holism, as, in principle, the correct view but that, nevertheless, for practical reasons, we proceed as if rigid rods and regular clocks were real, because, as he had argued in "Geometrie und Erfahrung," we cannot yet derive the structures of rods and clocks from our physical theories. Put differently, Einstein is arguing that the Duhemian view is the fundamentally correct one, but that, since we do not have a complete theory of everything, we are compelled, as a pragmatic matter, to proceed as if the right-wing logical empiricist, verificationist position, derived from Poincaré, were correct.

We do not know whether Frank ever read Einstein's review of Elsbach, nor whether he would have seen Einstein's subtle balancing of the Duhemian view and the Reichenbach-Poincaré view as a retreat from or a serious compromise on the Duhemian view that he defended. It would appear that Einstein did not regard it as a retreat or a compromise, however, because, in later years, he regularly defended the Duhemian view.

Einstein's continuing defense of the Duhemian point of view is most clearly seen in an exchange that he had with Reichenbach in 1949. Reichenbach contributed an essay on "The Philosophical Significance of the Theory of Relativity" (Reichenbach 1949) to the previously mentioned collection *Albert Einstein: Philosopher Scientist* (Schilpp 1949), a volume to which Frank was also a contributor. In his contribution, Reichenbach outlined the view on the empirical interpretation of relativity that he had been defending since 1924, the aforementioned view that individual concepts acquire determinate empirical meaning through conventional coordinating definitions:

> Another confusion must be ascribed to the theory of conventionalism, which goes back to Poincaré. According to this theory, geometry is a matter of convention, and no empirical meaning can be assigned to a statement about the geometry of physical space. Now it is true that physical space can be described by both a Euclidean and a non-Euclidean geometry; but it is an erroneous interpretation of this relativity of geometry to call a statement about the geometrical structure of physical space meaningless. The choice of a geometry is arbitrary

only so long as no definition of congruence is specified. Once this definition is set up, it becomes an empirical question *which* geometry holds for a physical space. . . . The combination of a statement about a geometry with a statement of the co-ordinative definition of congruence employed is subject to empirical test and thus expresses a property of the physical world. The conventionalist overlooks the fact that only the incomplete statement of a geometry, in which a reference to the definition of congruence is omitted, is arbitrary; if the statement is made complete by the addition of a reference to the definition of congruence, it becomes empirically verifiable and thus has physical content.

(1949, 297)

In his reply, Einstein went straight to the heart of the matter:

If, under the stated circumstances, you hold distance to be a legitimate concept, how then is it with your basic principle (meaning = verifiability)? Must you not come to the point where you deny the meaning of geometrical statements and concede meaning only to the completely developed theory of relativity (which still does not exist at all as a finished product)? Must you not grant that no "meaning" whatsoever, in your sense, belongs to the individual concepts and statements of a physical theory, such meaning belonging instead to the whole system insofar as it makes "intelligible" what is given in experience? Why do the individual concepts that occur in a theory require any separate justification after all, if they are indispensible only within the framework of the logical structure of the theory, and if it is the theory as a whole that stands the test?

(1949, 678)

Notice how, in this one brief passage, all of the threads that we have so far been following come together: a defense of Duhemian holism, a critique of right-wing logical empiricist verificationism, and a nod to the fact that we still lack the complete theory of everything. Frank, who surely read this, was sure to have felt vindication for the way in which he had been telling the story of Einstein's philosophy of science.

In one respect, however, Einstein's reply to Reichenbach takes the Duhemian view even further than he had before, by now explicitly extending theory holism into semantic holism. It is not just that individual propositions are never tested in isolation; there is the further fact that individual concepts lack empirical meaning, only whole theories possessing empirical content. This is exactly the point that Quine (1951) made famous two years later in "Two Dogmas of Empiricism." It is well to remember that, at this very time, 1949 to 1951, Frank was Quine's colleague at Harvard, and it is hard to imagine that these very issues did not arise in conversation between them, though Quine disavowed any acquaintance with Einstein's reply to Reichenbach (personal communication).[7]

5.4. Frank as Einstein's Champion

When Neville Chamberlain signed the Munich agreement on September 30, 1938, ceding control of the Sudetenland to Germany, it was clear that it would only be a matter of time before the Nazis took control of all of Czechoslovakia, as they did on March 15, 1939. Whether by luck or foresight – we do not know – Frank had arrived in the United States in the late summer of 1938 for an extended lecture tour. Einstein supplied a couple of letters of recommendation to help Frank with the funding and organization of the tour. In one of these, to Norman Kent of the Boston University physics department, he wrote:

> I know Herr Dr. Frank very well, all the more so as I successfully recommended him as my successor at the University of Prague. He is a sharp-minded theoretical physicist who, as an epistemologist, is close to the Vienna school and has also functioned successfully in the philosophical domain. The clarity and elegance of his lectures are widely praised.
> (Einstein to Norton A. Kent, April 27, 1938, EA 11–090)

Eventually, with the help of Percy W. Bridgman and Harlow Shapley, he obtained a half-time appointment in physics and mathematics at Harvard, where he taught thermodynamics and relativity along with two very popular courses on the philosophy of science in Harvard's general education program (Holton et al. 1968, 4).[8] Frank was not just a refugee physicist. He was the designated legatee of the Vienna Circle's *Verein Ernst Mach* [*Ernst Mach Society*], bringing not only the spirit of Vienna Circle logical empiricism to North America but also the modest financial resources of the *Verein*, which, eventually, were made available to the nascent Philosophy of Science Association. Frank worked hard, though unsuccessfully, to build North American philosophy of science after the expansive model of the Vienna Circle's left wing rather than the detached, narrowly formal conception of the philosophy of science that came to dominate the discipline in the 1950s (see Howard 2003; Reisch 2005). Not content just to teach, Frank eagerly undertook the challenge of transplanting the ideas and values of the Vienna Circle in a new North American home and nurturing its continuing development.

Shortly after his arrival in the United States, Frank was drawn into another struggle, this the distinctively American debate over the roles of science, philosophy, and religion in American public life. This debate had flared and subsided many times, as with the enormous public fascination with the Scopes trial in 1925 that had pitted religious skeptics about the Darwinian evolutionary story of human origins against progressive defenders of a scientific world view. The debate had once again erupted when, in 1940, Bertrand Russell's appointment at City College in New York was rescinded after a public outcry over his atheism and such things as his defense of sex outside of marriage. The attack was spearheaded by Episcopal Bishop William T. Manning and eventuated in a lawsuit that made its way to the New York Supreme Court, where the presiding judge, John F. McGeehan, who was Roman Catholic, found that Russell's works constituted an offense against religion and morals and ordered that the appointment be cancelled (see Kallen 1941). It was a great embarrassment for City College, and the case engendered considerable upset within both scientific and religious communities.

One year earlier, prominent religious leaders and other public intellectuals concerned about the rise of Fascism in Europe and the fate of democracy had convened a meeting at the Jewish Theological Seminary in New York at the invitation of Seminary President Louis Finkelstein to explore the role of science and religion in confronting these challenges. The Russell case only reinforced their concern about a culture in crisis. Out of their efforts grew what became a remarkable, nearly three-decade series of meetings under the general title of the "Conference on Science, Philosophy and Religion," that met for the first time in 1940 (see Symposium 1941). Among those participating in that first meeting were Mortimer Adler, Van Wyck Brooks, Harold D. Lasswell, Kenneth Sills, Paul Tillich, Jacques Maritain, Henry Slone Coffin, Sydney Hook, Paul Weiss, Harlow Shapley, Enrico Fermi, Arthur H. Compton, I. I. Rabi, Hermann Weyl, and Philipp Frank, who, in a sense, acted as Einstein's surrogate (see Frank 1950, xiii–xvi; Riddle 1941; Birmingham 1941). Einstein did not attend, but he contributed an essay on "Science and Religion" to the resulting published volume (Einstein 1941). Frank's lecture and his contribution to the volume were entitled "Science and Democracy" (Frank 1941).

Einstein's paper rehearses some of the central ideas that make up what he elsewhere termed "cosmic religion" (Einstein 1930, 1). Chief among these ideas is that the divine is the universe in its law-governed aspect, and that the doing of science, the quest to uncover those laws, is, in a sense, a religious act. Einstein went on to say more about the relationship between science and religion:

> Now, even though the realms of religion and science in themselves are clearly marked off from each other, nevertheless there exist between the two, strong reciprocal relationships and dependencies. Though religion may be that which determines the goal, it has, nevertheless, learned from science, in the broader sense, what means will contribute to the attainment of the goals it has set up. But science can only be created by those who are thoroughly imbued with the aspiration

towards truth and understanding. This source of feeling, however, springs from the sphere of religion. To this there also belongs the faith in the possibility that the regulations valid for the world of science are rational, that is comprehensible to reason. I cannot conceive of a genuine scientist without that profound faith. The situation may be expressed by an image: science without religion is lame, religion without science is blind.

(1941, 210–211)

A principal theme in Frank's essay for the conference volume is the work that a scientific education does in fortifying the mind against suasion by propaganda, and he thinks that it is pure science, especially, that has this effect, having noticed in his own experience that technicians and engineers were more prone than scientists to be duped:

> We can see quite clearly that the power of resistance to totalitarian ideologies is all the stronger, the more a given student is impregnated with the ideas of pure science and the more undiluted the form in which he has absorbed them. It is precisely the circumstance that the technician is not so capable of resistance that gives us material for thought, for it is in the technician that the purely scientific mode of thought is always mixed up with considerations of the application of his knowledge in economic life. And then thoughts always creep in of maintaining in all circumstances a certain economic structure which is favorable to technical activity and of being ready to accept any ideology which facilitates this.
>
> (1941, 219)

Is it mere coincidence that Frank penned this sharp critique of the implications for democracy from the technician's mindset in the very same year when Herbert Marcuse first introduced the similarly critical concept of "technological rationality" (Marcuse 1941)?

One of the specific, prominent issues taken up in this first and later meetings in the conference series was later remembered by Frank in this way:

> The precise aim of this Conference was to establish a common understanding of democratic principles that would help to overcome the high-pressure propaganda of totalitarian values. The members were anxious to prove that the danger of "relativism," which was, in a certain way, a frequent concomitant of liberalism and democracy, could be avoided by democratic methods.
>
> (1950, xiii)

A surprising but common misconception, then as now, is that relativity in physics somehow contributes to, legitimates, or encourages relativism in morals and culture more generally. Inspired by his participation in that first meeting in the series and his regular participation over the next decade, Frank resolved to focus his efforts on patiently but firmly exposing the errors that lay behind this confusion. The result was a sadly too-little-known book, *Relativity: A Richer Truth* (Frank 1950), to which Einstein contributed a foreword, "The Laws of Science and the Laws of Ethics" (Einstein 1950). In his preface, Frank describes his motivation and aim:

> In the beginning I soon saw that by the fight against "relativism" and its "totalitarian affiliations," the Conference was about to jump "out of the frying pan into the fire." I saw that the fight against "relativism" could easily degenerate into a fight against the spirit of modern science. And the denial of that spirit has always seemed to me to be a kind of ostrich truculence – for the leading role of science in our century cannot be denied. Still less can it be "restrained" by any political power – may it call itself a dictatorship, a democracy, or a religion. My principal goal in all my addresses and discussions at the Conference has been, therefore, to interpret the real spirit of modern science to the audience. I have not presented any excuses for the "relativism" of science, but rather have launched a frontal attack against its enemies. I have attempted to show that so-called "relativism" has not the slightest thing to do with agnosticism or skepticism, that it is in no way hostile to the belief in ethical or democratic values, that it is accompanying every advance in science and is nothing but a significant representation of the enrichment of human expression which is inseparably connected with our gradually increasing experience.
>
> (1950, xiv–xv)

Relativity in several guises has become an integral part of modern science, argues Frank, but that has no implications whatsoever for ethics.

Early chapters of Frank's book lay the scientific and philosophical foundations for his argument in a language suited to a general audience. The brief chapter on the theory of relativity concludes with these thoughts:

> We see that even Einstein's celebrated Relativity has nothing to do with any brand of "subjectivism" or "skepticism," and still less with a despair of the human mind to explore "the truth."
>
> Just the opposite is true. "Relativism" means the introduction of a richer language which allows us to meet adequately the requirements of an enriched experience. We are now able to cover these new facts by plain and direct words and to come one step nearer to what one may call the "plain truth about the universe."
>
> (Frank 1950, 18)

The fundamentals of logical empiricism are surveyed in such chapters as "How Can We Test the Meaning of a Statement? Operationalism and Semantics," "Why Have Sentences No Meaning Except in a System? Logical Empiricism," and "What Does All This Mean for Morality and Politics?" The argument grows more interesting when Frank turns to the core problems addressed by the whole conference series in chapters such as "Why Is Fundamental Science Often Embarrassing to Totalitarians and Heresy Hunters?", "Is Metaphysics the Language of Obsolete Science?", How Can an Anti-Metaphysical View of Science Help Democracy?", and "How Does Philosophical 'Idealism' Support Dictatorship?" The main message is straightforward and had been central to the Vienna Circle's scientific world conception for a quarter century: Science, properly understood, is inherently emancipatory and subversive of dogma in all of its forms because it demands that ideology be judged not by its power to move people but by whether it gets the facts right:

> Indifference to the words in which the highest principles are formulated proves to be a powerful weapon against all totalitarian ideologies. For those always rest on the assertion that all good lies in a definite principle, and that this principle must be maintained in all cases regardless of what follows from it in real life. But then this principle is reduced from a scientific principle to a mere banner. It is essential only for an army that the identity of the banner be preserved. The content of the practical objectives which are being sought can be changed if only the color of the flag remains the same. Only in this way could the Nazi government compel their "subjects" to work for a group of power-thirsty people, to obey all their commands, and to persecute or torture other men. This role of a banner is played in the totalitarian states by the "ultimate racial or national goals." They serve merely as the symbol for the gathering of troops and war material. The facts that follow from these principles lose all significance.
> (Frank 1950, 113)

Should anyone have ever doubted that the Vienna Circle was as much a social and political, as well as a purely philosophical, movement, such remarks should settle the case. One should also not mistake this as the kind of naive scientism that we have all been taught to deride since the post-modern turn in science studies, because this is the view of the democratic socialist left wing of the Vienna Circle, whose leader, Neurath (a fan of Duhem like Frank), mocked as "pseudorationalism" the idea that "auxiliary motives," including social, political, and cultural values, play no role in science. Frank, like Neurath, understood perfectly well that even social and political values infuse theory choice but in a manner that in no way compromises scientific objectivity, especially if we are honest with ourselves and others about the role of those values, allowing them to be exposed to critical scrutiny (see Howard 2003, 2009).[9]

Included in the letter in which Frank asked Einstein to write the foreword to *Relativity: A Richer Truth* was a now very hard-to-find short essay, "The Relativity of Truth and the Advance of Knowledge", that Frank had first published in 1946 in an obscure Italian journal (1946). Frank describes it as having been inspired by the "Finkelstein Conference" and characterizes the main thesis in words that closely match his previously quoted remarks from the later book about relativity and a richer experience:

> It is a very widespread misunderstanding that relativity has something to do with skepticism or even nihilism. In the accompanying essay, I have attempted to show what is actually meant by the increasing relativization of the language of physics: An adaptation of language to a richer experience. But it has absolutely nothing to do with skepticism. I also believe that the so-called "ethical relativism," insofar as it is scientific, plays a similar role. There is nothing "subversive" in it, but, rather, an adaptation to an increasing complexity of life or, at least, an increasing understanding of the complexity of life.

Frank added, with respect to the foreword:

> I do not at all want to suggest that you write something about my book, the reading of which would not be worth the trouble for you. The main idea is to be found in the published essay from the Italian journal that I am sending you with this letter. But if you are even only in the least bit unsympathetic with this idea and, in the moment, not so disposed, please forget immediately what I have written.
> (Frank to Einstein, n.d., CPAE 11–081)

Of course, Einstein did, gladly, write the foreword, because he agreed more or less totally with Frank in all such matters. In it, Einstein wrote:

> It is the privilege of man's moral genius, expressed by inspired individuals, to advance ethical axioms which are so comprehensive and so well founded that men will accept them as grounded in the vast mass of their individual emotional experiences. Ethical axioms are found and tested not very differently from the axioms of science. Die Wahrheit liegt in der Bewärung. Truth is what stands the test of experience.
> (1950, vii–viii)

In all of this, Frank was much more than just a voice for Einstein, with whom he shared not only scientific and philosophical interests but also a deep commitment to social justice and the critical role of science in promoting human emancipation and flourishing. Frank brought to the task his keen understanding of social, political, and cultural issues as well as an uncommon ability to articulate a vision of the scientific world view in language accessible to all and the patience and perseverance needed to engage in sustained, constructive dialogue even with those who propose a dramatically different vision of the world.

5.5. Frank as Einstein's Biographer

In these years, Frank did much more for Einstein than acting as his able second in public debate. By far his greatest service was his publishing in 1949 the best biography of Einstein that had so far been written, a biography that remains to this day one of the best, *Einstein: His Life and Times* (Frank 1947). It was written in close collaboration with Einstein, who supplied Frank with documents, photographs, and reminiscence, as well as detailed commentary on Frank's presentation of technical issues, as evident from the only surviving letter between Einstein and Frank from the time of the drafting of the biography, a letter in which Einstein answers a few direct questions from Frank, concluding with this:

> I do not want to give you the diaries, because there are far too personal things in them. Anyway, there is nothing adventurous in it. But, otherwise, there was enough adventure in my life, since I had a sense for personal considerations but not for conventional status, valuations, and formalities and no understanding for that which one commonly and thoughtlessly regards as "moral." That is why I have been more popular among simple people and children than among people of status – and I *also*, for the most part, prefer them.
> (Einstein to Frank, n.d., CPAE 123–032)

Happily, this did not make its way into Frank's biography, but it shows clearly the extent to which Einstein was willing to share with Frank thoughts and feelings that were otherwise not part of the public Einstein.

Frank's biography of Einstein is distinctive in several respects, quite beyond the unique personal access that he had to Einstein. Most noteworthy for our purposes is the space that Frank devotes to the philosophy of science as part of Einstein's work and legacy. Whole chapters are devoted to "Ernst Mach: The General Laws of Physics Are Summaries of Observations Organized in Simple Forms," "Henri Poincaré: The General Laws of Physics Are Free Creations of the Human Mind," "Positivistic and Pragmatic Movements," and "Einstein's Philosophy of Science." The chapter on Einstein's philosophy of science is interesting because it focuses on issues not highlighted in Frank's other writings on the topic and because it relies heavily on Frank's personal memories of conversations

with Einstein. It begins, curiously, with a discussion of the philosophical assumptions underlying Niels Bohr's complementarity interpretation
of quantum mechanics, which Frank characterizes as "positivistic." He
recalls a talk that he gave at a meeting of the German Physical Society
in Prague in 1929, where he commended both what he wrongly took to
be Bohr's positivistic interpretation of quantum mechanics and Einstein's
endorsement of a positivist point of view.

Another unnamed speaker is reported by Frank to have disputed that
characterization of Einstein's views and to have held that Einstein thoroughly endorsed Max Planck's well-known realist rejoinder to positivism.[10] Frank recalls his explaining to the unnamed interlocutor his more
nuanced interpretation of Einstein's views, according to which Einstein still
followed Mach in his insistence that physical theory always be grounded
in experience, but that, like Poincaré, he also insisted that physical concepts and theories were free creations of the human intellect. Frank then
reminisces about a conversation that he had with Einstein in Berlin in
1932, where they once again discussed Mach's positivism and the work
that it did in the theory of special relativity, as with Einstein's declaring
meaningless notions such as absolute distant simultaneity. Frank writes
about this exchange:

> Consequently I said to Einstein: "But the fashion you speak of was
> invented by you in 1905?" At first he replied humorously: "A good
> joke should not be repeated too often." Then in a more serious vein
> he explained to me that he did not see any description of a meta-
> physical reality in the theory of relativity, but that he did regard an
> electromagnetic or gravitational field as a physical reality, in the
> same sense that matter had formerly been considered so. The theory
> of relativity teaches us the connection between different descriptions
> of one and the same reality.
>
> (1947, 216)

After briefly rehearsing Frank's own view of Einstein as having reconciled
Mach and Poincaré and summarizing Einstein's account of his philosophy
of science in his 1933 Herbert Spencer Lecture at Oxford (Einstein 1934),
Frank concludes this chapter with an interesting observation about how
people sometimes misperceived Einstein's philosophical views:

> As in so many aspects of his life and thought, we also note a certain
> internal conflict in Einstein's attitude toward the positivistic concep-
> tion of science. On the one hand, he felt an urge to achieve a logical
> clarity in physics such as had not previously been attained, an urge to
> carry through the consequences of an assumption with extreme radi-
> calism, and was unwilling to accept any laws that could not be tested
> by observation. On the other hand, however, he felt that even *Logical*

Positivism did not give sufficient credit to the role of imagination in science and did not account for the feeling that the "definitive theory" was hidden somewhere and that all one had to do was to look for it with sufficient intensity. As a result, Einstein's philosophy of science often made a "metaphysical" impression on persons who are unacquainted with Einstein's positivistic requirement that the only "confirmation" of a theory is its agreement with observable facts.

(1947, 218)

Frank might have added that another source of confusion was the failure of most scientists and philosophers outside of the Vienna Circle to understand the crucial differences between verificationist right-wing logical empiricism and holist left-wing logical empiricism, the latter being, as Frank so often tried to explain, wholly compatible with Einstein's insistence on concepts and theories being free creations of the human intellect.

Yet another distinctive feature of Frank's Einstein biography, one that likewise betrays Frank's own interests and commitments, is its devoting an entire section to the topic of "Einstein's Theories as Political Weapons and Targets." Of course Frank discusses the horrible anti-Semitic attacks on Einstein and relativity in Germany in the 1920s and 1930. Led even by prominent physicists, like the Nobel Prize winners Johannes Stark and Philipp Lenard, the critics charged that relativity was a form of "Jewish physics" that had to be replaced by proper "German physics." However, he also devotes a separate chapter to a discussion of the curious way in which some other pro-Nazi physicists, like Pascual Jordan, tried to spin relativity as being supportive of Nazi ideology. But Frank was the first biographer to detail the critique of Einstein and relativity coming out of the Soviet Union in the 1920s and 1930s, with the charge that relativity as Einstein interpreted it was antithetical to dialectical materialism, inasmuch as its privileging the role of the subjective observer and its dispensing with the concept of the electromagnetic ether, understood by mechanical models of the ether, was idealist in its tendencies. Frank was more alert to this line of criticism than many would have been at the time because, he himself, as a defender of Mach and Poincaré, had come under attack from none other than Lenin, who, in *Materialism and Empirio-Criticism* wrote the following:[11]

For Poincaré . . . the laws of nature are symbols, conventions, which man creates for the sake of "*convenience.*" "The only true objective reality is the internal harmony of the world." By "objective," Poincaré means that which is generally regarded as valid, that which is accepted by the majority of men, or by all; that is to say, in a purely subjectivist manner he destroys objective truth, as do all the Machians. And as regards "harmony," he categorically declares in answer to the question whether it exists outside of us – "undoubtedly, no."

It is perfectly obvious that the new terms do not in the least change the ancient philosophical position of agnosticism, for the essence of Poincaré's "original" theory amounts to a denial (although he is far from consistent) of objective reality and of objective law in nature. It is, therefore, perfectly natural that in contradistinction to the Russian Machians, who accept new formulations of old errors as the latest discoveries, the German Kantians greeted such views as a conversion to their own views, i.e., to agnosticism, on a fundamental question of philosophy. "The French mathematician Henri Poincaré," we read in the work of the Kantian, Philipp Frank, "holds the point of view that many of the most general laws of science (e.g., the law of inertia, the law of the conservation of energy, etc.), of which it is so often difficult to say whether they are of empirical or of a priori origin, are, in fact, neither one nor the other, but are purely conventional propositions depending upon human discretion. . . ." "Thus [exults the Kantian] the latest *Naturphilosophie* unexpectedly renews the fundamental idea of critical idealism, namely, that experience merely fills in a framework which man brings with him from nature."

(1909/1947, 166; quoting from Frank 1907/1949)

Frank also displays a subtle and detailed understanding of the course of subsequent debates in the Soviet Union, culminating in a significant change in attitude toward relativity theory, when, in the 1940s, leading figures, like the physicist and long-time president of the USSR Academy of Sciences, Sergey I. Vavilov, advanced a different view, according to which mechanistic models of an electromagnetic ether were held to be themselves insufficiently dialectical. Vavilov rightly emphasized the fundamental role of matter, in the form of the stress-energy tensor, in general relativity.

This section of the biography concludes with a chapter "Einstein's Theories as Arguments for Religion," in which Frank assesses several attempts to enlist Einstein and relativity as affording evidence for one or another religious view. For example, he quotes Wildon Carr as follows:

The adoption of the principle of relativity means that the subjective factor, inseparable from knowledge in the very concept of it, must enter positively into physical science. . . . Hitherto the scientific problem has been to find a place for mind in the objective system of nature. . . . Now when reality is taken in the concrete, as the general principle of relativity requires us to take it, we do not separate the observer from what he observes, the mind from its object, the agent from his activity, the subject from the object, and then dispute as to the primacy of the one over the other.

(Frank 1947, 263, quoting Carr 1922, 340–346)

Frank comments:

> According to this, the achievement of the relativity theory for religion
> is simply that it provided a place for mind in nature, which during
> the period of mechanistic physics had been regarded as completely
> "material and mindless."
>
> If the reader will recall Einstein's physical theories, he will easily
> see that this interpretation is more closely related to the wording
> than to the content of these theories.
>
> (1947, 263)

What Frank means with his point about the "wording" is, of course,
that talk of frames of reference in relativity is often glossed in the
language of "observers," when, in fact, subjective, individual, human
perspective has absolutely nothing to do with the essential physical
content of the theory, a theory that would be just as true in a world
with no human beings. In the decades since Frank's Einstein biography,
a whole genre of literature on science and religion has come almost
to dominate the field that makes the same mistake that Frank here
calls out, namely mistaking merely verbal similarities to points of religious doctrine for matters of physical fact. Would that more physicists
like Frank would today take the trouble of explaining, patiently and
respectfully, why, for example, relativistic cosmology does not confirm
intelligent design or a religious origins narrative. Both science and religion would be the better.

Einstein had eight more years to live after the publication of Frank's
biography, so there would be more to the story. But, with this splendid
biography, as with his other services to Einstein, Frank more than repaid
the debt that he owed for Einstein's crucial role in launching his career
in 1912. Let us let Frank have the last word regarding the Einstein who
meant so much to him and to the world:

> The world around Einstein has changed very much since he published
> his first discoveries. He began work during the time of the German
> Kaiser in the environment characteristic of the German and Swiss
> petty bourgeoisie; he lived during the second World War in the last
> bulwark of democracy, the United States of America. He was able to
> make a substantial contribution toward an earlier conclusion of the
> war than had been expected, and is now anxious to help in making
> the peace a lasting one. But his attitude to the world around him has
> not changed. He has remained a bohemian, with a humorous, even
> seemingly skeptical approach to the facts of human life, and at the
> same time a prophet with the intense pathos of the Biblical tradition.
> He has remained an individualist who prefers to be unencumbered

by social relations, and at the same time a fighter for social equality and human fraternity. He has remained a believer in the possibility of expressing the laws of the universe in simple, even though ingenious mathematical formulae, but at the same time doubting all ready-made formulae that claim to be the correct solution for human behavior in private and political life.

(1947, 297)

5.6. Conclusion

In the years after the Einstein biography, Frank published two additional important books in addition to the collection of his own papers, *Modern Science and Its Philosophy* (Frank 1949c), which included the long and lovely historical memoir parts of which were quoted previously. The first was an edited volume, *The Validation of Scientific Theories* (Frank 1956a), that is recognized today as a landmark in the debates over the role of values in science. It opens with an essay by Frank himself on "The Variety of Reasons for the Acceptance of Scientific Theories" (1956b) and includes a classic paper by Richard Rudner, "Value Judgments in the Acceptance of Theories" (1956). The other was Frank's own magnum opus, *Philosophy of Science: The Link Between Science and Philosophy* (1957), that remains to this day a provocative challenge to more orthodox post-World War Two analytic philosophy of science.

Einstein died in Princeton, New Jersey, in April of 1955. Frank died in Cambridge, Massachusetts, in July of 1966. One year before Frank's death, a *Festschrift* was presented to him that commenced with reminiscences by friends and colleagues. Frank's former student, Peter Bergmann, who later became Einstein's assistant in Princeton, recalled the atmosphere created by Frank at the theoretical physics institute in Prague:

> There were five full professors close to each other, L. Berwald and K. Löwer in mathematics, you in theoretical physics, R. Fürth in experimental physics, and R. Carnap in philosophy of science. There were frequent joint seminars and other common undertakings. But the spirit of the unity of science and of scientific inquiry was taken so much for granted that I, for one, did not fully appreciate it until later, when I found compartmentalization the rule in many places elsewhere.
>
> You, Professor Frank, were as much interested in the problems of philosophy, and of epistemology in particular, as in those of theoretical physics. I remember your own lectures, at our seminars and at the Urania, as well as those by distinguished visitors, such as Neurath. Concurrently, you were putting the finishing touches on the second edition of "Frank-and-Mises," the standard encyclopedia of mathematical physics of the twenties and thirties. . . .

Two years after I left Europe, Czechoslovakia was sacrificed on the altar of "peace in our time." Those who were fortunate enough to have contacts abroad left, you after almost thirty years of professional service at Prague. At Harvard University you were able to translate into contemporary American two aspects of your lecturing at Prague, your concern for the philosophical foundations of science, and your efforts to acquaint non-scientists with the beauty and the charm of scientific inquiry.

(1965, ix–x)

The textbook to which Bergmann refers was the masterful *Die Differential- und Integralgleichungen der Mechanik und Physik* [*The Differential and Integral Equations of Mechanics and Physics*] (Frank and von Mises 1930– 1935). Frank's longtime Prague colleague Carnap began his remarks by recalling what he understood to be Frank's two great aims:

His first aim was to bring together science and philosophy, to bridge the gap that had separated them in the last century. . . . The reform of philosophy which Frank welcomed and at which he himself actively cooperated, was initiated and developed by men in two different fields. On the one hand, scientists like Duhem, Poincaré and Einstein tried to come to a better understanding of the foundations of their own scientific activity. On the other hand, in the field of modern logic more exact conceptual tools were developed which helped scientists and philosophers to improve their methods of concept formation and theory formation.

To bring science and philosophy into closer contact is useful but not sufficient. Frank recognized more clearly, I think, than most other philosophers and scientists that it is of greatest importance that those who work in theoretical fields be aware of the role of their work in the wider context of life, of the whole of society and culture. Therefore Frank, both in his own thought and in his teaching activities, paid close attention to the historical development in order to show how currents of thinking are motivated not only by striving for knowledge but to a great extent also by practical or emotional needs and social situations. He showed that this holds for theoretical work just as much as for work in other fields like art or religion. And further he regarded it as important that those working in theoretical fields, even in fields which seem quite remote from practical life, like theoretical physics or analytical philosophy, should be aware of their responsibilities for the development of humanity.

(1965, xi–xii)

It is noteworthy that so many people remember Frank as caring first and foremost about the integration of science and philosophy and the embedding of both in social and historical context.

Precisely those two points are also stressed by Frank's Harvard colleagues, Gerald Holton, Edwin Kemble, W. V. Quine, S. S. Stevens, and Morton White, in their moving obituary for Frank in the journal that he once edited, *Philosophy of Science*:

> He . . . saw it as a misfortune that science and philosophy are widely regarded as unrelated and incongruous pursuits. But it was also his conviction that this breach between a scientific and a humanistic orientation toward life – a breach that he thought to be of relatively recent origin – could be diminished, if not overcome, by an adequate philosophy of science. To be adequate, a philosophy of science in his view must also include certain socio-historical considerations – for example, analyses of circumstances (both internal and external to a science) under which changes in scientific doctrine and practice may occur; and of the grounds on which standards of scientific validity may differ for different times and in different subject-matters. Frank was not a purist in his conception of the philosophy of science, and he did not hesitate to "thicken" its content.
>
> (1968)

A Note on Einstein Citations

Correspondence and manuscripts that have already been published in *The Collected Papers of Albert Einstein* (Princeton, NJ: Princeton University Press, 1987–present) are cited by volume number, document number, and page, after the model "CPAE-x, Doc. yyy, z." Unpublished items from Einstein's correspondence are cited by their control index numbers in the Einstein Archive, after the model "EA xx-xxx."

Notes

1. In this volume, Einstein's role and ideas are discussed further by Thomas Ryckman and Marco Giovanelli.
2. Frank's philosophy of science is discussed further in Adam Tamas Tuboly's chapter in this volume.
3. About the role and significance of "simplicity" in Frank's works, see Wuest (2017).
4. Banks (2003) provides the most thoroughgoing debunking of this far-too-widespread, mistaken understanding of Mach's view, which is better characterized as a variety of neutral monism.
5. The relation of Neurath's ideas to Poincaré's conventionalism is scrutinized further by Katherine Dunlop's chapter in the volume.
6. The history of Einstein's acquaintance with Mach and Poincaré is reviewed in Howard (2014, 2020). See also Norton (2004) for more on Mach's influence on Einstein.
7. In fact, Frank followed Quine's work in the early 50s closely, as he discussed some of his ideas in his previously unpublished book manuscript. See Frank (2021).

8. For the details of Frank's arrival to the States and his socio-political struggles, see Reisch and Tuboly (2021).
9. The different roles and influence of values on science is the major theme of Frank's last book manuscript. There Frank shows that one path of humanizing the sciences is to make explicit the roles values could and actually play in scientific inquiries and the science-society role. See Frank (2021).
10. The unnamed speaker was, almost certainly, Arnold Sommerfeld. See Stöltzner (2020).
11. In fact, in his last book manuscript, Frank discussed many Russian encyclopedia-articles from the Great Soviet Encyclopedia, translated and quoted many of them (he even contributed two entries). Thus, his knowledge of Russian philosophy (of science) is well represented in various chapters of the book. See Frank (2021).

References

Banks, E. C. (2003), *Ernst Mach's World Elements: A Study in Natural Philosophy*. Dordrecht and Boston: Kluwer.

Bergmann, P. G. (1965), 'Homage to Professor Philipp G. Frank', in R. S. Cohen and M. Wartofsky, (eds.) *In Honor of Philipp Frank: Proceedings of the Boston Colloquium for the Philosophy of Science, 1962–1964*. New York: Humanities Press, pp. ix–x.

Birmingham, E. W. (1941), 'American Views on Modern Culture', *Nature*, 147: 367–369. Review of Symposium 1941.

Carnap, R. (1928), *Der logische Aufbau der Welt*. Berlin: Weltkreis.

——— (1965), 'A Few Words to Philipp Frank', in R. S. Cohen and M. Wartofsky, (eds.) *In Honor of Philipp Frank: Proceedings of the Boston Colloquium for the Philosophy of Science, 1962–1964*. New York: Humanities Press, pp. xi–xii.

Carr, W. (1922), *A Theory of Monads: Outlines of the Philosophy of the Principle of Relativity*. London: Macmillan.

Comte, A. (1830–1842), *Cours de philosophie positive*. 6 vols. Paris: Bachelier.

Duhem, P. (1906), *La Théorie physique, son objet et sa structure*. Paris: Chevalier & Rivière.

——— (1908), *Ziel und Struktur der physikalischen Theorien*. Translated by Friedrich Adler and Foreword by Ernst Mach. Leipzig: Johann Ambrosius Barth.

Einstein, A. (1905), 'Zur Elektrodynamik bewegter Körper', *Annalen der Physik*, 17: 891–921.

——— (1910–1911), 'Lecture Notes for Course on Electricity and Magnetism, University of Zurich, Zurich, Winter Semester 1910–1911', *CPAE*, 3, Doc. 3.

——— (1916), 'Ernst Mach', *Physikalische Zeitschrift*, 17: 101–102.

——— (1921), *Geometrie und Erfahrung. Erweiterte Fassung des Festvortrages gehalten an der Preussischen Akademie der Wissenschaften zu Berlin am 27. Januar 1921*. Berlin: Julius Springer.

——— (1924), 'Review of Elsbach 1924', *Deutsche Literaturzeitung*, 45: 1685–1692.

——— (1930), 'Religion and Science', *New York Times Sunday Magazine*, November 9: 1.

——— (1934), 'On the Methods of Theoretical Physics', *Philosophy of Science*, 1: 163–169.

——— (1941), 'Science and Religion', in Symposium (1941), pp. 209–214.

———— (1949), 'Remarks Concerning the Essays Brought Together in This Co-Operative Volume', in Schilpp (1949), pp. 665–688.

———— (1950), 'The Laws of Science and the Laws of Ethics', Foreword to Frank (1950), pp. v–viii.

Elsbach, A. (1924), *Kant und Einstein. Untersuchungen über das Verhältnis der modernen Erkenntnistheorie zur Relativitätstheorie*. Berlin and Leipzig: Walter de Gruyter.

Frank, Ph. (1907/1949), 'Experience and the Law of Causality', in Frank (1949c), pp. 53–60.

———— (1909), *Die Stellung des Relativitätsprinzips der Mechanik und Elektrodynamik*. Vienna: Hof- und Staatsdruckerei.

———— (1917/1949), 'The Importance for Our Times of Ernst Mach's Philosophy of Science', in Frank (1949c), pp. 61–78.

———— (1941), 'Science and Democracy', in Symposium (1941), pp. 215–228.

———— (1946), 'The Relativity of Truth and the Advance of Knowledge', *Analisi*, 1: 18–24.

———— (1947), *Einstein: His Life and Times*. New York: Alfred A. Knopf.

———— (1949a), 'Einstein's Philosophy of Science', *Reviews of Modern Physics*, 21: 349–355.

———— (1949b), 'Einstein, Mach, and Logical Positivism', in Schilpp (1949), pp. 271–286.

———— (1949c), *Modern Science and Its Philosophy*. Cambridge, MA: Harvard University Press.

———— (1949d), 'Introduction: Historical Background', in Frank (1949c), pp. 1–52.

———— (1950), *Relativity: A Richer Truth*. Boston: Beacon Press.

———— (ed.) (1956a), *The Validation of Scientific Theories*. Boston: Beacon Press.

———— (1956b), 'The Variety of Reasons for the Acceptance of Scientific Theories', in Frank (1956a), pp. 3–18.

———— (1957), *Philosophy of Science: The Link between Science and Philosophy*. Englewood Cliffs, NJ: Prentice Hall.

———— (2021), *The Humanistic Background of Science*. Edited by G. Reisch and A. T. Tuboly. New York: SUNY Press.

Frank, Ph. and von Mises, R. (1930–1935), *Die Differential- und Integralgleichungen der Mechanik und Physik*. 2nd ed. Braunschweig: F. Vieweg & Sohn.

Haller, R. (1991), 'The First Vienna Circle', in Th. Uebel (ed.), *Rediscovering the Forgotten Vienna Circle: Austrian Studies on Otto Neurath and the Vienna Circle*. Dordrecht: Kluwer, pp. 95–108.

Holton, G. (1968), 'Mach, Einstein, and the Search for Reality', *Daedalus*, 97: 636–673.

Holton, G. et al. (1968), 'In Memory of Philipp Frank', *Philosophy of Science*, 35: 1–5.

Howard, D. (1984), 'Realism and Conventionalism in Einstein's Philosophy of Science: The Einstein-Schlick Correspondence', *Philosophia Naturalis*, 21: 618–629.

———— (1990), 'Einstein and Duhem', *Synthese*, 83: 363–384.

———— (1993), 'Was Einstein Really a Realist?', *Perspectives on Science*, 1: 204–251.

———— (1994), 'Einstein, Kant, and the Origins of Logical Empiricism', in W. Salmon and G. Wolters (eds.), *Language, Logic, and the Structure of Scientific*

Theories. Pittsburgh: University of Pittsburgh Press and Konstanz: Universitätsverlag, pp. 45–105.

—— (2003), 'Two Left Turns Make a Right: On the Curious Political Career of North American Philosophy of Science at Mid-Century', in G. Hardcastle and A. W. Richardson (eds.), *Logical Empiricism in North America*. Minneapolis, MN: University of Minnesota Press, pp. 25–93.

—— (2009), 'Better Red Than Dead: Putting an End to the Social Irrelevance of Postwar Philosophy of Science', *Science and Education*, 18: 199–220.

—— (2014), 'Einstein and the Development of Twentieth-Century Philosophy of Science', in M. Janssen and Ch. Lehner (eds.), *The Cambridge Companion to Einstein*. New York: Cambridge University Press, pp. 354–376.

—— (2019), 'Otto Neurath: The Philosopher in the Cave', in J. Cat and A. T. Tuboly (eds.), *Neurath Reconsidered: New Sources and Perspectives*. Cham: Springer, pp. 45–56.

—— (2020), 'How General Relativity Shaped Twentieth-Century Philosophy of Science', in J. Z. Buchwald (ed.), *Einstein Was Right: The Science and History of Gravitational Waves*. Princeton, NJ: Princeton University Press, pp. 149–196.

Kallen, H. (1941), 'Behind the Bertrand Russell Case', in J. Dewey and H. Kallen (eds.), *The Bertrand Russell Case*. New York: Viking, pp. 13–53.

Lenin, V. I. (1909/1947), *Materialism and Empirio-Criticism: Critical Comments on a Reactionary Philosophy*. Moscow: Foreign Languages Publishing House.

Mach, E. (1894), 'Über das Prinzip der Vergleichung in der Physik', *Verhandlungen der Gesellschaft Deutscher Naturforscher und Ärzte*, 66, Part I: 44–56.

—— (1896), *Die Principien der Wärmelehre. Historisch-kritisch entwickelt*. Leipzig: Johann Ambrosius Barth.

—— (1897), *Die Mechanik in ihrer Entwickelung historisch-kritisch dargestellt*. 3rd impr. and enl. ed. Leipzig: Brockhaus.

—— (1900), *Die Analyse der Empfindungen und das Verhältnis des Physischen zum Psychischen*. 2nd ed. Jena: G. Fischer; 3rd enl. ed. 1902; 4th enl. ed. 1903.

—— (1905), *Erkenntnis und Irrtum. Skizzen zur Psychologie der Forschung*. Leipzig: Johann Ambrosius Barth.

—— (1906), *Erkenntnis und Irrtum. Skizzen zur Psychologie der Forschung*. 2nd ed. Leipzig: Johann Ambrosius Barth.

—— (1908), 'Vorwort', in Duhem (1908), pp. iii v.

Marcuse, H. (1941), 'Some Social Implications of Modern Technology', *Studies in Philosophy and Social Science*, 9: 414–439.

Natorp, P. (1910), *Die logischen Grundlagen der exakten Wissenschaften*. Leipzig and Berlin: B. G. Teubner.

Norton, J. D. (2000), '"Nature Is the Realisation of the Simplest Conceivable Mathematical Ideas": Einstein and the Canon of Mathematical Simplicity', *Studies in History and Philosophy of Science Part B*, 31: 135–170.

—— (2004), 'How Hume and Mach Helped Einstein Find Special Relativity', in M. Domski and M. Dickson (eds.), *Discourse on a New Method: Reinvigorating the Marriage of History and Philosophy of Science*. Chicago: Open Court, pp. 359–386.

Poincaré, H. (1902), 'L'expérience et la géométrie', in *La science et l'hypothèse*. Paris: Ernest Flammarion, pp. 92–109. Originally published as: 'On the Foundations of Geometry', *The Monist*, 9 (1898): 1–43.

—— (1904), *Wissenschaft und Hypothese*. Translated by Ferdinand Lindemann and Lisbeth Lindemann. Leipzig: B. G. Teubner.

Quine, W. V. O. (1951), 'Two Dogmas of Empiricism', *Philosophical Review*, 60: 29–43.

Reichenbach, H. (1920/1965), *The Theory of Relativity and A Priori Knowledge*. Berkeley and Los Angeles: University of California Press.

—— (1924/1969), *The Axiomatization of the Theory of Relativity*. Berkeley and Los Angeles: University of California Press.

—— (1928/1958), *The Philosophy of Space and Time*. New York: Dover.

—— (1949), 'The Philosophical Significance of the Theory of Relativity', in Schilpp (1949), pp. 289–311.

Reisch, G. (2005), *How the Cold War Transformed Philosophy of Science: To the Icy Slopes of Logic*. New York: Cambridge University Press.

Reisch, G. and Tuboly, A. T. (2021), 'Philipp Frank: A Crusader for Scientific Philosophy', in Frank (2021), Forthcoming.

Rudner, R. (1956), 'Value Judgments in the Acceptance of Theories', in Frank (1956a), pp. 24–28.

Riddle, D. W. (1941), 'The Conference on Science, Philosophy and Religion', *The Journal of Religion*, 21: 54–55. Review of Symposium (1941).

Schilpp, P. A. (ed.) (1949), *Albert Einstein: Philosopher-Scientist*. Evanston, IL: The Library of Living Philosophers.

Stadler, F. (2001), *The Vienna Circle: Studies of the Origins, Development, and Influence of Logical Empiricism*. Vienna and New York: Springer.

Stöltzner, M. (2020), 'Scientific World Conception on Stage: The Prague Meeting of the German Physicists and Mathematicians', in R. Schuster (ed.), *The Vienna Circle in Czechoslovakia*. Cham: Springer, pp. 73–95.

Symposium (1941), *Science, Philosophy, and Religion: A Symposium*. New York: Conference on Science, Philosophy, and Religion in Their Relation to the Democratic Way of Life.

Wuest, A. (2017), 'Simplicity and Scientific Progress in the Philosophy of Philipp Frank', *Studies in East European Thought*, 69 (3): 245–255.

6 On the Empirical Refutation of Epistemological Doctrine in Hans Reichenbach's Early Philosophy

Alan Richardson

> Beginning from an analysis of the concept of knowledge, we will inves-
> tigate what presuppositions are included in the epistemology of Kant,
> and when we confront these with the results of our analysis of relativity
> theory, we will determine in what sense Kant's theory has been refuted
> by experience.
>
> (Hans Reichenbach 1920, 4–5)

6.1. Introduction

One curious feature of the recent emergence of history of philosophy of
science as a research specialty is this: while logical empiricism has a large
role in the literature, and Hans Reichenbach is surely one of the most
influential of the logical empiricists on the development of twentieth-
century philosophy, Reichenbach does not have a particularly important
role in the literature that sets about to re-evaluate logical empiricism. In
this he contrasts with Rudolf Carnap, Moritz Schlick, and Otto Neurath –
around each of these other figures a voluminous secondary literature has
emerged. One thing certainly lacking in the recent secondary literature
is a well-established framework for providing a general interpretation
of Reichenbach's overall philosophical project.[1] I believe that he had
such a project and that in the history of logical empiricism in North
America, at least, his project was very influential for the topics central to
philosophy, the methods used to deal with them, and indeed the general
organization of philosophical work. But I do admit that this project is
harder to characterize than is, for example, Carnap's. Moreover, it might
be seen as more clearly an example of the more philosophically naïve
version of logical empiricism, especially if one concentrates on his more
popular philosophical writings, such as his *The Rise of Scientific Philosophy* (Reichenbach 1951) – although this, I would suggest, would be
a bad mistake.

This chapter is not an attempt to argue for a general characterization
of Reichenbach's project, but it is an invitation to a reassessment of that

project. I will offer a general claim about Reichenbach's *wissenschafts-analytisch* project in the period from 1920 to the mid-1930s and then use that claim to offer a new reading of his 1920 book on relativity theory, *The Theory of Relativity and A Priori Knowledge*.[2] My reading of that work will suggest that a common understanding in the recent literature of that book's significance for the subsequent development of logical empiricist philosophy of science is substantially at odds with Reichenbach's own understanding of his project there. I will end by suggesting that the book lands Reichenbach in an unsatisfactory philosophical place and that two of the central aspects of Reichenbach's work for the next 20 or so years thereafter – his acceptance, at Schlick's insistence, of the language of conventionalism and his pragmatic vindication of induction – were efforts to finding more satisfactory philosophical resolution to the problems of a suitably empiricist scientific philosophy.

My general view is that Reichenbach had two main goals for his overall philosophical project, one philosophical and epistemological and one metaphilosophical and methodological. His methodological goal for philosophy was the institution of a properly scientific philosophy – a philosophy that used the methods and standards of the exact sciences in its own work. His substantive philosophical claim was that such a scientific method in philosophy would show that attention to the formal features of knowledge in the exact sciences of nature requires that philosophy of science move from scientific neo-Kantianism to a theory of knowledge that is closer to empiricism. This was the main argument of the 1920 book. However, the form empiricism took in that book turned out to be quite fragile and potentially incoherent. What Schlick did in his conventionalist criticisms was offer Reichenbach a different strategy for distancing his empiricism from neo-Kantianism. The pragmatic vindication of induction, meanwhile, provided him with a more stable empiricist alternative to neo-Kantianism than did his 1920 account of induction.

6.2. Reichenbach and the History of Logical Empiricism

A history of logical empiricism that locates its significance mainly within the history of analytic philosophy does not make much room for Reichenbach. Allow me to take only the most prominent contemporary example: Volume 2 of Scott Soames' *The Analytic Tradition in Philosophy* (Soames 2017) devotes much space to a consideration of logical empiricism. Schlick and, especially, Carnap are discussed in several chapters. Reichenbach is discussed only in one chapter and that only in connection to his (alleged) attempt to analyze truth in terms of high probability. Features that philosophers of science would understand to be central to Reichenbach's work – his accounts of causal processes and the time-order, his philosophy of space and time, his vindication of induction, and so on – are presumably for Soames not central to the importance of logical empiricism for the analytic tradition and may safely be left undiscussed.

Now, there are several reasons for this lack of interest in Reichenbach in the official histories of logical empiricism as a chapter of analytic philosophy. One might, as indeed Soames does, think of analytic philosophy largely as a story of philosophical accounts of meaning. Then all the action in logical empiricism is its account of meaning, and the most action falls not to Reichenbach's "truth theory of meaning" in *Experience and Prediction* (Reichenbach 1938), say, but to the well-known logical positivist account of meaning as verifiability in principle. This is in line with a variety of different accounts, including those by Quine (1951) and Ayer (1936), that find the most exciting project in the history of logical empiricism to be the hard-core verificationism and strict translational reductionism of Vienna circa 1928, as exemplified (so they believe) in Carnap's 1928 *Logischer Aufbau der Welt*. If, like Quine, you take everything after the *Aufbau* to be a retreat from the failure of the *Aufbau* and a betrayal of the lessons of proper empiricism, Reichenbach is not going to be an interesting figure in your account of philosophy, since Reichenbach never endorsed anything like the *Aufbau* project. This systematic predilection is combined with a sort of Vienna-centrism right through the histories of logical empiricism. For example, A.J. Ayer's 1963 account of logical empiricism in the BBC series and volume *The Revolution in Philosophy* is titled "The Vienna Circle", which rather excludes Reichenbach right out of the gate on geographical grounds. The passage in which Ayer introduces Reichenbach – and after which Reichenbach is never discussed again – is telling for other reasons:

> From the beginning the members of the Circle met regularly to discuss philosophical problems among themselves, but it was not until 1929 that they, as it were, registered themselves as a philosophical party. . . . In 1930 they took over a journal called *Annalen der Philosophie*, renamed it *Erkenntnis*, and used it as the principal vehicle for the diffusion of their ideas. Its editors were Carnap and Hans Reichenbach, the leader of a similar though less important movement in Berlin.
>
> (1963, 71)

The reader of Ayer would be forgiven for not realizing after such an account that the *Annalen* was a prominent Berlin philosophical journal, that Reichenbach organized the takeover (which the previous editor did not view as friendly), that Reichenbach had chosen the new title for the journal (one Schlick, on Wittgensteinian grounds, hated), and that this brought to fruition a plan for a journal that Reichenbach had been pressing upon Carnap since they met in 1923.[3] Whether Ayer himself ought to be forgiven is another question.

It would be hard to expunge Reichenbach from a history of 20th-century empiricism – indeed, he has a rather prominent place in van Fraassen's (2002) history of empiricism – unless one thought that the crucial service to empiricism performed in logical empiricism came from semantics, as accords with Soames's account noted above. A related narrative is one that finds that the crucial feature of analytic philosophy is the method of analysis and that logical empiricism crucially followed the founders of analytic philosophy in providing for a specific and technical form of logical analysis.[4] You might think this even if you did not think analysis reveals meaning. This move gets me closer to one of my own interests in Reichenbach, because a history of analytic philosophy that features logical analysis as the key notion should be interested in various forms of philosophical analysis. Moreover, it is clear that in his early works Reichenbach understood himself to be doing analysis.

Reichenbach's favored terms in 1920 for his overall project were, after all, "*Erkenntnisanalyse*" – the analysis of knowledge – and "*Wissenschaftsanalyse*" – the analysis of science. In 1936, now writing for an Anglophone audience and seeking to make room for a distinctively German as opposed to Viennese form of logical empiricism, it is precisely this commitment to a method in the analysis of knowledge that Reichenbach stressed as the difference between the Berlin and the Vienna branches of logical empiricism. In 1936, Reichenbach had fixed on "*wissenschaftsanalytisch*" as his favored term and often renders it in German. (A wise choice since such German adjectives have no good counterparts in English; the best we can do is something like 'pertaining to the analysis of science'.) Here is a passage in which he seeks to distance the work of the Berlin Group from the work of the Vienna Circle:

> The system of logistic positivism marks a stage in the development of recent epistemology. . . . To say that it is one of the most important theories of knowledge ever propounded is an indication of its significance. But this calls attention at the same time to the weak point of the construction. It is a system – and nothing is so dangerous as the construction of a system at a time when analysis of the details has not yet been achieved. . . . The method of scientific analysis is the method of examining details. The adherents of the "*wissenschafts-analytische Methode*" – i.e., the members of the Berlin group – had always insisted that systematic construction must be foresworn until all details have been analyzed. They concentrated on minute work; and hoped to advance the work of the whole step by step. A great part of this work, indeed, confirmed the positivistic view that nothing but tautological constructions are allowed in science; for the disappearance of the synthetic a priori was a detailed confirmation of the positivistic principle. But there remained other points of the positivistic system which could not be sustained.
>
> (Reichenbach 1936, 149–150)

In Reichenbach's view, philosophy done according to the *wissenschaft-sanalytisch* method stood in a dual relation to science: science was the *topic* of epistemology, and also scientific ways of thinking were importantly *exemplified* in epistemology. The final section of the 1936 paper stresses precisely this point, which is also his gesture toward international community in philosophy and his rejection of "a certain modern nationalism" – a nationalism that had driven him from Berlin to Istanbul in 1933:

> But the German empiricist movement never intended to launch a uniform system of philosophy. Uniformity of opinion is an aim, but not a presupposition of joint action. What it aimed to do was to gather together a group of men working with empiricist methods and fully conscious of their intellectual responsibility. It was the common acceptance of this aim which enabled discussions to go on, even when opinions varied widely. The use of words with ambiguous meaning is a permissible device for the poet, if he wants to arouse certain feelings in his auditors; but in science, and scientific philosophy as well, it must be absolutely forbidden. It seemed to us that this was overlooked by much contemporary philosophy. We therefore set as our program the clarification of every scientific concept of the human understanding. As opposed to the era in which essays and enquiries "concerning human understanding", or "critiques of pure reason" were written, we understood "understanding" not in the abstract sense of a general human power, but in the concrete form of positive science; and so, our method developed into an analysis of science.
> (Reichenbach 1936, 159)

It is worth attending then to the project of *Erkenntnisanalyse* as Reichenbach explained it in the 1920s. I will not attempt to assess the success or failure of his project specifically with respect to the theories of relativity. Others are far more capable of evaluating that than I am. I am interested in the philosophical project that Reichenbach argues his work on relativity exemplified. I want particularly to attend to that project in its relations to scientific neo-Kantianism, where I draw conclusions quite different from those associated with the account of that work given by Michael Friedman. In particular, I wish to argue that Reichenbach's desire to distance himself from Kantianism and endorse empiricism in the 1920 book bequeaths to him two of his central concerns right through the 1920s and 1930s.

6.3. *Wissenschaftsanalyse* and the Relativized A Priori

The first thing to note about the project of *Erkenntnisanalyse* is that the term was quite specifically chosen by Reichenbach to distance himself from Kant's own project – he stresses this in 1920 and not just retrospectively in 1936. He says so specifically in the passage in which he introduces the contrast between his project and Kant's:

> If [Kant] searched for the conditions of knowledge, he should have ana-
> lyzed *knowledge*; but what he analyzed was *reason*. He should have
> searched for *axioms* instead of *categories*. It is correct that the nature
> of knowledge is determined by reason; but how this influence of reason
> manifests itself can be expressed only in knowledge, not in reason.
> (Reichenbach 1920, 69, cf. 1965, 72, original emphases)

Much is going on in this passage, but among the things most important to note is the specificity of his position compared to Kant: he grants that the Kantian question "how is knowledge possible?" is the proper question of the theory of knowledge; he raises a methodological objection to Kant's strategy for answering this question. Kant's strategy is his Copernican Revolution: to know how knowledge is possible is to know how synthetic a priori knowledge is possible, and to know that is to know how the fac- ulties condition the possibility of objective knowledge. Thus, Kant's own answer to the question requires understanding the forms of sensibility and understanding and the ideas of reason and their joint operation in the edifice of knowledge. This whole project is a mistake, according to Reichenbach. If you want to understand the formal conditions of the pos- sibility of knowledge, you need to know how knowledge works. Thus, you need to know how our best scientific theories work as knowledge systems, and this is the project of *Erkenntnisanalyse*. As Reichenbach says in the very last sentence of Chapter 6: "However illuminatingly the critical question 'How is knowledge possible?' stands before all theory of knowledge – it cannot lead to valid answers before the method of answer- ing it has been freed from the narrowness of psychological-speculative insight" (1920, 70; cf. 1965, 73). Obviously, this here means "transcen- dental psychological-speculative insight".

We will return to Reichenbach's relations to Kant shortly, but first we need to say some things about "axioms, rather than categories". That is, we must understand why axiomatization is, for Reichenbach, the proper project of *Wissenschaftsanalyse* even in the empirical realm. Roughly speaking, the answer is this: to understand how a physical the- ory is knowledge, you must understand the formal principles by which it constitutes its objects of knowledge and also how these principles are confirmed or refuted by experience. That sounds strange, since a priori principles ought not be refuted or confirmed by experience, you might think. But that is how Reichenbach thinks of the a priori. Here, roughly, is the structure of Reichenbach's thinking in 1920:

In the realm of pure mathematics, axioms are implicit definitions of concepts. As long as they are not inconsistent, they (partially) define con- cepts, and they present relations between concepts that can be used to deduce theorems.[5] Axioms in physics are not like that exactly. Some axioms in physics present the relations of physical concepts to one another (these are axioms of connection); these are typically in modern physics expressed mathematically. Other axioms are axioms of coordination – they coordinate the theoretical axioms of connection to "physical reality". (What is physical reality? We cannot answer that properly yet. Pretend you understand it.) To understand how a physical theory is knowledge is to understand how its axioms create a well-defined mathematical manifold and how that manifold is hooked by other axioms onto the physical world. His project in 1920 and then more fully in 1924 is to show that for the Special and General Theories of Relativity. (In 1924, he is also much more clear in the relations of axioms to definitions.)

This is enough detail for us to see why Reichenbach thought of this method as analysis. It is analysis in the most fundamental sense: he is taking theories that are presented and used as corporate bodies in physics and analyzing them into various components in order to show how they function as systems of physical knowledge – how they interact with the empirical data, how they connect to mathematical structures, how they relate to one another in the historical development of physics, and so on. He stresses these elements of analysis in 1936:[6]

> The appearance of Einstein's theory of relativity and the heated controversy about its interpretation (the degree of heat being commonly in an inverse ratio to the scientific education of the opponents) set new tasks for analysis; it produced an initial confusion, due in part to the inadequate exposition of the logical foundations of their subject by the physicists. The necessary analytical work in this field was done by our German group. It resulted in the distinction between the definitional and empirical components of the theory of relativity; and for the first time an adequate construction of space-time-doctrine was developed satisfying all philosophical requirements. In this construction, the empirical basis of space-time concepts was formulated, independently of definitions; space-time-order was shown to be the general order of the causal structure of the world. Incidentally, it turned out that the expositions of the theory at the hands of the physicists needed some essential corrections. The subjectivity of the observer, for instance, was eliminated and replaced by the definitional character of certain stipulations, called definitions of correspondence (*Zuordnungsdefinitionen*). In this way the untenable idea of a manifold of individual times was replaced by a manifold of definitions of the measurement of time (an idea which logically is entirely independent of the relativity of motion), and the term "relativity" came to be interpreted as the eligibility of any one of a set of definitions of correspondence.
>
> (Reichenbach 1936, 145–146)

To discover that theories face the tribunal of experience as a corporate body – which is indeed Reichenbach's view – is not to discover that there are no distinctions that can be drawn in the ways specific portions of theory function in making that possible. This delineation and differentiation of functional role is precisely what Reichenbach thinks forms the project of the analysis of science. Of course, the techniques of the new formal logic are not really in point here – this is epistemological analysis that takes the original mathematical formulation of the theory for granted.

As was already mentioned, in the recent scholarship on early analytic philosophy, Reichenbach's 1920 book has pride of place and thus serves as the key text for his current reputation. This is due to a pair of argumentative moves that he makes in the book and that have been stressed in Michael Friedman's historical and systematic work (Friedman 1994, 2001): Reichenbach distinguishes between two meanings of Kantian synthetic a priori principles; he then argues that these two understandings can be separated and argues we should set one understanding aside and should endorse a suitably retheorized version of the other. What are the two meanings of the synthetic a priori? On the one hand, they are understood to be universal and necessary principles. The second meaning is perhaps the deeper methodological meaning in the Kantian project – they are principles that constitute the objects of empirical knowledge. For Kant, of course, these understandings come together: the principles constitutive of the objects of empirical knowledge are discovered by investigating the permanent forms of the faculties of the mind.

In accordance with *Wissenschaftsanalyse*, which sets aside transcendental idealism and the forms of the mind in favor of the analysis of the actual knowledge systems themselves, Reichenbach makes two further moves. He argues that the history of natural scientific theorizing – especially the then recent history of the physics – showed that there were no permanent, universal, and necessary principles of objective knowledge. Thus, the first understanding of the synthetic a priori could safely be set aside. But, *Wissenschaftsanalyse* both endorsed and clarified the second notion of the synthetic a priori. You do not understand how scientific theorizing works if you do not see that they have formal principles that constitute the objects of empirical knowledge. However, in the history of physical theorizing, we, again, see that these principles change. Thus, the principles are relativized to the theories they are embedded in; this is the relativized constitutive a priori as emphasized by Michael Friedman.

It is worth drilling down a bit here, since the specific setting within which Reichenbach endorses the "constitutive of the object of knowledge" understanding of the synthetic a priori is particular to the project of *Wissenschaftsanalyse* and reveals quite a bit of Reichenbach's thinking about both axiomatization and how scientific theorizing relates to the empirical world. It is a fairly deep debt to Kantian thinking and also connects his work to themes in both neo-Kantians like Cassirer and to formalism in mathematics à la Hilbert and in science à la Duhem.

The setting, as we already began to enunciate previously, is this: the nature of mathematics has been fundamentally clarified by axiomatization. In the formal sciences, the axioms implicitly define conceptual structures. There is no additional question as to whether there are objects corresponding to those structures. We wish to bring the clarity that comes from axiomatization to empirical science as well. But we must respect the difference between empirical and formal science. For Reichenbach, this requires "coordination with the physical world". But he is neither a radical empiricist nor a phenomenalist nor a naïve realist. Thus, there are some serious questions about what exactly we are coordinating our theories to and what coordination here might mean.

On these issues, Reichenbach bites the bullet. Theories are not coordinated to perception or experience: perceptions are not the proper sorts of things to be the independently real objects to which theories are coordinated. Nor is there a world of everyday objects that we can straightforwardly coordinate with the theories of physics. Instead, the coordination itself defines or creates the set of objects to which the equations are coordinated. Here are some characteristic ways Reichenbach (1920, 38, cf. 1965, 40) speaks at this point: "Thus we are faced with the strange fact that in knowledge we accomplish a coordination of two sets, one of which not only attains its order through this coordination, but whose elements are *defined by means of this coordination*" (original emphasis). Or again:

> Since perceptions do not define the elements of the overarching set, one side of the knowledge process contains a completely undefined class. Thus it happens that physical law first defines the individual things and their order. The coordination itself creates one of the sequences of elements to be coordinated.
> (Reichenbach 1920, 39–40, cf. 1965, 42)

And, really, how could it be otherwise: if you take completely seriously the objectivity-conferring role of synthetic a priori principles, then these principles cannot be coordinated with *independently existing objects*.

Having bitten the bullet on the ontological matter, Reichenbach makes the move you would expect: if the coordination is not successful by virtue of connecting two independently existing sets, it must be successful (or not) in some other way. It is successful if it defines a consistent class of objects. Reichenbach expresses this in the language of univocality or uniqueness – *Eindeutigkeit*.[7] The coordination should yield the same properties for the empirically constituted objects no matter how it is deployed. He illustrates this with an imagined case of a disconfirmation of the Einsteinian prediction of the deflection of light in the gravitational field as empirically investigated by Eddington. Reichenbach imagines a theoretical prediction of a 1.7" deflection and a measurement of a 10" deflection. The crucial issue here is that experimental measurement is, according to Reichenbach and fully in line with Duhem's philosophy, itself a highly theoretical calculation:

> The value 1.7" has been obtained on the basis of equations and experiences of other material; but the value of 10" has in principle not been ascertained in a different way since it is not read off directly. . . . It can be maintained therefore that *one* chain of reasoning and experience coordinates the value 1.7 to the physical event, the *other* the value 10, and here lies the contradiction. That theory which continuously leads to consistent coordination is called true.
>
> (1920, 41, cf. 1965, 43)

All of this, as has been emphasized repeatedly, is indeed in Reichenbach (1920). It forms the center of the current understanding of the significance of the book. But it is the business of Chapters Four and Five of an eight-chapter book. So, it is neither the first nor the last thing that Reichenbach argues for. Attending to the larger setting in which these arguments are given can substantially alter our sense of the project Reichenbach is engaged in in the book. The book comes to look both more neo-Kantian in its argumentative structure and less Kantian in its philosophical aims.

6.4. The Larger Philosophical Project of *Wissenschaftsanalyse*

Where are we before we arrive at the question of Chapter Four and the account of knowledge as unique coordination? After a brief introductory chapter, Chapters Two and Three take us through "the contradictions asserted" by the special and general theories of relativity. These are unfortunate chapter titles because Reichenbach's point is not that the special and general theories of relativity contradict themselves. What they assert, rather, are contradictions to the principles that Kant argued were synthetic and a priori. Reichenbach wishes to establish not only that the constitutive principles of the theories of relativity differ from those Kant argued were synthetic and a priori but also that history of physics in this period reveals how Kantian a priori principles had been empirically refuted.

The language of unique coordination in Chapter Four helps express why Reichenbach has what sounds like an absurd view – a priori principles can be empirically refuted. His claim is precisely that the principles that Kant claims to be a priori (supplemented by a few of his own suggestions) cannot be uniquely coordinated with the world of experiment and experience. Thus he states that:

According to the special theory of relativity, we assert that the following principles in their totality are incompatible with experimental observations:

> the principle of the relativity of uniformly moving coordinates;
> the principle of irreversible causality;
> the principle of action by contact;
> the principles of the approximate ideal;
> the principle of normal induction;
> the principle of absolute time.
> (Reichenbach 1920, 15, cf. 1965, 15)

Taken together, Reichenbach argues, these Newtonian principles entail the possibility of arbitrarily high and physically attainable velocities. He understands that that entailment was experimentally refuted by evidence that in real physical processes, the speed of light was the highest attainable velocity.

In this argument, one particular a priori principle has a special place.[8] That principle is the principle of normal induction. This is the principle that allows findings contrary to high-level principles to be asserted:

> Even the Michelson experiment is a proof only if very ingenious theories for the rescuing of the old theorem concerning the addition of velocities are rejected. The extrapolation therefore has only a certain degree of probability. Let us call the principle of using the most probable extrapolation of experiential material the *principle of normal induction*.
> (Reichenbach 1920, 14; cf. 1965, 14, original emphasis)

It is the principle of normal induction that grounds all experimental results and thus experimental testing. But its power is sufficient to bring down even systems of allegedly a priori principles due to their inability to achieve unique coordination. Indeed, he is clear that this discussion from Chapter Two that grants the special epistemological status of the principle of normal induction is what motivates the discussion in Chapter Four about knowledge as unique coordination.

The argumentative setting of the discussion in Chapter Four and the unique place for specifically the principle of normal induction have not been sufficiently stressed in the recent literature on Reichenbach. The principle of normal induction is a clearly epistemologically distinct sort of a priori principle for Reichenbach in 1920; indeed, it was arguably already in his dissertation (1916/2008).[9] The way he deploys it underscores how far his philosophy had moved toward empiricism from the very beginnings, even as it denied experience or perception a privileged place as the basis or content of knowledge. The principle of normal induction is what grounds his fallibilism about a priori principles.[10]

There is another argument in the 1920 book that should also change our appreciation of the philosophical project Reichenbach was there engaged in. This is the argument that the distinction he draws between the two notions of the a priori directly subserves. The argument in Chapter Two was meant to argue that the specific a priori principles that Kant claims to give transcendental grounding to have been empirically refuted. There we see a highly familiar account of Kant as attempting to transcendentally ground the principles embedded in Newtonian physics and the argument that these principles provide too much content to be uniquely coordinated with the world revealed in experiment. The project of Chapters Five and Six is different from this. There Reichenbach wants to argue not against the specific principles Kant took to be a priori but rather with an epistemological principle he thinks Kant is committed to. This epistemological principle is interesting and strange.

We must approach this argument in a couple of steps. First, we need to get clearer on what role experience or "perception" plays in the unique coordination that constitutes knowledge for Reichenbach. We have already seen that experience is not that to which theoretical structures are coordinated. There is no way independently of the coordination even to characterize the objects to which the theoretical structures are coordinated. But experience does not play no role, for then all empirical confirmation and disconfirmation seem impossible. Reichenbach argues that perceptions form "the criterion of uniqueness of the coordination":

> We always call a theory true when all chains of reasoning lead to the same number for the same phenomenon. This is the only criterion of truth; it is that criterion which, since the discovery of exact empirical science by Galileo and Newton and of its philosophical justification by Kant, has been regarded as the absolute judge. And we notice that we can now point out the role played by perception in the process of knowledge. *Perception furnishes the criterion of uniqueness of the coordination.* We saw previously that it cannot define the elements of reality. But it is always able to lead to the determination of uniqueness.
>
> (1920, 41–42, cf. 1965, 43–44, original emphasis)

By this he does not mean that the mere having of perceptions is somehow the test of coordination. The perceptions he discusses are perceptions interpreted according to scientific theory and principles of measurement, as the original relativity example showed and the reliance on the principle of normal induction underscores. By means of often highly recondite theory, we reach the conclusion that we can through certain operations actually measure physical quantities. Thus, in addition to the theoretical predictions of what value a quantity has, we have a prediction that thus and so a method measures that value. Having experience that is interpreted as a measurement value that accords with a theoretical prediction that also depended upon experience is the criterion of uniqueness or failure of uniqueness of the coordination of the measured value and the theoretically predicted value. As he says "[e]very perception has this property, and this is its only epistemological significance" (Reichenbach 1920, 44).

Second, we need to understand the significance Reichenbach places precisely in Kant conflating the necessary principles of reason with the principles constituting the object of empirical knowledge. Beginning in the final pages of Chapter Five, Reichenbach argues that Kant needs to show, rather than simply presume, that the principles of reason are knowledge-constituting principles, that is, that they allow of unique coordination. Since Kant cannot rely on pre-established harmony between reason and reality and cannot show a priori that his system does uniquely coordinate with empirical reality, Kant needs, according to Reichenbach, to show that it must uniquely coordinate through some alternative means of argument. The only remaining alternative, so thinks Reichenbach, is that Kant needs to show that the principles of reason uniquely coordinate with empirical reality because *any* system of axioms of connection allows of unique coordination. As Reichenbach (1920, 57, 1965, 60) says "Kant's theory contains the hypothesis *that there are no implicitly contradictory systems of coordinating principles for the knowledge of reality*" (original emphasis); he calls this "the hypothesis of the arbitrariness of coordination".[11]

Now this argument is astonishing. As Kant interpretation, it is scarcely credible. Our robustly Newtonian Kant who built Euclidean geometry and Newtonian laws of motion into the very possibility of experience now stands accused of requiring that any system of axioms allows of unique coordination to experience. The problem as Kant interpretation is, of course, Reichenbach's view that perceptions furnish the criterion of uniqueness of the coordination and thus that this uniqueness can only be determined a posteriori via the axiom of normal induction. (It is important in this regard that Reichenbach places this axiom precisely at the center of his disagreement with Kant. He thinks Kant tries and fails to establish that the axiom cannot ever properly be deployed in such a way as to contradict a system of axioms.) Reichenbach thinks the only alternatives to his view of the situation are the hypothesis of preestablished harmony, which he rightly sees Kant as rejecting, and the arbitrariness of coordination (which must therefore be Kant's position). Of course, to view these as the only options is precisely to beg the question against transcendental idealism and the view that "object" is an empty intellectual concept unless informed by the formal conditions under which objects become objects of knowledge for us. That is, the very transcendental project that Reichenbach has already set aside is Kant's answer to precisely this question.

Even stranger in some ways is that Reichenbach is here bringing up a central issue for early twentieth-century scientific neo-Kantianism and yet fails either clearly to specify or to argue against the views taken by the neo-Kantians. The most salient example of a scientific neo-Kantian who had raised this issue is Ernst Cassirer. As Reichenbach makes clear, he had not read the Einstein pamphlet by Cassirer when he published his book. But he had certainly read Cassirer's 1910 *Substance and Function* (to which he refers in the very footnote in which he says he hadn't read the Einstein pamphlet before his book went to print), and it is there where Cassirer covers this whole issue in terms very close to Reichenbach's but where he takes a much more strictly constructivist Kantian line.[12]

Cassirer discusses these issues in conjunction with what he called "the core logical problem of physics", which he expressed as follows:

> The naïve view that the measures of physical things and processes inhere in them like sensible qualities and, like those qualities, only need to be read off, recedes ever more with the progress of physics. With this, however, and at the same time, the relation between law and fact changes. For the explanation that we attain to laws insofar as we compare and measure individual facts now reveals itself as a logical circle. The law can only emerge from measurement because we have put it into the measurement in hypothetical form. However paradoxical this mutual relation might appear, it indicates exactly the core logical problem of physics.
>
> (1910, 193–194)

Cassirer is here, I claim, interested in exactly the issue raised by Reichenbach: both the theoretical prediction of a measurement value and the physical processes of measurement are guided by theory. But Cassirer does not make the individual perceptual fact (as intellectually interpreted) the criterion of unique coordination. Rather, he views the whole process as holistic and dialectic: science seeks to construct a world of "experience in its entirety". Where contradictions exist, they preclude a completed world of experience. The harmonizing of theoretical prediction and the processes of measurement is the construction of the world of experience in its entirety.[13]

There is another feature of Reichenbach's overall argument in the book that perhaps reveals his debt to Cassirer. The relativized constitutive a priori that Reichenbach is thought to be endorsing in 1920 is often contrasted with what Cassirer in *Substance and Function* calls "the universal invariant" theory of experience. In this latter account, there are necessary and universal a priori forms for experience but, in quasi-Hegelian fashion, we only know what they are at the end of inquiry, since they are revealed as the constants across changes of theory. Reichenbach's insistence that Newton, the special theory of relativity, and the general theory of relativity all have different a priori principles seems to count in favor of his endorsement of the relativized a priori over the universal constitutive a priori. But things are not so simple.

Reichenbach's Chapter Eight suggests a very different narrative embedded in the history of physics. In Chapter Eight, Reichenbach stresses again the object-constituting function of a priori principles but, by again stressing the possibility (and actuality) of systems of a priori principles being refuted by experience, he now sets up a progress of physics across changes of principles: he stresses in the final chapter a progressive history of changes in the concept of the object of physical knowledge. Thus, he sums up his argument as follows:

> We understand that today's conditions of knowledge are no longer those of Kant, *because the concept of knowledge has changed, and the changed object of physical knowledge presupposes different logical conditions.* The change could occur only in connection with experience, and therefore the principles of knowledge are also determined through experience. But their validity does not depend only upon the judgment of particular experiences, but also upon the possibility of the whole system of knowledge: this is the sense of the a priori.
> (Reichenbach 1920, 99–100, cf. 1965, 104,
> original emphasis)

This directionality in the progress of science draws out some aspects implicit in his Chapter Seven account of a priori principles, in which it is clear that Reichenbach views the direction of changes in the constitutive from the more specific to the more general, without positing that there is a most general form of them. "For all conceivable principles of coordination the following statement holds: for every principle, however it may be formulated, a more general one can be indicated that contains the first as a special case" (Reichenbach 1920, 78, cf. 1965, 81).

Indeed, Reichenbach goes on to make a similar argument at the meta-level and to argue that the very idea that the object of knowledge depends on unique coordination is a principle that could be generalized. In fact, he castigates Schlick for not allowing for this possibility, accusing him of engaging in an overly (transcendentally) psychological version of epistemology and making unique coordination a permanent principle of epistemology.[14] Nonetheless, he maintains a directionality to history even here, so even in the history of epistemology itself:

> Whereas a change in particular laws produces only a change in the relations between particular things, the progressive generalization of the principles of coordination represents a development in the *concept of object* in physics. Our view differs from Kant's philosophy as follows: whereas in Kant's philosophy only the determination of

the *particular concept* is an infinite task, we contend *that even our concepts of the object of science in general, that is, of the real and its determination can only meet with a gradual precisification.*
(Reichenbach 1920, 84, cf. 1965, 88, original emphasis)

I do find it odd that this aspect of Reichenbach's views in 1920, which is very much in line with Cassirer's universal invariant theory of experience, has been submerged in recent accounts of the book. His account here depends on two things: the idea that objective scientific knowledge is unique coordination and his view that the principle of normal induction allows systems of (relativized) a priori principles to be refuted by experience (because they lead to ambiguous coordination). It is worth noting in a bit more detail how close this all is to Cassirer in 1910. Cassirer's account of the "universal invariant theory of experience" occurs in Chapter Five of *Substance and Function*, which is given over to the problem of induction. He introduces the universal invariant theory of experience in a passage in which he discusses the need in physics sometimes to revise not only "the facts" and "purely empirical 'rules'" but indeed "principles and axioms" and uses the overthrow of Newtonian physics as an example (Cassirer 1910, 355). His discussion ends with these claims:

Since we never compare the totality of the hypotheses as such with the naked facts, but rather always only compare one hypothetical system of axioms with another, more general and more radical one, we require of this progressive comparison a final constant measure in the highest axioms that are valid for all experience in general. Thought demands the identity of this logical measurement system despite all change in that which is measured. In this sense the critical theory of experience forms in fact at the same time the universal invariant theory of experience and thus fulfills a demand toward which the inductive procedure itself ever more clearly urges.
(Cassirer 1910, 355–356)

The connection is even stronger than this: Reichenbach argues that the evidence for the new constitutive principles needs to be adduced in accordance with the earlier principles. One uses, for example, classical optics in one's understanding of the optical telescope used in the Eddington expedition, for example. This connects to Cassirer's claim in this same passage that "[t]he new form should contain the answer to questions that were framed and formulated in the older form" (1910, 355).

So, what is our take-home message here? I think it is clear that the overall argument of Reichenbach's 1920 book is far more informed by scientific neo-Kantianism than is typically understood. Nonetheless, he still wishes ultimately to argue that, carried through to the end, the *wissenschaftsanalytisch* method leads to empiricism. The problem here for Reichenbach is that Cassirer's project for scientific neo-Kantianism was to attempt to colonize the entire field – he wanted to show that neo-Kantianism subsumed what was right in empiricism, in realism, and so on. Reichenbach needed to find a way to distinguish his position from Cassirer's. So what did he do at this point in the dialectic? He argued that the principle of normal induction also was fallible and might be rejected in the course of science and that the principle of uniqueness of coordination might be rejected in the course of epistemological thinking. While these moves do distinguish him from Cassirer and carve out a sort of through-going fallibilist empiricism, they leave him with a very fragile form of empiricism. The rejection of normal induction and the rejection of uniqueness both threaten to change empiricism into a form of skepticism or anarchism. It is unclear how any form of contact with the world of experience is left for theory if normal induction is rejected – given the central role it has played in Reichenbach's argument so far. Similarly, to reject uniqueness is to leave it unclear what the epistemological problem would look like that would drive any further changes in scientific theory. Reichenbach's ways to distinguish himself from Cassirer in this framework offer science and scientific epistemology a less constrained future but at the cost of there being any coherent future direction for science or indeed any *Erkenntnisproblem* to be solved.

6.5. The Lasting Legacy in Reichenbach's Work of the Problems of 1920

These problems are internal issues for Reichenbach's project as framed in 1920. I end by suggesting that they are at least in part responsible for some important changes to his view later on. First, these considerations provide an internal reason for him ultimately to accept Schlick's criticisms of the language of constitution and his adoption of the language of conventionalism.[15] For the language of conventionalism gives Reichenbach an alternative way to distance himself from Kantianism. By the late 1920s, Reichenbach's preferred language critical of Kantianism is not a simple criticism of the psychological language of Kant's faculties of the mind per se but a criticism that Kant presupposes that such faculties have constant forms that structure knowledge: in the realm of knowledge, Kant constrains the will to the permanent forms of intuition, understanding, and reason. Reichenbach, now speaking the language of conventionalism, reverses the order here – he argues that while knowledge needs structure to be knowledge, this structure is not imposed by the forms of the mind but is freely chosen by the rational will.[16] This theme is rung throughout his work in the late 1920s and early 1930s. Often the stress in such passages is on the emancipatory aspect of this view of the nature of scientific knowledge. Here is one such passage from 1930:

> And perhaps one may view as the most significant result of modern natural knowledge that the picture of the world to which it has led

has at the same time brought to light a new picture of humans as thinking minds: for natural science has shown us that reason is not a rigid scaffolding of logical pigeonholes, that thinking does not signify the endless repetition of norms that have been superseded, but has taught instead that humans grow with their knowledge and carry within themselves the capacity for forms of thought that they could not have conceived at earlier stages.

(Reichenbach 1930, 71)

Friedman has repeatedly argued that Reichenbach made a mistake in adopting the language of the conventional and dropping the language of the constitutive. This is because Friedman grants the Quinean-Einsteinian argument that holism with respect to theory testing does not induce any non-arbitrary distinction between the conventional and the empirical. Here we need not concern ourselves with whether Friedman should grant this argument, because I do not see any evidence that Reichenbach's adoption of the language of the conventional was due to a misunderstanding he shared with Schlick of the Duhemian point. *Wissenschafts-analyse* never proceeded from holism with respect to theory testing to the conventional/constitutional vs. empirical split. Throughout, the ability of physical theory to be tested in experience at all depended on our ability to distinguish the formal and conventional (and object-constituting) features of physical theories from their particular empirical claims and laws. I am arguing here that what the language of the conventional did for Reichenbach was not at the level of the analysis of science but at the level of his anti-Kantian metaphilosophy: rather than arguing for a thoroughgoing fallibilism of principles and axioms, Reichenbach's epistemology of science could stress a different epistemological status for constitutive principles – they are not formal principles of objective thought forced upon us by the nature of thought but are rather constitutive principles adopted by the will. Reichenbach's empiricism is grounded in the free choice of the rational will.

Second, the quite specific place for the principle of normal induction in the arguments of the 1920 book leads Reichenbach into a dilemma, neither horn of which should quite satisfy him. He must either maintain the stance that this principle is robustly a priori – indeed, the a priori precondition of all contact of theory with reality – or he must adopt fallibilism with respect to it. This latter move might be more robustly empiricist but at the potential cost of rendering it unclear what the knowledge-producing enterprise without normal induction would look like. It is unclear what knowledge would look like and what a scientific question or answer would consist in. I submit that the pragmatic vindication of induction is Reichenbach's ultimate answer to this question: in the structure of the pragmatic vindication of induction, the principle of induction is not an a priori principle that must be followed for even the concept of knowledge to make any sense. But it is a principle that is not quite subject to empirical refutation. Rather it has the status of a principle that will lead to the construction of predictive knowledge of the world if such knowledge is possible at all.

Notes

1. Of course, there is a voluminous literature on Reichenbach, and I don't mean to undervalue the work of many scholars – including, importantly, Kamlah (1977, 2013), Salmon (1991), Friedman (1994), Howard (1994, 2010), Ryckman (2005), Milkov (2011, 2013), Eberhardt (2011), Glymour and Eberhardt (2016), Padovani (2011, 2015a, 2015b, 2017), or others in saying this. There is certainly much important work on Reichenbach's philosophy; my own work here is substantially influenced especially by Padovani's work, in which she emphasizes the larger setting of Reichenbach's constitutive principles. But compared with, for example, the sheer amount and the basic agreement on where the interpretative issues are with respect to Carnap's work or Neurath's, the Reichenbach literature is far behind.
2. I have found it necessary slightly to alter most translations from the canonical English edition, so I provide both the page numbers to the German (Reichenbach 1920) and a pointer to the relevant passage in the English edition (Reichenbach 1965).
3. I do not believe the full history of the start of the journal *Erkenntnis* has been written yet, although a start was made in Hegselmann and Siegwart (1991). For Carnap and Schlick's reactions to the title and the draft of Reichenbach's initial editorial, see Dewulf (2018, 69). It is clear from their letters that Reichenbach was trying to convince Carnap that they should start a journal right back to their first contact in 1923. See, for example, Reichenbach's letter to Carnap of March 29, 1923 (HR 15–50–01).
4. Perhaps the person who has most vigorously pursued this line in recent work is Michael Beaney (2007, 2013)
5. In our current technical environment, this idea that concepts are partially defined through axioms is under dispute – see, for example, the last three sections of §2 of Hodges (2013); but, of course, Reichenbach was not privy to these technical developments in 1920, and his view here is not unusual.
6. In the quoted passage, Reichenbach is guilty of (at least) exaggeration: space-time theories were also important topics among the members of the Vienna Circle; certainly at least the work of Schlick and Carnap needs to be mentioned here.
7. Howard (1996) is a long meditation on the importance of *Eindeutigkeit* in early analytic philosophy but does not take up Reichenbach (1920).
8. It is of interest that the two papers that more than any others established this new understanding of the significance of Reichenbach's book, Friedman (1994) and Howard (1994), never once mention the principle of induction.
9. I read Padovani (2011), especially section 2.2, as providing precisely this argument.
10. Interestingly, in another publication from 1920, Reichenbach (1920/1978) claims that the law of (probabilistic) distribution and causality are necessary presuppositions of the possibility of knowledge. His view here is very close to the view he castigates Schlick for in the book; see footnote 14.
11. Howard (1994, 59–60) is a rare exception to the silence in the literature about this strange argument.

12. See footnote 20 to Reichenbach (1920, 108–109, cf. 1965, 114–115). See Padovani (2011) for an account of Cassirer's influence on early Reichenbach.
13. I have discussed Cassirer's views on this issue at length in Richardson (2015).
14. Reichenbach's criticism of Schlick on this matter is perhaps the sharpest in the entire book (Reichenbach 1920, 110, cf. 1965, 116): "It is characteristic of Schlick's psychological method, that he believes himself to have refuted with many proofs the correct part of the Kantian theory, namely, the constitutive meaning of co-ordinating principles, and to have, without noticing it, taken on the mistaken part. The characterization of knowledge as unique co-ordination is Schlick's analysis of reason, and uniqueness is his synthetic judgment a priori."
15. On this matter, see Coffa (1991), Friedman (1994), Howard (1994), Padovani (2015a).
16. I discuss the volitional aspects of Reichenbach's position in the late 1920s in Richardson (2005, 2017).

References

Ayer, A. J. (1936), *Language, Truth, and Logic*. London: Gollancz.
——— (1963), 'The Vienna Circle', in A. J. Ayer (ed.), *The Revolution in Philosophy*. London: Macmillan, pp. 70–87.
Beaney, M. (ed.) (2007), *The Analytic Turn in Philosophy: Analysis in Early Analytic Philosophy and Phenomenology*. London: Routledge.
——— (2013), 'The Historiography of Analytic Philosophy', in M. Beaney (ed.), *The Oxford Handbook of the History of Analytic Philosophy*. Oxford: Oxford University Press, pp. 30–60.
Carnap, R. (1928), *Der logische Aufbau der Welt*. Berlin: Weltkreis.
Cassirer, E. (1910), *Substanzbegriff und Funktionsbegriff*. Berlin: Bruno Cassirer.
Coffa, A. (1991), *The Semantic Tradition from Kant to Carnap: To the Vienna Station*. Cambridge: Cambridge University Press.
Dewulf, F. (2018), *A Genealogy of Scientific Explanation*. Ph.D. Dissertation. Faculteit Letteren en Wijsbegeerte, Gent.
Eberhardt, F. (2011), 'Reliability via Synthetic A Priori: Reichenbach's Doctoral Thesis on Probability', *Synthese*, 181: 125–136.
Friedman, M. (1994), 'Geometry, Convention, and the Relativized A Priori: Reichenbach, Schlick, and Carnap', in W. Salmon and G. Wolters (eds.), *Logic, Language, and the Structure of Scientific Theories*. Pittsburgh: University of Pittsburgh Press, pp. 21–34.
——— (2001), *The Dynamics of Reason*. Stanford, CA: CSLI Publications.
Glymour, C. and Eberhardt, F. (2016), 'Hans Reichenbach', in *Stanford Encyclopedia of Philosophy*, https://plato.stanford.edu/entries/reichenbach/.
Hegselmann, R. and Siegwart, G. (1991), 'Zur Geschichte der "Erkenntnis"', *Erkenntnis*, 35 (1/3): 461–471.
Hodges, W. (2013), 'Model Theory', *Stanford Encyclopedia of Philosophy*, https://plato.stanford.edu/entries/model-theory/.
Howard, D. (1994), 'Einstein, Kant, and the Origins of Logical Empiricism', in W. Salmon and G. Wolters (eds.), *Logic, Language, and the Structure of Scientific Theories*. Pittsburgh: University of Pittsburgh Press, pp. 45–105.
——— (1996), 'Relativity, Eindeutigkeit, and Monomorphism: Rudolf Carnap and the Development of the Categoricity Concept in Formal Semantics', in R. N.

Giere and A. W. Richardson (eds.), *Origins of Logical Empiricism*. Minneapolis, MN: University of Minnesota Press, pp. 115–164.

—— (2010), '"Let Me Briefly Indicate Why I Do Not Find This Standpoint Natural", Einstein, General Relativity, and the Contingent A Priori', in M. Domski and M. Dickson (eds.), *Discourse on a New Method: Reinvigorating the Marriage of History and Philosophy of Science*. LaSalle, IL: The Open Court, pp. 333–355.

Kamlah, A. (1977), 'Hans Reichenbach's Relativity of Geometry', *Synthese*, 34 (3): 249–263.

—— (2013), 'Everybody Has the Right to Do What He Wants: Hans Reichenbach's Volitionism and Its Historical Roots', in N. Milkov and V. Peckhaus (eds.), *The Berlin Group and the Philosophy of Logical Empiricism*. Dordrecht: Springer, pp. 151–175.

Milkov, N. (2011), 'Hans Reichenbachs wissenschaftliche Philosophie', in N. Milkov (ed.), *Hans Reichenbach: Ziele und Wege der heutigen Naturphilosophie*. Hamburg: Felix Meiner Verlag, pp. vii–xliv.

—— (2013), 'The Berlin Group and the Vienna Circle: Affinities and Divergences', in N. Milkov and V. Peckhaus (eds.), *The Berlin Group and the Philosophy of Logical Empiricism*. Dordrecht: Springer, pp. 3–32.

Padovani, F. (2011), 'Relativizing the Relativized A Priori: Reichenbach's Axioms of Coordination Divided', *Synthese*, 181: 41–62.

—— (2015a), 'Measurement, Coordination, and the Relativized A Priori', *Studies in History and Philosophy of Modern Physics*, 52: 123–128.

—— (2015b), 'Reichenbach on Causality in 1923: Scientific Inference, Coordination, and Confirmation', *Studies in History and Philosophy of Science*, 53: 3–11.

—— (2017), 'Coordination and Measurement: What We Get Wrong about What Reichenbach Got Right', *European Studies in Philosophy of Science*, 5: 49–60.

Quine, W. V. O. (1951), 'Two Dogmas of Empiricism', *Philosophical Review*, 60: 20–43.

Reichenbach, H. (1916/2008), *The Concept of Probability in the Mathematical Representation of Reality*. Translated and Edited by F. Eberhardt and C. Glymour. Chicago: Open Court.

—— (1920/1978), 'A Philosophical Critique of the Probability Calculus', in M. Reichenbach and R. S. Cohen (eds.), *Hans Reichenbach: Selected Writings, 1909–1953, Volume 2*. Dordrecht: Reidel, pp. 312–327.

—— (1920), *Relativitätstheorie und Erkenntnis A Priori*. Berlin: Springer.

—— (1930), 'Die philosophische Bedeutung der modernen Physik', *Erkenntnis*, 1: 49–71.

—— (1936), 'Logistic Empiricism in Germany and the Present State of Its Problems', *The Journal of Philosophy*, 33: 141–160.

—— (1938), *Experience and Prediction*. Chicago: University of Chicago Press.

—— (1951), *The Rise of Scientific Philosophy*. Berkeley and Los Angeles: University of California Press.

—— (1965), *The Theory of Relativity and A Priori Knowledge*. Translated by Maria Reichenbach. Berkeley and Los Angeles: University of California Press.

Richardson, A. W. (2005), '"The Tenacious, Malleable, Indefatigable, and Yet, Eternally Modifiable Will": Hans Reichenbach's Knowing Subject', *Proceedings of the Aristotelian Society*, 79: 73–87.

———— (2015), 'Holism and the Constitution of "Experience in Its Entirety": Cassirer Contra Quine on the Lessons of Duhem', in J. T. Friedman and S. Luft (eds.), *The Philosophy of Ernst Cassirer: A Novel Assessment*. Berlin: De Gruyter, pp. 103–122.

———— (2017), '"Neither a Confession nor an Accusation": Michael Polanyi, Hans Reichenbach, and Philosophical Modernity after World War One', *Historical Studies in the Natural Sciences*, 47: 423–442.

Ryckman, T. (2005), *The Reign of Relativity: Philosophy in Physics 1915–1925*. Oxford: Oxford University Press.

Salmon, W. C. (1991), 'Hans Reichenbach's Vindication of Induction', *Erkenntnis*, 35: 99–122.

Soames, S. (2017), *The Analytic Tradition in Philosophy, Volume 2: A New Vision*. Princeton, NJ: Princeton University Press.

van Fraassen, B. (2002), *The Empirical Stance*. New Haven, CT: Yale University Press.

Part 2

The Philosophy of Physical Theories

7 Carnap, Einstein, and the Empirical Foundations of Space-Time Geometry

Robert DiSalle

7.1. Introduction

One of the chief failings of logical empiricism, in the view of the late 20th century, concerned one of its central aims: to explain the relation between scientific theories and experience. Logical empiricists' accounts of theoretical and observational languages, with associated distinctions between theoretical and observational concepts and statements, were intended to connect theories as formal axiomatic structures with the experiments and observations that provided their empirical meaning and support. Realizing this intention turned out to be surprisingly difficult, however, and the consensus of subsequent philosophers of science seemed to be that such a project could not succeed in the way in which it was framed. Indeed, this failing came to be regarded as indicating some misconception at the heart of the logical empiricists' view of theories. The rise of the semantic view of theories, led by Suppes, was meant to replace their view with one less preoccupied with axiomatic structures and more in keeping with the practice of science.

This was a somewhat ironic outcome. The logical empiricists, after all, distinguished themselves by their efforts to absorb the philosophical lessons of contemporary scientific work – in particular, the work of Einstein in developing special and general relativity. Unlike most philosophers who commented on relativity in the early 20th century, their leading figures, Schlick, Reichenbach, and Carnap, were well versed in technical details of the theories and in close intellectual contact with the physicists involved. Of course, this was not the first time that philosophers had found profound lessons in the most advanced science of their time. For Hume and Kant, especially, Newton's theory of universal gravitation, and his avowed method of arriving at it, were indispensable examples for progress in philosophy. To the logical empiricists, however, Einstein's work was more than an exemplary scientific methodology. It was a kind of philosophical work in itself. Einstein had arrived at both special and general relativity by a particular effort of philosophical analysis, and central to his success was his philosophical investigation of the empirical meaning of theoretical concepts. The ideas behind this analysis had antecedents in the work of Helmholtz, Poincaré, and Hilbert, but it was Einstein who had showed that their use could lead to profound theoretical transformations in physics. In short, the logical empiricists' goal of a "scientific philosophy" was motivated in general by decades of profound reflection on the connections between mathematics and experience and in particular by Einstein's appeal to that tradition in creating special and general relativity.

That the logical empiricists' program would be judged unsuccessful, having so informed and being so motivated by the actual practice of science in their time, is a fact that requires a complex explanation, and this chapter will not attempt to find a complete one. At least one part of it must lie in the misconceptions of the foundations of general relativity, and its philosophical significance, that colored their philosophy of physics generally. Much philosophical work has since been devoted to addressing those misconceptions, and this chapter has little to add.[1] Another part of the explanation must lie in preoccupations and presuppositions of the late 20th century that led to a different conception of the aims of the philosophy of science and that are not addressed here.

This chapter starts, rather, with some sympathy with the logical empiricists' aims and examines some central challenges to their fulfillment. Central among these, I suggest, was the problem of connecting theoretical structures with their empirical content. Carnap, in particular, explained this connection in a manner intended, in part, to characterize the sense in which physical theories make genuinely synthetic claims about the empirical world. It was also intended to capture characteristic elements of then-contemporary science and especially Einstein's insights into the relation between geometry and experience. That view, according to Einstein, was central to the motivating arguments for the general theory of relativity, which included not only the familiar arguments about generalizing the relativity of motion, but also the more complicated argument about the empirical content of space-time geometry. On this matter, Einstein's views, broadly speaking, reflected the influence and insight of Poincaré and Hilbert regarding the empirical interpretation of formal structures. More specifically, however, Einstein offered a reductive analysis of the empirical foundation of geometry, that is, the argument reducing geometrical measurements to observations of "point-coincidences". This reductive argument was an inspiration, in turn, to Carnap's conception of the empirical content of formal theories.

This chapter offers a critical analysis of the Carnapian picture as an account of the empirical, synthetic character of physics as distinct from the analytic character of its mathematical formalism. By considering the case of space-time theory, in particular general relativity, we can see the contrast between Carnap's picture and the way in which the theory actually determines its characteristic theoretical magnitudes, such as the curvature of space-time. It suggests that the reductive analysis ultimately obscures the empirical significance of general relativity as a novel theory of space-time structure and the nature of the evidentiary basis for this dramatic conceptual shift. I suggest an alternative account of how general relativity connects with spatial and temporal measurement, based in the history of the epistemology of geometry, by extending historical analyses of spatial measurement, and of the empirical character of non-Euclidean geometry, to the analysis of curved space-time. This account suggests, more generally, an account of scientific representation, and of the "coordination" between empirical descriptions and mathematical structures, that avoids the characteristic difficulties of standard recent approaches and connects with themes previously seen in the work of Stein (1994) and Demopoulos (2013).

7.2. Carnap on Mathematics, Physics, and Experience

Regarding the foundations of geometry, Carnap certainly shared a philosophical inheritance from the 19th-century work of Helmholtz and Poincaré on certain points that are obvious but worth recalling here. One was replacing the Kantian idea that geometrical postulates are synthetic a priori principles, founded in pure intuition, with the idea that they are empirical principles based on experience with the motions of rigid bodies and lines of sight. Another was the acknowledgement of the role of convention in the formation of these principles. For Helmholtz, the free mobility of rigid bodies and the rectilinear propagation of light were, at least approximately, empirical regularities established by long experience; for Poincaré, they were essentially stipulations guiding the empirical use of our concepts of congruence and straightness. Indeed, their stipulative character was essential to their role in providing an empirical foundation for geometry. By the same token, however, it gave rise to the possibility of conventional choice among possible alternative geometries determined by alternative stipulations. The stipulation that light travels in a straight line, for example, allows the paths of light rays in space to be treated as measuring the curvature of space. But we could equally stipulate that space is flat and adjust our physical hypotheses concerning light propagation accordingly. The foundational role of these two physical principles, as furnishing criteria for the application of geometrical concepts, prevents us from regarding them as empirical generalizations. They are "definitions in disguise"; that is, they seem to have the grammatical form of statements of laws of nature. They are strongly suggested by experience but not determined by it.

This background is familiar to all students of logical empiricism. Evidently, perhaps unfortunately, it guided Schlick and Reichenbach in their philosophical accounts of special and general relativity (cf. Schlick 1917; Reichenbach 1928). For these 19th-century analyses of the physical geometry of space, assumed to be homogenous, did not easily lend themselves to an understanding of the inhomogeneous space-time geometries of general relativity. For that purpose, the work of Riemann, Levi-Cività, Ricci-Curbastro, and Cartan, little appreciated by the logical empiricists, was essential to Einstein's work.[2] Carnap's early work, especially his doctoral thesis (1922/2019) and some papers on causal structure in space and time (1924/2019, 1925/2019), reflected his interest in the empirical foundations of geometry as developed in the thought of Helmholtz, Poincaré, and Einstein. In most of his work, however, he did not focus on the preoccupations of Schlick and Reichenbach with specific foundational questions regarding special and general relativity or with broader issues in the philosophy of space and time; when he did discuss the those topics, he generally referred to Poincaré's conventionalism and deferred to Reichenbach's account of its contemporary relevance (for example, cf. Carnap 1995, Part III).

Carnap's main interest in these topics was for their value as examples, for his account of the relations between mathematics and physics, and between physical theory and observation. Among the central philosophical achievements of the 19th century, in his view, was drawing a proper distinction between pure and applied mathematics. For Kant and his predecessors, pure mathematics, including most notably geometry, concerned ideally precise relations, conforming to the forms of pure spatial and temporal intuition, and applied geometry was distinguished by its imprecision. In the 19th century, Poincaré and Hilbert, among others, realized that mathematics in itself represents purely formal structure, whereas applied mathematics involves additional principles of interpretation.

Regarding the latter topics, a distinctive feature of Carnap's development was the influence of Russell. From his interest in the work of Frege and Russell on logic, he developed a further interest in Russell's ideas about the theory of knowledge and in particular the role of logical structure in scientific knowledge. Russell's influence on Carnap is the subject of much literature and scarcely needs to be summarized here (see, for example, Demopoulos 2011; Coffa 1991). For our purposes, it suffices to consider Russell's and Carnap's accounts of how mathematical structures apply to world of experience and to compare it with Carnap's.

A consistent aspect of Russell's view, among many changes throughout his philosophical career, was his dissatisfaction with Kant's conception of synthetic a priori knowledge. Kant's view of synthetic a priori knowledge was inseparable from his distinction between appearances and things in themselves: synthetic a priori knowledge is possible only insofar it is founded on, nature of our own cognitive faculty. For if it depended on the nature of things as they are in themselves, we would have no certainty of its truth; the apodeictic certainty of mathematical sciences, including geometry and natural philosophy, derives from their conformity to the forms of our spatial and temporal intuition and the categories of our understanding. The nature of our cognitive faculty is therefore the true subject of metaphysics. For metaphysical principles are by definition synthetic a priori principles: if not synthetic, they are merely analytic principles that say nothing about what there is; if not a priori, then they are no more than hypotheses such as scientific specula-tions. Russell insisted that, on the contrary, our knowledge could extend to the nature of things in themselves. Kant's failure to grasp this was explained, according to Russell, by his failure to understand the impor-tance of structure. The latter failure was natural, given how little was understood, in Kant's time, of structure as the notion developed in 19th-century mathematics. By Russell's time, he thought, the understanding of structure enabled him to grasp the relation of experience to the way that the world is itself as essentially a structural relation. The structure of reality need not be geometrically similar to the world as we experience it in space and time, but it can be structurally similar in a more abstract sense.

> There has been a great deal of speculation in traditional philosophy which might have been avoided if the importance of structure, and the difficulty of getting behind it, had been realized. For example, it is often said . . . that phenomena are subjective, but are caused by things in themselves, which must have differences inter se corre-sponding with the differences in the phenomena to which they give rise. Where such hypotheses are made, it is generally supposed that we can know very little about the objective counterparts. . . .
>
> [I]f the hypotheses as stated were correct, the objective counter-parts would form a world having the same structure as the phenom-enal world, and allowing us to infer from phenomena the truth of all propositions that can be stated in abstract terms and are known to be true of phenomena. In short, every proposition having a commu-nicable significance must be true of both worlds or of neither.
>
> (Russell 1919, 61)

According to Russell's version of structuralism, in short, mathematical theories can apply to the world because the objective world has an under-lying structure to which a correct theoretical account, based on the phe-nomenal world, is isomorphic (cf. Russell 1927, Ch. XX).

Russell's structuralism thus proposed a metaphysical claim, assert-ing a structural similarity between our mathematical representation of the phenomena and the underlying structure of the world in itself. One might consider Russell's thesis not as a substantive claim but as a defi-nition: the true structure of the world is by definition the isomorphism class of our mathematical theory of its structure. But Russell's thesis was evidently meant as a substantive claim; the modern theory of structural relations would explain what the world in itself could possibly have in common with our picture of it. Contrary to Kant's view, structural relations could account for a kind of residue when our picture is purged of whatever depends on the nature of our cognitive faculty. Soon after it was articulated, however, Russell's view came under severe criticism from the famous argument of Newman (1928) that Russell's thesis is not the substantive claim that it was intended to be. As Newman pointed out, whether a given formal structure is satisfied by a given set of elements depends only on the cardinality of the set. To know that the world is structurally similar to our theoretical picture is only to know that the set of real elements has the cardinality appropriate to the relevant mathematical structure.

> Any collection of things can be organized so as to have the structure W, provided there is the right number of them. Hence the doctrine that only structure is known involves the doctrine that nothing can be known that is not logically deducible from the mere fact of existence, except ("theoretically") the number of constituting objects.
> (Newman 1928, 144; cf. Demopoulos and Friedman 1985)

Evidently we can't directly compare the world's underlying structure to with any mathematical structure that our theories attribute to it, as we might compare the structure of a subway system with its representation on a map; Russell's more abstract conception of structural relations was meant to overcome precisely this difficulty. Newman's argument shows that this conception of pure structural similarity makes Russell's thesis so abstract as to be nearly empty.

Russell's form of structuralism, again, profoundly influenced Carnap's conception of the structure of scientific theory and its connection to experience. This was especially obvious in Carnap's early work on "the logical structure of the world" (1928). He soon abandoned this Russellian effort to construct a representation of the world on the basis of the immediately observable. In lifelong investigation of "the logic of science", however, he retained certain elements of Russell's view. Above all, he retained the view that scientific theory is a formal structure, or calculus, whose application to the world of sense requires empirical interpretation. In Carnap's understanding of interpretation, however, lay a profound difference from Russell that had a profound bearing on the metaphysical questions that Russell had emphasized. For Carnap, as for the other logical empiricists, the distinction between formal structure and its interpretation was essentially bound to an anti-metaphysical point of view. Such a structure was capable of generating empirical claims only by means of rules (of "correspondence") connecting its formal principles to empirical facts. The metaphysical question whether such a structure somehow captured the structure of the world in itself, as Russell required, could not arise. In Carnap's terms, the formal theory should be regarded as a language or, more precisely, as a theoretical language, enabled by correspondence rules to yield statements in an observation language.

All of this, too, will be familiar to the reader. Perhaps less obvious, in a period influenced by Quine's criticisms of Carnap, is how naturally Carnap's view accorded with the scientific understanding of his own period. Mathematics had shown its power to generate arbitrarily many formal structures, in themselves not connected with any factual content yet capable of being interpreted through the association of mathematical objects with physical principles as more or less useful tools for describing physical phenomena. Einstein himself had repeatedly emphasized this conception of the nature of mathematics, and the way in which it is incorporated in physics, as essential to his own theoretical progress:

> The progress achieved by axiomatics consists in its having neatly separated the logical-formal from its objective or intuitive content. . . . It is clear that the system of concepts of axiomatic geometry alone cannot make any assertions as to the behaviour of real objects of this kind. . . . To be able to make such assertions, geometry must be stripped of its merely logical-formal character by the coordination of real objects of experience with the empty conceptual schemata of axiomatic geometry. . . . I attach particular significance to the view of geometry just set forth, because without it I should have been unable to formulate the theory of relativity.
>
> (1921, 5–6)

7.3. Formal Structure and Experience

From the foregoing, one might conclude that the progress of the mathematical sciences, up to Carnap's time, encouraged at least some form of distinction between the analytic and the synthetic. The use of non-Euclidean geometry in physics, for example, had no bearing on the truth or falsity of Euclidean geometry as an axiomatic structure – the question is not even clearly posed – but only on the value of the physical interpretation of Euclidean geometry as a description of the world. To see the question in this light, it is not necessary to take any metaphysical position on realism or anti-realism; even a claim such as, "the geometry of our real space is not Euclidean", in this understanding of formal science, has no bearing on the standing of Euclidean geometry as a structure.

Nonetheless, the task of explaining how such structures actually do yield synthetic claims about the world was not as straightforward as it might have seemed, as may be seen from the logical empiricists' long struggle to characterize the empirical basis of science. It seemed straightforward enough to say, for example, that "light travels in a straight line" provides an empirical interpretation for Euclidean geometry sufficient for empirical measurement of, and thus synthetic claims about, the curvature of the space that we inhabit. But the logical empiricists sought to exhibit the significance of such claims for the experience of observers. This is not the place to rehearse the history of their efforts (see, for example, Oberdan 1993; Coffa 1991, Ch. 19). Some recent questions about role of observation, however, have an important bearing on Carnap's larger picture of the connection between structure and experience.

Russell's structuralism, as we saw, faced the objection that its central metaphysical claim is vacuous. Carnap's use of formal structure ought to avoid such an objection simply by its anti-metaphysical starting point: the fundamental question is not how the structure of science relates to some epistemically inaccessible structure but whether the observational basis of science can bear such a structure. Recently, however, Demopoulos (2011) argued that Carnap's account also faced the objection of triviality. For Newman's argument did not crucially depend on the metaphysical claim in Russell's original version of structuralism but rather on the more general notion of a structure as instantiated in a set of objects. That, in Carnap's case, the structure was to be satisfied in the phenomenal world rather than in an underlying reality, according to Demopoulos, provided no defense against Newman's objection. Carnap's aim of metaphysical neutrality made essential use of the "Ramsey sentence" (cf. Ramsey 1929/1990) as a reduction of a theory to its factual content and therefore neutral between realist and instrumentalist interpretations of the theory.

> [The Ramsey sentence] is obviously a factual sentence. It says that the observable events in the world are such that there are numbers, classes of such, etc. which are correlated with the events in a pre-scribed way and which have among themselves certain relations; and this assertion is clearly a factual statement about the world.
>
> (Carnap 1963, 963)

Demopoulos argued that this application of a relational structure to the set of observable events has the same arbitrariness, and hence triviality, as Russell's application of structure to the set of unobservable events. The difficulty begins with Carnap's "understanding of the truth of a theoretical claims is one according to which their *truth* is the same as their *satisfiability* in a such a model" (Demopoulos 2011, 198, original emphases). By the Newman argument, such a requirement for truth is altogether too easy to meet for Carnap's "factual statement" to have genuine factual significance (see also Psillos 2011).

It would be unfortunate if this argument were a sort of final verdict on Carnap's idea not only for Carnap's general project but also for the philosophical insights that his work had been attempting to draw from the sciences. From the later 19th century, as he knew – and, indeed, at least since the work of Newton – physicists had been able to admit the most abstract theoretical terms by explicating them through their observable consequences. Concepts such as the electromagnetic field and the curvature of space-time had thus been introduced and thereby enabled to take their place in the scientific picture of the observed world. As Eddington put it, in defending the then-novel general theory of relativity,

> The reader may not unnaturally suspect that there is an admixture of metaphysics in a theory which thus reduces the gravitational field to a modification of the metrical properties of space and time. . . . There is nothing metaphysical in the statement that under certain circumstances the measured circumference of a circle is less than π times the measured diameter; it is purely a matter for experiment. We have simply been studying the way in which physical measures of length and time fit together – just as Maxwell's equations describe how electrical and magnetic forces fit together. The trouble is that we have inherited a preconceived idea of the way in which measures, if "true," ought to fit . . . and we certainly ought not to be accused of metaphysical speculation, since we confine ourselves to the geometry of measures which are strictly practical, if not strictly practicable.
>
> (1918, 29)

Without a plausible reconstruction of such reasoning, it would be difficult or impossible for the "logic of science" to give a plausible account of how 20th-century physics took the form that it did. Naturalism certainly shed no light on this matter. It would hardly help to know that such "practical measures" should be encompassed within the structure of general relativity not because they provide a special empirical significance to its theoretical terms but simply by virtue of the general features of formal structures. It is fortunate, then, that Carnap's reconstruction can be successfully defended against Demopoulos' use of Newman, as Lutz (2017) has shown.

Even if the charge of triviality can be set aside, however, there is another challenge to Carnap's program of reconstruction that strikes at the entire question of connecting theory with observation. It concerns the question of the logical connection between theoretical statements and observational ones, and was posed most cogently by Stein:

> [Carnap] always assumed that the "observation language" is more restricted than, and included in, a *total* language that *includes an observational part and a theoretical part, connected by deductive relations*. And this, I think . . . is not the case: there is no department of physics in which it is possible, in the strict sense, to *deduce* observations, or observable facts, from data and theory.
>
> (1994, 638, original emphases)

This raises the question: if the connection of theory with observation in Carnap's account is indeed a non-trivial matter of fact, how exactly is that

connection made? Stein considered this question in the context of Newtonian mechanics. The logical empiricists, especially Schlick (1917) and
Reichenbach (1928), tended to treat general relativity not only as a novel
theory but as the embodiment of a novel philosophical approach to theoretical physics, abandoning the metaphysical aspects of previous theories
such as Newton's. Yet Newtonian physics was, if anything was, an instructive and convincing example of a theoretical framework in Carnap's sense.
Quine's objections notwithstanding, it can be treated entirely analytically,
as an axiomatic system that remains the subject of formal development,
independently of its theoretical replacement by 20th-century physics and
its continuing use as a practical tool (cf. Stein 1994, 650).

At the same time, it has obviously been a successful tool for measurement and prediction, and it has done so by playing something very much
like the role of a Carnapian framework: the laws of motion define, via
the propositions derived from them in the *Principia*, procedures for treating observations of relative motions as measures of forces. More broadly,
it is, as Newton himself emphasized, a "method" for detecting what we
might call "forces of nature" acting among particles that, left to themselves, would do nothing but move uniformly or remain at rest. To put
it in Carnapian terms, which nonetheless are consistent with Newton's
own understanding, the laws of motion constitute a framework within
which "internal" questions of ontology can be raised and answered by
empirical methods that the framework defines. This is illustrated in his
distinction between what he called the "mathematical" part of his trea-
tise, based on purely mathematical derivations from the laws of possible cases of centripetal forces, and the "philosophical" part in which
specific empirical phenomena permitted the application of the laws to
actual cases (Newton 1687, 401). That Newton could treat those phenomena as a "neutral" observation basis, with respect to his theory, was
ensured by their origin in astronomical measurements that presupposed
nothing about Newton's laws of motion. Interpreting those observations
within Newton's dynamical framework was a project that not only gave
natural philosophy a previously unimagined level of empirical precision
and scope, but also introduced an ontological novelty – the gravitational
field – that remains part of theoretical physics.

7.4. The Schematic Observer

From the empirical success of Newton's framework, one might be sure
that its theoretical principles had some secure connection with observation. But this connection is not so easy to cast in Carnapian terms. As
Stein put it, applying his general point to this case,

> there can be no thought of deducing observations within [the New-
> tonian] framework. To do so in the strict sense, one would need to

have a physical theory of the actual observer, and to incorporate it into the Newtonian framework. I certainly do not want to say that there is a reason "in principle" why such a thing can never be done, for any possible (future) physical framework; but everyone knows that Newton could not do it, and that we – in the best versions of our own physics – cannot do it.

(1994, 649)

Indeed, in the history of physics up to now, the question of deducing an observation has never seriously arisen and so was never seen as a prerequisite for applying mathematical theories. The predictions of the exact sciences have relied, rather, on what Stein called a "schematic" conception of the observer – not as a sentient being predicted to have certain sensations, but as a kind of mathematical abstraction (Stein 1994, 650). In the case of traditional astronomy, a prediction from a theory such as Ptolemy's would assert that from a given configuration of the planets with respect to the stars at a given time, one could infer certain positions on the celestial sphere; from this, one could infer that optical lines from these positions would converge at a certain place on the earth at given angles at that time. This is a schematic representation of the claim that an observer would observe certain relative positions at a certain time. Deduction from Ptolemy's theory could go no further than this toward the prediction that a human observer, at that time, would have a certain experience. Ptolemy's planetary theory involves, moreover, not only the schematic picture of positions on the (geocentric) celestial sphere, but also a theoretical picture of the motion of any planet as lying in a plane and composed of some number of geocentric or eccentric circles and their epicycles.

The mathematical reasoning that connects this picture with the geocentric picture of the heavens, and thereby with observations at times and places on the earth, is intricate but philosophically straightforward. Its only physical assumption is the usual one, the straightness of lines of sight, and in this respect it is no more or less straightforward than the reasoning by which Kepler connected positions observed from Earth with calculations of the area swept out by the mean radius of a planet's orbit around the sun. This reasoning provided a sufficient basis in observation for Newton's effort to interpret those relative motions as accelerations that must, by the laws of motion, measure the magnitudes and the directions of the forces that are responsible for them.

Why should a particular problem of connecting theory with observation arise for Carnap's theory, given physics' successful history of bringing its principles into increasingly detailed accord with observation and measurement? The explanation lies in two aspects of the scientific context of Carnap's time and the philosophical perspectives that they seemed to encourage. Both conspired, I suggest, to shift focus from the traditional conception of measurement. Carnap and his fellow logical empiricists had clearly seen the importance of the view of axiomatic structure, and its connection with physical science, that Einstein had used to such remarkable effect. But they blurred an important aspect of that view when they extended the emphasis on purely formal structure from pure mathematics to physical theories. A physical theory is not a pure formal calculus. Indeed, Einstein had acknowledged this when he pointed out that, by interpretive principles, geometry becomes "the oldest branch of physics". A physical theory, in other words, is already an interpreted formalism; Minkowski geometry may be seen as a pure formalism, for example, but Minkowski space-time is "about'" the invariance of the velocity of light. Perhaps Newton confused this issue, with some enduring effect, by characterizing the first book of his *Principia* as "only" mathematical. But he was nonetheless fairly clear on the distinction between mathematical principles such as the laws of motion, which are interpreted principles about the motions of bodies in space and time and so based on experience, and principles that are genuinely pure mathematics. As he explained to Hooke, in defense of his "new theory of light and colours",

> I said indeed that the Science of Colours was Mathematical & as certain as any other part of Optiques; but who does not know that Optiques and many other mathematical sciences depend as well on Physicall Principles as on Mathematicall Demonstrations: And the absolute certainty of a Science cannot exceed the absolute certainty of its Principles.
>
> (Newton 1672, 5101)

It is their character as interpreted principles, moreover, that allows such theoretical principles to serve, themselves, as principles of interpretation. Newton's laws, as we already saw, provided the means to interpret celestial relative motions as measurements of forces of interaction. Indeed, Poincaré (1902, Ch. 6) had emphasized this aspect of Newton's laws as essential to their definitional aspect. They do not directly make empirical claims but rather define an interpretive program – what Carnap might call a proposal – to treat every observable acceleration as an instance of, and a possible measure of, the action of a force.

The other factor in Carnap's scientific context, tending to create difficulties regarding the relation between theory and observation, was the reductive conception of observation that influenced Carnap not only from the side of Russell, but also from Einstein's account of the observational basis of general relativity. This was Einstein's well known "point-coincidence" account, and I will not try to say much about its philosophical and historical dimensions than is strictly necessary (see, e.g., Giovanelli 2013a). Einstein explained it as the motivation for requiring generally covariant field equations:

All our space-time verifications invariably amount to a determination of space-time coincidences. If, for example, events consisted merely in the motion of material points, then ultimately nothing would be observable but the meetings of two or more of these points. Moreover, the results of our measurements are nothing but verifications of such meetings of the material points of our measuring instruments with other material points, coincidences between the hands of a clock and points on the clock-dial, and observed point-events happening at the same place at the same time.

The introduction of a system of reference serves no other purpose than to facilitate the description of the totality of such coincidences. . . . As all our physical experience can be ultimately reduced to such coincidences, there is no immediate reason for preferring certain systems of coordinates to others, that is to say, we arrive at the requirement of general co-variance.

(1916, 153)

It is not necessary to consider here whether Einstein's argument achieves its purpose. It is enough to note that the point-coincidence argument plays no role in the observation basis for his new treatment of the gravitational field. The actual argument of Einstein's paper, rather, concerns the measurement of accelerations in local coordinate systems that are themselves accelerating relative to each other and the new understanding of gravity (and space-time) that results from acknowledging that such coordinate systems are physically equivalent. The need to transform field-quantities (the gravitational potentials) into reference-frames that are equivalent, but not in uniform relative motions, leads to a radical new conception of the unity of inertia and gravity. But the observation basis is not fundamentally different in character from that of Newton's gravitation theory. The conceptual gap between the two theories is historically wide, yet the measurements of accelerations in particular coordinate systems are as schematic in the one case as in the other. It is only the demand to reduce the empirical content to what is immediately observable basis, and to interpret the formalism of the theory through this reductive basis, that makes it difficult to see the empirical continuity between the two ways of thinking.

7.5. The Interpretation of Physical Theory

These examples exhibit the profound philosophical difficulties that arise from treating a physical theory as an uninterpreted calculus and its interpretation as a matter of finding a set of objects in which the formal theory is satisfiable. They suggest that another understanding of interpretation might be useful. According to a more recent way of thinking about the problem of interpretation, to demand an interpretation is to pose the question, "what would the world be like if the theory were true?" The meaning of this question is not as obvious as it may seem. Consider the question posed with respect to quantum mechanics: what would the world be like if quantum mechanics were true? The question invites the simple retort, "like this world": quantum mechanics is true, as far as we know – in any sense in which we can ever say that any physical theory is true – and any world in which it is true must be like this one. Presumably, the question is meant to demand something more than this. It is usually taken as a demand for a clear ontological picture, that is, an account of the structures or entities that might produce such a phenomenal world as we observe. So taken, it might seem to be a trivial demand: the ontological picture underlying the theory is determined by just the theoretical entities that the theory postulates, and if the latter is true, then those entities or structures constitute "what the world is like". In the case of quantum mechanics, the demand is a serious one, evidently, because of seriously diverging views of how the theory should be formulated. In Carnap's terms, divergent interpretations, in the current sense, would correspond to divergent language-forms or linguistic frameworks with certain overlapping central principles.

There is another way of understanding the question – "what would the world be like?" – that sheds more light on the connection between theory and observation. The reformulated question is, what are the aspects of our experience that tell us that the world is such as the theory says it is? Consider the theory that the earth is a sphere of a certain size: Eratosthenes showed how the observation of certain relative positions and angles of celestial objects, at certain times of day on certain parts of the earth, give empirical content to his theory of the earth's shape and size. Evidently, such reasoning treats the observations in an entirely schematic manner. By the similar use of schematically described observations, Newton showed how the relative accelerations of all satellites known to orbit bodies in our system can be seen as the operations of the same force and that this force can be identified as terrestrial gravity. This is, in a perfectly reasonable sense, an answer to the question, what would the world be like if gravity was not merely a familiar terrestrial phenomenon, but a universal force?

In contrast, Einstein's view of the interpretation of space-time geometry rests on a reduction of measurement to the observation of space-time coincidences. Setting aside whether there is a sound epistemological argument for this view, it is easy to see that it does not provide a basis for understanding "what the world would be like if general relativity were true". For, as we have already seen, the actual basis on which Einstein argues for his new theory of gravitation of geometry is essentially continuous with the arguments of Newton for the former one. They emerge from the comparison of schematically characterized local measurements and the constraints on any more global picture into which they can be integrated. In that sense, the argument for the curvature of space-time is no more based in reduction to the epistemically immediate than is the argument for the curvature of the earth.

7.6. Conclusion

In the end, there were two obstacles to a persuasive Carnapian account of physical science. If "Carnapian" is now meant to convey what is most essential to Carnap's own account and most worth preserving and pursuing, it is perhaps best characterized by what is most essentially opposed to naturalism: that a physical theory provides a conceptual framework that defines a set of "internal" questions about the nature and existence of its objects, along with empirical methods for answering them; that such a framework is characterized by rules that explicate its fundamental concepts and their application to observation and measurement; and that there is, after all, an important difference between the evaluation of such a framework from without and the evaluation of its internal theoretical and empirical claims. That such an account illuminates the structure of modern physical theory, at least since Newton, should by now be sufficiently obvious from its historical evolution.

One stumbling block in the way of this account was Carnap's tendency to remove the evaluation of a framework altogether from theoretical and empirical concerns, placing it entirely in the realm of the pragmatic; this made it more difficult to articulate – in response to Quine, for example – how such frameworks are subject to the pressure of empirical evidence and how such pressure influences in their historical evolution. The other stumbling block, closely related to the first, was an insufficient account of how evidence provides empirical content to physical theories. He rightly emphasized the non-factual character of uninterpreted mathematical structures and the role of empirical interpretation in their use as theoretical structures in physics. But in treating a physical theory, too, as an uninterpreted formal calculus, Carnap started off on the wrong foot, as it were, to account for empirical content of physical theory. He thereby assimilated the question of interpreting a physical theory with the question of interpreting a formal structure by providing a set in which the structure may be satisfied.

As a result, he did not provide very clear picture of the functions of actual observation and measurement in constituting a physical theory as an interpretation of the observed physical world. In the development of his approach, some of the insights that originally inspired him, derived from Poincaré and others, were left behind. This lacuna does not imply that Carnap's general program for analyzing and exhibiting the logic of science, in broad outline, had to fail. Rather, it reveals it as an unfinished project and not always moving in a clarifying direction. Important steps were taken afterwards – as I hope to have shown – to connect a broadly Carnapian conception of theories with a more realistic conception of empirical content. I also hope that this chapter will be seen as a small further step.

Notes

1. See, for example, Stein (1967, 1977), Earman (1989), DiSalle (1995), Friedman (1999), Giovanelli (2013b).
2. For an illuminating account of this tradition, its importance to the development of general relativity, and its comparative unimportance to the thinking of the logical empiricists, see Giovanelli (2013b). For more general discussion of the connection between the logical empiricists and Poincaré, see Friedman (1999), DiSalle (2002), and Ben-Menahem (2006). See also the chapter of Katherine Dunlop in this volume.

References

Ben-Menahem, Y. (2006), *Conventionalism*. Cambridge: Cambridge University Press.

Carnap, R. (1922/2019), 'Space: A Contribution to the Theory of Science', in A. W. Carus et al. (eds.), *The Collected Works of Rudolf Carnap: Early Writings, Volume 1*. Oxford: Oxford University Press, pp. 21–171.

——— (1924/2019), 'Three-Dimensionality of Space and Causality: An Investigation of the Logical Connection between Two Fictions', in A. W. Carus et al. (eds.), *The Collected Works of Rudolf Carnap: Early Writings, Volume 1*. Oxford: Oxford University Press, pp. 247–289.

——— (1925/2019), 'On the Dependence of the Properties of Space on Those of Time', in A. W. Carus et al. (eds.), *The Collected Works of Rudolf Carnap: Early Writings, Volume 1*. Oxford: Oxford University Press, pp. 297–325.

——— (1928), *Der logische Aufbau der Welt*. Berlin-Schlachtensee: Weltkreis-Verlag.

——— (1963), 'My Intellectual Development and Replies and Systematic Expositions', in P. A. Schilpp (ed.), *The Philosophy of Rudolf Carnap*. LaSalle, IL: The Open Court, pp. 3–84, 859–1013.

——— (1995), *An Introduction to the Philosophy of Science*. New York: Dover Publications.

Coffa, J. A. (1991), *The Semantic Tradition from Kant to Carnap*. Cambridge: Cambridge University Press.

Demopoulos, W. (2011), 'On Extending "Empiricism, Semantics, and Ontology" to the Realism-Instrumentalism Controversy', *Journal of Philosophy*, 108: 647–669.

——— (2013), *Logicism and Its Philosophical Legacy*. Cambridge: Cambridge University Press.

Demopoulos, W. and Friedman, M. (1985), 'Bertrand Russell's *The Analysis of Matter*: Its Historical Context and Contemporary Interest', *Philosophy of Science*, 52 (4): 621–639.

DiSalle, R. (1995), 'Spacetime Theory as Physical Geometry', *Erkenntnis*, 42: 317–337.

—— (2002), 'Conventionalism and Modern Physics: A Re-Assessment', *Noûs*, 36: 169–200.

Earman, J. (1989), *World Enough and Space-Time: Absolute versus Relational Theories of Space and Time*. Cambridge, MA: MIT Press.

Eddington, A. (1918), *Report on the Relativity Theory of Gravitation*. London: Fleetway Press.

Einstein, A. (1916), *Die Grundlage der allegemeinen Relativitätstheorie*. Leipzig: Johann Ambrosius Barth.

—— (1921), *Geometrie und Erfahrung*. Berlin: Julius Springer.

Friedman, M. (1999), *Reconsidering Logical Positivism*. Cambridge: Cambridge University Press.

Giovanelli, M. (2013a), 'Erich Kretschmann as a Proto-Logical-Empiricist: Adventures and Misadventures of the Point-Coincidence Argument', *Studies in History and Philosophy of Modern Physics*, 44: 115–134.

—— (2013b), 'The Forgotten Tradition: How the Logical Empiricists Missed the Philosophical Significance of the Work of Riemann, Christoffel and Ricci', *Erkenntnis*, 78: 1219–1257.

Lutz, S. (2017), 'Armchair Philosophy Naturalized', *Synthese*, 197 (3): 1099–1125.

Newman, M. (1928), 'Mr. Russell's "Causal Theory of Perception"', *Mind*, 37: 137–148.

Newton, I. (1672), 'Mr. Isaac Newton's Answer to Some Considerations Upon His Doctrine of Light and Colors', *Philosophical Transactions of the Royal Society*, 88: 5084–5103.

—— (1687), *Philosophiae naturalis principia mathematica*. London: The Royal Society.

Oberdan, T. (1993), *Protocols, Truth and Convention*. Amsterdam: Rodopi.

Poincaré, H. (1902), *La science et l'hypothèse*. Paris: Flammarion.

Psillos, S. (2011), 'Carnap's Ramseyfications Defended', *European Journal for Philosophy of Science*, 1: 71–87.

Ramsey, F. P. (1929/1990), 'Theories', in D. H. Mellor (ed.), *Frank Ramsey: Philosophical Papers*. Cambridge: Cambridge University Press, pp. 112–139.

Reichenbach, R. (1928), *Philosophie der Raum-Zeit-Lehre*. Berlin: De Gruyter.

Russell, B. (1919), *Introduction to Mathematical Philosophy*. London: Allen and Unwin.

—— (1927), *The Analysis of Matter*. London: Kegan Paul, Trench, Trubner & Co. Ltd.

Schlick, M. (1917), *Raum und Zeit in der gegenwärtigen Physik*. Berlin: Springer.

Stein, H. (1967), 'Newtonian Space-Time', *Texas Quarterly*, 10: 174–200.

—— (1977), 'Some Philosophical Prehistory of General Relativity', in J. Earman, C. Glymour, and J. Stachel (eds.), *Foundations of Space-Time Theories*. Minneapolis, MN: University of Minnesota Press, pp. 3–49.

—— (1994), 'Some Reflections on the Structure of Our Knowledge in Physics', in D. Prawitz, B. Skyrms, and D. Westerstahl (eds.), *Logic, Methodology, and Philosophy of Science IX*. Amsterdam: Elsevier Science, pp. 633–655.

8 Einstein, General Relativity, and Logical Empiricism

Thomas Ryckman

> At that time (i.e., 1905) my mode of thinking was much nearer positivism than it was later on . . . my departure from positivism came only when I worked out the general theory of relativity.
>
> Einstein to D.S. Mackey, 26 April and 28 May 1948
> (cited in Fine 1996, 86)

8.1. "Geometry and Experience": Exemplar of Logical Empiricism?

Einstein's paper "Geometry and Experience" ("Geometrie und Erfahrung") was initially given as a public address in Berlin on 27 January 1921 at the Prussian Academy of Sciences on its annual founder's day (*Friedrichstag*) celebration (Einstein 1921/2002). The lecture ostensibly revisited the 19th-century dispute between geometric empiricism and geometric conventionalism, set now in the new context provided by the general theory of relativity. Published in the *Proceedings of the Prussian Academy* on February 3, an expanded version of "Geometry and Experience" appeared as a separate booklet and sold out within weeks. Within the year it had been translated into French, English, Russian, Italian, and Polish. It would become one of the widely reprinted and influential essays in 20th-century philosophy of science, one of three Einstein texts included in the 1953 anthology of papers in philosophy of science edited by Herbert Feigl and May Brodbeck, the first such collection to be published in the United States.

Already in 1929 it was deemed an essential text of the "Scientific World Conception" (*Wissenschaftliche Weltauffassung*) in the manifesto of the Vienna Circle, written (mostly) by Otto Neurath, Rudolf Carnap, and Herbert Feigl. In a summary section of that document (in Stadler and Uebel 2012, 108), the philosophical significance of "Geometry and Experience" was epitomized in four points: i) the need for a clear distinction between pure mathematical (axiomatic) geometry and applied geometry, a branch of physics; ii) the definition of the latter as 'practical geometry' characterized by the possible positions of a rigid body in space; iii) an insistence that the definition of rigid body has an 'empirical basis', namely 'the preservation of coincidences (equality of spatial intervals)'; and finally the positivists' conclusion iv) that from the standpoint of 'practical geometry', the question regarding the spatial structure of the universe (concerning both metrical structure as well as finite or infinite extension) had a clear empirical answer.

Fundamental tenets of logical empiricism (or logical positivism; I consider the terms interchangeable, following Uebel's [2013, 85] conclusion that "even among the participants" there was "no systematic unity of usage") are implicated in this brief and rather tendentious *précis*. Above all, there is an implied rejection of any Kantian metaphysics of space. In declaring the question concerning the geometry of physical space to have an unambiguous empirical answer, Einstein was understood not only to invalidate Poincaré's geometric conventionalism but also to ride the positivist hobbyhorse of renouncing all metaphysics, the ideological cornerstone of the "Scientific World Conception". In particular, Einstein had shown the error of any *a priori* foundation for geometry, in particular in the Kantian doctrine that Euclid's postulates express the "necessity and universality" of the form of outer intuition and so are synthetic *a priori* conditions of possible experience. To logical positivism, the collapse of this bastion of the *a priori* in the theory of space signaled more generally the triumph of empiricism over idealist and other metaphysical philosophies; Einstein's "clear formulations brought order into a field where confusion often prevailed" (Frank 1947, 177).

Furthermore, the latter's sharp distinction between empirical 'practical' physical geometry and purely formal axiomatic geometry went in tandem with the logical empiricist account of pure mathematics as grounded ultimately in logic. Inspired by Whitehead and Russell's *Principia Mathematica*, logical empiricism considered mathematical statements reducible to logic and purely mathematical truths a species of logical truth. As logical truths are paradigmatically analytic, true in virtue of the meanings of the terms they contain, the account of mathematics as logic supported the core logical empiricist thesis that any 'cognitively meaningful' statement is either analytic or an *a posteriori* synthetic statement, confirmable or refutable by experience. In this way, the logical positivists would point to "Geometry and Experience" as the illustrious precursor of their dictum that the synthetic *a priori* statements of metaphysics were literally meaningless. Two decades later, this sharp dichotomy between analytic and synthetic statements was attacked as one of the "dogmas" of empiricism in Quine's (1951) famous critique of Carnap.

Less than apparent in the previous clipped summary is just how the empiricist conception of the geometry of physical space (-time) ostensibly presented by Einstein actually responds to, and defeats, Poincaré's geometric conventionalism. Seeing this requires unpacking what was stated with considerable compression in iii). "Practical geometry" explicitly rests on Einstein's concept of the "practically rigid body". Now the rigid body is a problematic concept (a "child of sorrow" – *Schmerzenskind*, according to Einstein) already in special relativistic physics. Nonetheless, iii) states that the concept has an "empirical basis" in that observed coincidences between the end points of two rigid bodies are preserved when the bodies are translated in space. That statement entails that whenever the end points of two "practically rigid" measuring rods are found to coincide in one region of space, they always will be found to do so when the rods are brought together in any other region. Is this really an empirical statement?

A brief reflection should convince that it is not, at least not in any straightforward sense. And so logical empiricism emphasized the "empirical basis" of Einstein's definition of a rigid body in iii) required a *stipulation* of "preservation of coincidence". This states that if two bodies (e.g. measuring rods), whose endpoints are in coincidence at one time at location *A*, are then separated and translated to a distant location *B*, their endpoints again will be found in coincidence when compared at *B* at a later time. The definitional nature of this statement is readily appreciated if one considers that the bodies may travel from *A* to *B* at different velocities along distinct, possibly circuitous paths. That the application of pure mathematics in natural science requires stipulations of this sort, here investing a physical object (measuring rod) with the meaning of "practically rigid body" (and of "equality of spatial intervals") would be enshrined as a central facet of logical empiricism's account of scientific methodology. The necessary first step in the application of any formal mathematical theory (e.g., pure axiomatic geometry) to empirical phenomena required similar stipulations or "coordinative definitions" associating certain concepts or relations of mathematics with observable objects or processes. That empirical determination of the geometry of physical space rested upon postulation of definitional linkages between formal geometric concepts ("distance", "straight line", etc.) and physical objects ("measuring rod", "light ray", etc.), became the logical empiricists' paradigmatic example of how formal expressions of a mathematized theory acquire empirical meaning. Rudolf Carnap in 1927 provided an early, and certainly most graphic, illustration of the significance of the methodology of coordinative definitions. Only subsequent to implementing definitions coordinating formal concepts with concrete empirical objects can the blood of empirical reality enter through these touch points to flow upward into the most diffuse veins of a hitherto empty theory-schema (see Carnap 1927).

Einstein's lecture does indeed suggest something of this kind. Yet "Geometry and Experience" is not really concerned with the methodological issues emphasized by logical empiricism. Its message is considerably more tempered, endorsing geometric empiricism only in the guise of a *pro tem* strategy. Practically rigid measuring rods and ideal clocks of geometric empiricism play, at least provisionally, an epistemologically privileged but in principle logically objectionable role in the general theory of relativity. It is no coincidence that in January 1921, the privileged epistemic role of rods and clocks in Einstein's general theory of relativity had become a live issue of contention between Einstein and mathematician Hermann Weyl, a friend and colleague who sought, though in a mathematically speculative way, to reconstruct general relativity without it.

What the logical positivists did not, nor, for the most part, would not, mention is the existence of this controversy and the threat it posed to their methodology of coordinative definitions and to the ensuing conception of empiricism in physical science more generally. Einstein's paper is then only superficially a pointed intervention on behalf of geometric empiricism in its storied confrontation with geometric conventionalism *à la* Poincaré. Instead "Geometry and Experience" appropriates that earlier debate, a conflict in any case now outmoded in the different context opened up by the variably curved space-times of general relativity, effectively carrying over the no-longer-suitable terms of the earlier discussion into a new, and considerably more intricate, setting. Interestingly, rather than insisting upon what is essential to the geometry of general relativistic spacetimes, territory firmly occupied in 1921 by Weyl, Einstein choose to largely mute the controversy, emphasizing the *pro tem* benefits of a pragmatic justification of the "practical geometry" of rods and clocks while admitting that Weyl (though cloaked in the *persona* of Poincaré) is correct, in principle (*"sub specie aeterni"*).[1] In the end, what is really at issue in the new situation – unless one knows the backstory – is only dimly perceptible. Like the masked actor in a Nōh play, Einstein relates a largely symbolic drama, seasoned with stylized elegance (and memorable quotes), directed at an audience that was unlikely to read between and behind the lines of ritualized presentation.[2]

8.2. Einstein the "Believing Rationalist"

According to physicist, historian, and Einstein scholar Gerald Holton, a significant shift occurred in Einstein's philosophical views as a result of his work on the general theory of relativity. In Holton's (1968/1988, 244) influential account, Einstein began his "philosophical pilgrimage . . . starting on the historic ground of positivism", heavily under the influence of Mach.[3] Yet following the completion of general relativity in late 1915, his "apostasy" from Mach became more and more apparent. The end point of Einstein's philosophical odyssey lay in the latter's conversion to what Holton termed a "rationalistic realism", that is, the conviction that

> there exists an external, objective, physical reality which we may hope to grasp – not directly, empirically, or logically or with fullest

certainty, but at least by an intuitive leap, one that is only guided by experience of the totality of sensible "facts".

(1968/1988, 263)

Documenting this trajectory, Holton pointed to a letter of 24 January 1938 to mathematical physicist Cornelius Lanczos in the course of which Einstein identifies himself as "a believing rationalist",

> Coming from skeptical empiricism of somewhat the kind of Mach's, I was made, by the problem of gravitation, into a believing rationalist, that is, one who seeks the only trustworthy source of truth in mathematical simplicity.

(1968/1988, 259)

Lanczos, a Berlin collaborator on unified field theory in 1928–1929, was not only a colleague but also a philosophical soul mate. In a 1931 monographic report on what was then the latest Einstein unified field theory (the so-called "distant parallelism theory"), Lanczos observed that the distinction between the general theory of relativity and quantum mechanics cannot be captured in a simple opposition of "classical" and "quantum" but signals a far wider divide, a difference "in the whole intellectual orientation which lies at the basis of modern research", namely that between the "metaphysical-realistic" and "positivistic". He continued,

> Something unique, and perhaps even uniquely lasting, happened with the theory of relativity – we are thinking here of primarily the general theory. It happened that "metaphysical" thinking, or logical-constructive imagination (*Phantasie*), won insight into the mysteries of Nature that arguably would never have been attained along purely empirical paths.

(Lanczos 1931, 99)

Small wonder Einstein wrote a 1942 letter to Lanczos stating that Lanczos was the only other physicist who held "the same orientation towards physics as I", namely "belief in the comprehensibility of reality through something logically simple and unified".[4] Einstein's rationalist orientation received full public expression for the first time in the Herbert Spencer lecture, "On the Method of Theoretical Physics", delivered at Oxford on 10 June 1933. Lecturing in English, also for the first time, in no uncertain terms Einstein expressed a conviction that the "right method" of theoretical physics consisted in seeking laws of nature with the simplest mathematical formulation.

> Our experience up to date justifies us in feeling sure that in Nature is actualized the ideal of mathematical simplicity. It is my conviction

that pure mathematical construction enables us to discover the concepts and the laws connecting them which give us the key to the understanding of the phenomena of Nature. Experience can of course guide us in our choice of serviceable mathematical concepts; it cannot possibly be the source from which they are derived; experience of course remains the sole criterion of the serviceability of a mathematical construction for physics, but the truly creative principle resides in mathematics. In a certain sense, therefore, I hold it true that pure thought is competent to comprehend the real, as the ancients dreamed.

(1934, 167)

The striking declarations at Oxford, echoing Platonism in apparently identifying fundamental physical reality with the abstract reality of ideal mathematical structures, was immediately recognized as a fundamental change in what had been regarded to be Einstein's philosophy. Appearing a few months later in the second issue of the new journal *Philosophy of Science*, Einstein's text was widely discussed and criticized, a shock to many who viewed his scattered philosophical remarks as congenial to positivism or logical empiricism. Closer scrutiny of Einstein's writings, particularly on unified field theory, give a very different picture. As early as 1923, Einstein unambiguously proclaimed his adherence to a method of mathematical speculation:

The search for the mathematical laws which shall correspond to the laws of nature then resolves itself into the solution of the question: What are the formally most natural conditions that can be imposed upon an affine relation?

(1923, 448)

In 1923 and for the next few years, Einstein followed the path laid out by A.S. Eddington in believing that a unified field theory encompassing gravitation and electromagnetism might be more simply erected on a differential geometric basis by beginning from an "affine relation" (i.e., affine connection) rather than with the metric tensor; the former is a much more general notion than the latter, providing needed additional mathematical degrees of freedom in which the unification might be formulated (for details, see Ryckman 2005, Ch. 7). For Einstein, employing mathematical simplicity as a criterion in finding the fundamental laws of nature appears to have been the principal means of addressing "the Promethean element of scientific experience", that is, the hubris of a theoretician who would seek to know whether the Creator had any choice:

We wish not only to know *how* Nature is (and *how* her processes transpire), but also to attain as far as possible the perhaps utopian and seemingly arrogant goal, to know why Nature is *thus and not otherwise.*

(1929a, 126)

To explain according to the Principle of Sufficient Reason, the ultimate aim of explanation, is perhaps the rationalist motive *par excellence.* Even earlier expressions of Einstein's rationalistic pronouncements are not difficult to find, going back to at least 1918 (see subsequently). Of course, to any variety of positivism or empiricism, these declarations are difficult to assimilate. But to logical empiricists, who had revered Einstein as a patron saint in their campaign to rid both philosophy and science of metaphysics, they were heretical. The discomfort is palpable in physicist and logical positivist Phillip Frank's 1947 biography of Einstein. Commenting on the Oxford lecture, Frank attempted to dilute the objectionable remarks with irrelevancies, noting that Einstein felt that "even *Logical Positivism* did not give sufficient credit to the role of imagination in science". The overtly metaphysical tenor of the address had to be diminished in significance; this was done by reminding that Einstein still clung to what might be presented as an empiricist notion of truth:

As a result Einstein's philosophy of science often made a 'metaphysical' impression on persons unacquainted with Einstein's positivistic requirement that the only 'confirmation' of a theory is its agreement with observable facts.

(Frank 1947, 218)

Just a few years later, Frank (1949, 354) was less circumspect; Einstein's avowal that mathematical thought could lead to the discovery of the true fundamental laws of nature was a "belief in the rationality of nature" of a "religious nature".[5] By this time Frank, an old friend of Einstein, did not simply dismiss such metaphysical statements for being meaningless as logical positivism advocated in the 1930s. Rather, after 1947 or so, Frank argued for their meaningfulness by connecting them with empirical statements regarding their effects on human behavior, presumably in terms of human motivation. In a lengthy unpublished manuscript drafted post-1947 (forthcoming as *The Humanistic Background of Science*; see Frank 2021), this conception of metaphysical statements received a clear articulation:

[M]etaphysical propositions about the physical universe are actually meaningful propositions about human behavior or, in other terms, propositions of sociology . . . according to the opinions of the Vienna Circle, metaphysical propositions about the physical world are meaningless within the system of physical concepts, but have meaning with the 'universe of discourse' that embraces physical and sociological concepts.

(Frank 2021, P1. Chap. 5. Sect. 6., cf. Wuest 2015, 150)

One can only imagine a smile on Einstein's face upon learning that his conviction "in Nature is actualized the ideal of mathematical simplicity" might be given a sociological explanation. We will see immediately in the following that not sociology, but the claims of Reason as tempered within the Transcendental Dialectic of Kant, brings comprehensibility to Einstein's overt metaphysical utterances and perhaps even gave him the courage to make them.

In any case, by the end of the 1920s, Einstein was quite forthcoming about his conviction that a methodology coupling mathematical speculation with a belief in "the uniformity of natural laws and their accessibility to the speculative intellect" had not only yielded general relativity but might well do the same in the project of unified field theory.[6] Post-1930, ever more deeply immersed in a paradigm that viewed unified field theory as the necessary completion of the problem of gravitation as well as the fundamental basis of a theory of matter, from which the empirical successes of quantum mechanics might be seen as having merely statistical validity, Einstein repeatedly stressed that pure mathematics, not physical insight or physical requirements, had been the winning ticket back in 1912–1915. The message that mathematical simplicity provided the needed key to unlock the door shielding general relativity was reiterated many times in his last years, including to de Broglie, on 15 February 1954 just a little more than a year before his death:

> The equations of gravitation were only found on the basis of a purely formal principle (general covariance), that is to say, on the basis of trust in the greatest logical simplicity of laws of nature thinkable.
> (Einstein to de Broglie, 15 February 1954, in *Annales de la Fondation Louis de Broglie*, v. 4, no. 1, 1979, 56)

Yet Einstein's "believing rationalist" is neither a Platonist nor really a rationalist at all, at least as that term is understood within the tradition stemming from Descartes, Spinoza, and Leibniz. Notwithstanding well-known railings against Kant's unsustainable doctrine of synthetic *a priori* concepts and judgments, the later Einstein affirmed that while experience alone remained the ultimate arbiter of any theory, a "belief in the comprehensibility of reality through something logically simple and unified" was a methodologically legitimate heuristic, whose general form Kant had thoroughly probed in the Transcendental Dialectic of *The Critique of Pure Reason*. There Kant had given a tempered account of the necessary, although "hypothetical", cognitive role of reason in science as an injunction to seek systematic unity in nature. The message that Einstein came to understand as "the truly valuable in Kant" lay in the doctrine of the regulative principles of reason, according to which the very concept of an order of nature presupposes a decision to seek unity through systematic connections, a synthesis that, in the absence of detailed microphysical knowledge unavailable at that time, could only be projected by mathematical construction. It is readily documented that the further Einstein went down the road of unified field theory, the more he stressed his "belief in the comprehensibility of reality through something logically simple and unified", legitimation for which he found belatedly in Kant.[7]

8.3. Einstein's Kant

In the course of a celebrated discussion with French luminaries at the Collège de France in April, 1922, Einstein pointedly replied to philosopher Leon Brunschvicg's question concerning the bearing of Kant's philosophy upon the theories of relativity:

> As concerns Kant's philosophy, in my opinion, every philosopher has his own Kant.

After remarking that he remained uncertain of Brunschvicg's interpretation of Kant as manifested by the latter's lengthy intervention, Einstein went on to say,

> It seems to me that the most important matter in Kant's philosophy is that one speaks of *a priori* concepts in the construction of science. But here there are two opposing viewpoints: the *apriorism* of Kant, in which certain concepts preexist in our mind, and the conventionalism of Poincaré. These two points of view agree that science requires, for its construction, arbitrary concepts; with regard to whether these concepts are given *a priori* or are arbitrary conventions, I cannot say.[8]

It is somewhat surprising that in such an august public venue Einstein chose to sit on the fence, refusing to clearly voice an opinion previously affirmed in correspondence with Schlick and Born, among others,[9] that the concepts required for theoretical construction are conventions as well as to state his opposition to any Kantian doctrine holding that "certain concepts preexist in our mind". As is widely known, this dissent emphasized that the concepts of theoretical science ("categories") are not "unalterable"; they are by no means necessary ("conditioned by the nature of the understanding") but are "(in the logical sense) free conventions". Accordingly, what might be termed the negative moment of Einstein's Kant targets the doctrine of the Transcendental Analytic and the

Transcendental Aesthetic that it presupposes. To the extent that these sections are regarded as constituting the core of Kant's theory of cognition, as logical empiricists such as Schlick and Reichenbach affirmed, there could be but one attitude to Kant in the aftermath of general relativity: utter rejection (see Ryckman 2005, Ch. 2).

Still, the larger significance of Einstein's remarks on Kant in Paris lies in their implied dissent to pure empiricism or positivism, a criticism contained in the statement that "science requires, for its construction, arbitrary concepts"; here "arbitrary" is an ellipsis for "logically arbitrary", which, in Einstein's peculiar epistemological vocabulary, has the meaning "not derived from sense experience". That physical theory requires freely posited concepts or "free conceptual construction" is a long-recognized tenet of Einstein's epistemology. And, though not in Paris, it is in an insistence that "all concepts . . . are from the point of view of logic free conventions" that Einstein also locates his departure from Kant – that is to say, from the doctrine of *a priori* categories in the Transcendental Analytic. While Einstein was critical of Kant's account of the constitutive rules of the understanding, he resonated with the Transcendental Dialectic's emphasis on regulative principles, transcendental ideas, or concepts of the faculty of reason.

In response to the new vogue of positivism among philosophers and quantum physicists beginning in the late 1920s, an increasingly prominent feature of Einstein's later writings on the epistemology of science points to the insufficiency of the empiricist thesis that all knowledge rests solely on the deliverances of the senses. Einstein counters empiricism's shortcoming with an emphasis that "reason" or "pure thought" is an ineliminable, albeit "'metaphysical'" (scare quotes within quotations are always Einstein's), factor in cognition. But, as Einstein recognized, this presents an epistemological challenge: With what "justification" (*"Berechtigung"*) does the physicist employ concepts that are neither analytic nor derived from sense experience? Kant, in the opinion of the late Einstein, at least provided a partly correct statement of the problem:

> The following, however, appears to me to be correct in Kant's statement of the problem: in thinking we use, with a certain "justification" such concepts in thinking, to which there is no access from the material of sense experience, if one considers the matter from the logical standpoint.
>
> (1944, 287)

He went on to affirm once again that physico-mathematical concepts are "free creations of thought which cannot be inductively derived from sense experience". Now the mantra of concepts as "free creations of the human mind" is one of the most familiar components of Einstein's epistemology of science, one he repeatedly stressed against empiricists, from

Russell to Reichenbach. But then if concepts (certain ones, surely, more than others) do lack empirical justification, just what other kind of justification might Einstein have had in mind?

His answer, triangulating between realism and conventionalism, is perhaps most fully stated in the course of an extended discussion covering several pages in the 1949 "Reply to Criticisms". Ostensibly engaging once more in battle against positivist epistemology, Einstein articulated a viewpoint decidedly non-positivist, but also non-realist and non-conventionalist, by focusing on the justification of one such concept, that of "the real" or "being", whose extension includes the "not observable". Derided by positivists as a metaphysical excrescence on the fabric of science, Einstein proceeded to give reasons for considering the concept not merely convenient, but essential.

> "Being" (*Das "Sein"*) is always something intellectually constructed by us (*von uns gedanklich Konstruiertes*), hence freely posited statutes (*frei Gesetztes*) by us (in the logical sense). The justification of such posits (*Setzungen*) does not lie in their derivation from what is given to the senses. Never and nowhere is there a derivation of this kind (in the sense of logical deducibility), not even in the domain of prescientific thought. The justification of the constructs that represent the "real" (*das "Reale"*) for us alone lies in their more perfect, or less imperfect, suitability for making intelligible what is given in sensation (the vague character of this expression is forced here upon me by my striving for brevity).
>
> (1949, 669)

Einstein's striving for brevity should not occlude two important points stated here. First, "being" – in this context a clear reference to the portrayal of physical reality in theoretical physics – is always a "mental construction" (i.e., "conceptual construction for grasping the interpersonal"; see subsequently) freely posited by the theorist. This admission distances Einstein's realism from either a "rationalistic realism" or scientific realism more generally, since the core of any such realism far outstrips the mere avowal of a mind-independent external reality. Rather, any realist interpretation of physical theory worthy of the title must in addition claim *knowledge of the real* (without scare quotes!), holding firm to a conception that the best physical theories are at least approximately true. For such a realist, only *à la façon de parler* might true physical theories be considered "mental constructions", in the rather trivial sense that it takes a theorist to write down a theory, just as it takes a photographer to snap a portrait. But for the realist, the theorist/photographer is irrelevant except for "stylistic" details, while the significance of the ensuing portrait or image is not that it is a "mental construction" at all but that it (approximately) corresponds to, maps, or represents in literal fashion a mind-independent reality.

Second, the passage affirms that the only justification accruing to claims that conceptual constructs represent "being" or the "real" lies in their ability to make our sense experience "intelligible". It cannot be, as conventionalism prototypically maintains, that there is "no fact of the matter" about which concepts are correct or that justification is the subjective matter of choosing concepts that pragmatically prove most convenient or commodious in managing experience. It is "intelligibility" that matters, and to Einstein, intelligibility ranked supreme over other virtues of a good theory that have been adduced (e.g., fruitfulness, accuracy, simplicity, consistency, scope; see Kuhn 1977). Intelligibility mandates that the conceptual constructs of a fundamental physical theory manifest to the greatest extent possible the unity of systematic interconnection of an ideal order of nature, an ideal projected by the hypothetical employment of reason. It is just here that Einstein points to "the truly valuable in Kant", that the real is not "given" but "*aufgegeben*" – posed as a problem (for theoretical construction). He paraphrased this to mean:

> There is such a thing as a conceptual construction for the grasping of the interpersonal, the authority of which lies purely in its validation. This conceptual construction refers precisely to the "real" (by definition), and every further question concerning the "nature of the real" appears empty.
>
> (Einstein 1949, 680)

The expression "conceptual construction" as enabling a "grasp [of] experiences intellectually" is key, and in this sense, it can be regarded as "'knowledge of the real'"

> [i]nsofar as physical thinking justifies itself, in the more than once indicated sense, by its ability to grasp experiences intellectually, we regard it as "knowledge of the real".
>
> (Einstein 1949, 673–674)

At this point, he remarked that "the 'real'" in physics "is to be taken as a type of program, to which we are, however, not forced to cling *a priori*". Einstein's last documented discussion of Kant concludes with a reminder that he naturally dissents from the doctrine of *a priori* categories in the Transcendental Analytic.

> The theoretical attitude here advocated is distinct from that of Kant only by the fact that we do not conceive of the "categories" as unalterable (conditioned by the nature of the understanding) but as (in the logical sense) free conventions. They appear to be *a priori* only insofar as thinking without positing categories and concepts in general would be as impossible as is breathing in a vacuum.
>
> (1949, 674)

These late philosophical pronouncements stress the role of reason, that is, theoretical speculation based on seeking mathematical simplicity, in creating a unifying conceptual structure within which empirical phenomena find intelligible representation. They are not *ex cathedra* pronouncements but reflect a confluence of two philosophical currents. First, Einstein's assessment that progress in fundamental physical theory required new physical and mathematical ideas and that the most appropriate way of finding them was to use the beacon of mathematical-logical simplicity. Second, pursuit of such a path required a philosophical orientation appropriate to an ever-growing distance between observation and the basic concepts and relations of the theoretical structure, one that is "metaphysical-realistic" and not "positivistic".[10]

8.4. Einstein and Planck

Einstein will always deny the self-attribution "realist". On the other hand, in the full flush of success following the completion of the general theory of relativity, Einstein embraced Planck's attempt to define the "real" in terms of a striven for constant *world-image* (*Weltbild*) and of the task of the theoretician as that of advancing such a world-image through establishing the permanence of certain of its constituents and of its most general basic laws. This occurs in the text of Einstein's warm appreciation of Planck at the Berlin Physical Society's celebration of Planck's 60th birthday on 26 April 1918. Einstein did not take this responsibility lightly. Earlier that day, he wrote to Sommerfeld, "I'll be happy tonight if the gods grant me the gift to speak profoundly, because I am very fond of Planck, and he will certainly be pleased when he sees how much we all care for him and how highly we value his life's work". It cannot be a coincidence that the prominent message of Einstein's tribute is an endorsement of Planck's signature philosophical position, the characterization of the task of theoretical physics as the search for a unifying and universal '*Weltbild*'.

> Do the results [of the theoretical physicist] deserve the proud name '*Weltbild*'? . . . the proud name is well-deserved since the most general laws, upon which the thought-structure [*Gedankengebäude*] of theoretical physics is based, raise the claim of being valid for all natural occurrences. . . . Therefore the highest task of the physicist is the search for those most general elementary laws from which the world-image (*Weltbild*) is to be obtained by pure deduction.
>
> (1918/2002, 57, my own translation)

Einstein's audience undoubtedly understood that this embrace of Planck's conception of "the highest task of the physicist" is a deliberate public alignment with Planck and so with the latter's passionate polemic against

Mach and positivist conceptions of physical theory. The theoretical physicist is to be understood as engaged in building up a "physical world-
image", much of which is an admittedly mental construction. Employing
symbols for the metaphysically real, the *Weltbild* is not readily expressible in ordinary language, and it may posit elements that outstrip present
capacities for observational test. At the same time, it must yield consequences that are possible to observationally confirm, and in addition it
must be flexible enough to accommodate new phenomena. Though in
a sense a creation of mind, its implications purport to refer to a mind-
independent real external world. As it is always incomplete, it can never
be supposed in satisfactory agreement with mind-independent states of
affairs, and in any case, such agreement can never be directly ascertained
but at most indirectly inferred. Perhaps its most important functions are
to serve as a platform within which further thought may develop, as well
as to suggest experiments whereby it can be transcended or amended.[11] It
is worth mentioning that in 1933, Einstein allowed the text of this 1918
lecture to appear as a "Prologue" to a collection of Planck's (1933) philosophical articles in English.

Some dozen years after 1918, Einstein again expressed fundamental
philosophical agreement with Planck's anti-positivist realism, this time
with Planck's article "Positivism and the Real External World" ("*Positivismus und reale Aussenwelt*"). Planck's theme is the looming dual crisis of civilization, both material (the Great Depression) and "spiritual"
(by which Planck certainly included "political"). Though physics should
have claim to be a firm foundation on which to base the modern outlook on the world, Planck expressed his disquiet at the confusion and
contradiction now active among certain physicists. Though Mach was
long dead, quantum physicists resuscitated positivism in physical science,
fundamentally posing limits to how far and in what way the human mind
is capable of attaining knowledge of the external world and in challeng-
ing the law of causality. Once again in opposition to positivism, Planck
articulates the *Bild* conception of physical theory and the invariable tension between the presupposition of a real world in the metaphysical sense
and the realization that theories are images, never capable of completely
grasping its nature.

> Positivism always rigorously maintains that there are no other sources
> of knowledge except sense perceptions. Now, the two theorems (*Sätze*):
> (1) *there is a real external world independent of us* and (2) *the real
> external world is not directly knowable* form together the hinge point
> (*Angelpunkt*) of the whole of physical science. And yet they stand in
> a certain opposition to one another. In this way they at the same time
> disclose the purely irrational element that adheres to physics as to
> every other science. The result is that a science is never able to fully
> complete its task. We must accept that as an irrefutable fact, one that

cannot be removed, as positivism tries to do, by restricting the task of science at its very start. The work of science therefore poses itself to us as an incessant struggle toward a goal that will never be reached and fundamentally is unattainable. For this goal has a metaphysical character, lying beyond any experience.

(1934, 217)

Planck's lecture marks not so much an intervention in but a response to the debate over the meaning of the Heisenberg uncertainty relations and quantum indeterminism. Whereas Planck deplores the quantum physicists' abandonment of the "law of causality", Einstein's criticisms by 1930 had shifted to questioning whether the Ψ-function of wave mechanics could be considered a complete description of the states of individual systems. But the quantum physicists' resort to positivist doctrine, then under vigorous revival by philosophers, particularly in Vienna and Germany, spurred Planck into action. One of these philosophers was Moritz Schlick (1932/1979), Planck's former student and leader of the Vienna Circle. Indeed, Planck's lecture prompted a response from Schlick retorting that the previous statement (1) is meaningless.

Two documents show that Einstein was in full accord with Planck's conception of realism while opposed to Schlick's positivism. A handwritten draft note from 1931 in the Einstein Archives was probably intended as the preface to the published pamphlet of Planck's lecture published on 9 March 1931. For one reason or another, the preface never appeared. Lauding Planck's article, Einstein wrote:

I presume I may add that both Planck's conception of the logical state of affairs as well as his subjective expectation concerning the later development of science corresponds entirely with my own understanding.

Citing this passage, Gerald Holton (1968/1988, 262) observes that from this time forward, "Einstein's and Planck's writings on these matters are often almost indistinguishable from each other" (Holton reports the note was "written on or just before 17 April 1931"). Of course, by 1930, it had become increasingly apparent to Einstein that many of the leading quantum physicists sought to cloak what he considered the theory's shortcomings under the mantle of positivist strictures of meaningful expressions. This in turn led to more and more forceful counters of a realism *à la* Planck. In the autumn of 1930, Schlick, perhaps the leading logical empiricist, completed a lengthy essay purporting to attain "philosophical clarity" about the standing of the principle of causality in the new physics of quantum mechanics. It appeared in the scientific weekly *Die Naturwissenschaften* at the beginning of 1931. Upon finishing it, sometime in

November 1930, Schlick posted a typescript copy to Einstein. If Schlick had any awareness of Einstein's objections to quantum mechanics, he knew beforehand that Einstein was not disposed to favor the essay's conclusion, that "quantum physics teaches us ... that within the bounds established by the uncertainty relations the [causality] principle is *bad*, useless or idle, and incapable of fulfillment" (Schlick 1931/1979, 196; original emphasis).

But his transgression had gone much further, impugning the very conception of physical theory guiding both Planck and Einstein. For at the beginning of his essay, Schlick lampooned the "obstinacy" of the physicist's traditional belief that the explanation of nature required comprehensible models, a rejection of the *Bild* conception of physical theory. At the end, he proclaimed "the human imagination is incapable of conjecturing the world-structure revealed to us by patient research" (Schlick 1931/1979, 206). Einstein was quick to disagree with these two claims in particular. Responding in a letter to Schlick of 28 November 1930, two weeks after Planck's lecture, Einstein condemned the constraints of positivism's conception of physical theory. Alluding to the irrational bifurcation between real world and the world of experience that Planck viewed as the hinge point of natural science, Einstein freely admitted to the metaphysical character of the theoretician's realist impulses:

> I tell you straight out: Physics is the attempt at the conceptual construction of a model of the *real world* and of its lawful structure. To be sure, [physics] must present exactly the empirical relations between those sense experiences to which we are open; but only *in this way* is it chained to them. . . . In short, I suffer under the (unsharp) separation of Reality of Experience and Reality of Being.
>
> . . . You will be astonished by Einstein the "metaphysicist". But every four- and two-legged animal is *de facto* in this sense metaphysicist.
>
> (Einstein Archives document FA 21–603)

Without acknowledging a scientist's motivational realism, admitted even by descriptivists like Duhem, positivism can give only a misleading account, perhaps only a caricature, of scientific method. This would become an increasingly prominent theme in Einstein's response to quantum mechanics.[12]

8.5. Realism "As a Program"

In the immediate post-WWII period, Einstein coalesced two previously distinct lines of argument against quantum mechanics into the thesis of "macrorealism". Prior to then, Einstein wielded a two-tiered critique. First, a parade of heuristic examples of individual systems purported to show the incompleteness of quantum mechanical description without targeting Heisenberg's strictures on simultaneous measurement of exact values of canonically conjugate observables. The famous EPR paper is but one instance of this line of critique.

At the methodological/philosophical level, Einstein repudiated complementarity's positivist inference that since simultaneous measurements cannot reveal exact values of conjugate properties, such values cannot be said to exist and accordingly need not be described. By the late 1940s, Einstein began to emphasize that even positivists found it difficult to deny that *macroscopic* objects possessed "real states" independent of observation. An incompleteness argument could then be marshaled if the quantum mechanical characterization of the transition from micro-object to macro-object, such as is required by allowing the mass of single particle to increase to classical scale (e.g., one gram, a transition quantum mechanics, if considered universal, was obliged to describe) did not yield a "real state" of the corresponding macro-object, that is, did not produce the definite assignment of kinematical and dynamical properties that classical particle mechanics takes for granted.

A first step is to throw down the gauntlet to positivism, postulating what in 1953 Einstein termed the "thesis of reality":

> "thesis of reality": *there is such a thing as the "real" state* of a physical system existing independently of any measurement or observation that in principle can be described by the means of expression of physics.
>
> (1953a, 6)

The scare quotes around "real" serve to alert us that such a state (unobserved, unmeasured, independent of characterization by theory and so independent of the human mind) is a metaphysical hypothesis. The thesis of reality is a fundamental posit of what in 1949 is termed realism as a "program", that is, a program for complete description of "real states".

> The "real" in physics is to be taken as a type of program to which we are . . . not forced to cling *a priori*. No one is likely to give up this program within the realm of the "macroscopic". . . . But the "macroscopic" and the "microscopic" are so interrelated that it appears impractical (*untunlich*) to give up this program in the "microscopic" alone.
>
> (Einstein 1949, 674)

Three related points can be mentioned here. First, realism "as a program" is widely assumed, though perhaps implicitly, in classical physics. Objects or states of objects are regarded as having definite values of observable properties independently of observation or measurement. Second, the entwined "interrelation" between microscopic and macroscopic systems gives reason to think the previous operative assumption should not be restricted to the realm of the macroscopic. Third, adherence to the program is not compelled "a priori"; after all, the assumption of the real external world is a metaphysical assumption that may be consistently denied, as by solipsism. But such an assumption is not controversial. Even quantum theoreticians adhere to it "so long as they are not discussing the foundations of quantum theory" (Einstein 1953a, 6).

The thesis of realism is affirmed to be generally applicable and so to apply indifferently to all physical systems. Controversy ensues only when it is claimed that there are "real states" of micro-objects that are only incompletely described by the Ψ-function. But then the quantum theoretician has had to admit that the notion of "real state" is not univocal between classical and quantum and that there is a fundamental difference in kind between the character and method of the classical physical description of objects and the quantum mechanical description of quantum objects. This is, of course, a postulate, even a defining characteristic, of Bohr's complementarity. Einstein insists on viewing this postulate as a positivist inference and is prepared to be a "metaphysicist" for doing so.

In place of the incoherent EPR "criterion of reality", the "thesis of reality" is an attempt to reinstate the conception of physical theory outlined previously, in which the intended meaning of the concepts <physical reality>, <external world>, and <real state of (an individual) system> characterize what it is that physical theory attempts to describe. Realism as a "program" thus pertains to the intended target of description of any physical theory, the real external world as partitioned by physics into systems and subsystems. With the advent of quantum mechanics, positivism reappeared in physical theory in the guise of complementarity. And to the extent that complementarity enjoins a sharp conceptual separation between the respective realms of classical and quantum, it presents a vulnerable target to the "thesis of reality". Not because the thesis merely affirms what Bohr and quantum orthodoxy deny, the existence of definite states of quantum objects at all times: this is moot and merely begs the question. Rather the intent of the thesis is to highlight that, as a matter of consistency and as a *theoretical requirement of quantum mechanics*, there must be a faithful quantum mechanical characterization of the transition in the limit from the domain of quantum phenomena to macroscopic physics. This quantum mechanics does not do, then or arguably now, and so the "thesis of reality" points out a glaring gap in the case that complementarity can make to be an adequate and definitive philosophical interpretation of quantum mechanics.

Einstein's realism vis-à-vis the quantum physicists is also a dispute over the explanatory goals of physical theory. "Realism as a program" delineates in the first instance categories of understanding, not necessarily of nature. But it supports the characterization of quantum mechanics as incomplete. Realist theories, as Einstein conceived of them, share a common world picture, that is, reflect a cumulative *understanding* of nature as composed of relativistic (in the sense of the general theory) deterministic (causal) non-quantum continuous fields in which individual systems are spatiotemporally separable and can be characterized completely by properties having determinate values independent of any act of measurement or observation. The concepts of this realist program of description are not to be assumed either *a priori* valid or as impossible. They are not essential to the practice of science, but neither are they to be merely dismissed as excrescences of the metaphysically diseased mind. Like all other concepts, they are ultimately justifiable only by experience. Occasionally rivals (e.g., "purely algebraic physics") are mentioned. Characteristically, after laying out his vision, Einstein will immediately observe there is absolutely no compelling reason, conceptual or metaphysical, to regard it as correct. Justification for any choice of a physical theory and its attendant concepts and presumed ontology lies principally with the ability of such theories to implement coordination with experience (confirmation) "with advantage" (*mit Vorteil*) (Einstein 1953b, 34). The latter phrase is significant, for it is an oblique indication of the requirement that a theory must provide comprehension, that is, *understanding*,

> a conceptual model [*Konstruction*] for the comprehension of the interpersonal whose authority lies solely in its confirmation [*Bewährung*]. This conceptual model refers precisely to the 'real' (by definition), and every further question concerning the "nature of the real" appears empty.
>
> (Einstein 1949, 680)[13]

8.6. Tamed Metaphysicist

By contemporary lights, the unified field theory program that occupied Einstein for three decades was a hugely premature, bound-to-fail search for the cosmic world order. It produced no new physics but only a trail of equations too unwieldy or complex to be solved. But his final philosophical vantage point was the result of a long preoccupation with what he delicately termed "the present difficulties of physics" (Einstein 1944, 279). These "difficulties" are a reference to the two principal concerns of his last three decades: i) the critique of the quantum theory as incomplete and a related attempt to counter positivist conceptions of physical theory; ii) struggles with unified field theory, which, epistemologically speaking, presented new concerns on account of the tenuous and indirect connection between the fundamental mathematical concepts and possible observable evidence. Dovetailing philosophical commitments emerged from these twin battles.

As seen, Einstein stressed against positivism the legitimacy of a concept of "the real", a concept that carries the presumption of comprehensibility by the human mind, that is, that nature is intelligible. But a considerably more filled-out picture emerges by examining the several overt expressions of rationalist methodology accompanying his quest for unified field theory, the hypothetical projection of the intelligible-real, framed as a non-linear, continuum-based field theory whose laws are generally covariant.

Previously, it was seen that a famous instance of such a declaration, at Oxford in 1933, may perhaps be tempered by consideration of the context of its occurrence. But there are others that should be considered. A little-known Einstein text of 1929 featured another strident avowal of rationalist aspiration. Appearing in an obscure *Festschrift* for an old Bern teacher, Aurel Stodola, the paper opens by listing the "two ardent desires" of the theoretical physicist. The first is the wish to satisfy the requirement of "completeness" (*Vollständigkeit*), that is, to contain within one theory "all the relevant phenomena and their connections". But the second points in an orthogonal direction, an aspiration that is "the Promethean element of scientific experience".

> We wish not only to know *how* Nature is (and *how* her processes transpire), but also to attain as far as possible the perhaps utopian and seemingly arrogant goal, to know why Nature *is thus and not otherwise.*
>
> (Einstein 1929a, 126)

This demand, no less than an expression of the Principle of Sufficient Reason, is certainly the ultimate goal of explanation and, if satisfied, would be undeniable proof of the comprehensibility and intelligibility of Nature cherished by rationalism. But the full story of Einstein's rationalism requires introducing the "tamed metaphysicist". Writing for a lay audience in *Scientific American* in 1950, Einstein again returned to endorsing the method of mathematical speculation that drove the search for a "generalization" of the theory of gravity, a unified field theory:

> Time and again the passion for understanding has led to the illusion that man is able to comprehend the objective world rationally by pure thought without any empirical foundations – in short, by metaphysics. I believe that every true theorist is a kind of tamed metaphysicist, no matter how pure a "positivist" he may fancy himself to be. The tamed metaphysicist believes that not all that is logically simple is embodied in experienced reality, but that the totality of all sensory experience can be "comprehended" on the basis of a conceptual system build on premises of great simplicity.
>
> (1950/1954, 342)

Admitting this a "miracle creed", nonetheless Einstein claimed it "has been borne out to an amazing extent by the development of science". He made no attempt to justify such a highly controversial characterization of the history of physics, and perhaps none can be given. It may only be intended as an indirect reference to the perceived success of previous unification efforts, a narrative of physical theory related many times elsewhere. The story begins with classical mechanics as the fundamental basis of all physics, where the fundamental concepts are of particles and their laws of motion. The next stage is the Faraday/Maxwell concept of field and the unification of optics, electricity, and magnetism. A unification between mechanics and field theory is effected through the special and the general theories of relativity and points toward a future theory of the "total field" where particles and their properties are derived as solutions to non-linear field equations. The underlying theme is unity, whereas the underlying theme of unity is comprehension via logical simplicity. In this way, one can view "the grand aim of all science" in the attempt to "cover the greatest number of empirical facts by logical deduction from the smallest number of hypotheses or axioms" (quoted in Barnett 1950). Philosophers like Phillip Frank struggled to turn declarations of this kind into manifestations of Einstein's concordance with logical empiricism.

Yet in referring to "the *illusion* that man is able to comprehend the objective world rationally by pure thought", Einstein drops a clue that theory unification by the method of mathematical speculation may well be a rationalist pipe dream, that is, of *the illusion of reason* that comprehension and understanding, in precisely the rationalist sense of knowing why Nature *is thus and not otherwise*, is possible. It is clear that Einstein recognized the possible deceptive character of this rationalist faith by the two personae with which he frequently associated it, Hegel and Don Quixote.

Einstein was impressed by Émile Meyerson's treatment of the *a prioristic* tendency in science in his 1925 book *La Déduction Relativiste* and by the utility of an analogy Meyerson employed between the explanatory *pangeometrisim* of the general theory of relativity and Hegel's idealist narrative of the evolution of reason.[14] Einstein returned time and again to the figure of Hegel as exemplifying both the danger and the promise of the rationalist dogma that the human mind can force reality into the mold created by mind or, in terms relevant to Einstein, that mathematical speculation based on "logical simplicity" can be employed as a methodology of successful theory construction. For example, in the Stodola essay just mentioned, Einstein inserts a footnote after speaking of the theorist's goal of deductively capturing empirical laws as logical necessities:

> Meyerson's comparison with Hegel's aspirational goal (*Zielsetzung*) surely has a certain justification; he sharply illuminates the frightening danger here.

> (1929a, 128)

And in *The Times* (London) on 5 February 1929, attempting to explain his new "distant parallelism" unified field theory, Einstein noted that the characteristics

> distinguishing the general theory of relativity, and even more so the new third stage of [relativity] theory, the Unitary Field Theory, from other physical theories, are the degree of formal speculation, the slender empirical basis, the boldness in theoretical construction, and finally the fundamental reliance on the uniformity of the secrets of natural law and their accessibility to the speculative intellect. It is this feature which appears as a weakness to physicists who incline towards realism or positivism, but is especially attractive, nay, fascinating, to the speculative mathematical mind. Meyerson, in his brilliant studies on the theory of knowledge, justly draws a comparison of the intellectual attitude of the relativity theoretician with that of Descartes, or even of Hegel, without thereby implying the censure which a physicist would naturally read into this.
>
> (1929b, 114)

Einstein is nonetheless clear that "the speculative mathematical mind" of the theorist must be tamed by sober recognition that the very premise that such a methodology is possible at all may well be illusory. If so, the grandiose aspirations but utter failure of rationalist understanding, exemplified by Hegel, may then be linked to the madness afflicting Don Quixote, as in this letter to Max Born, responding to Born's 1943 attack on the speculative non-empirical "Hegelian physics" of Eddington and Jeans:

> I have read with much interest your lecture against Hegelianism (*"die Hegelei"*), which with us theoreticians amounts to the Don Quixotean element (*"das Don Quijote'sche Element"*) or, should I say, the seducer? But where this evil or vice is fundamentally missing, the hopeless philistine in on the scene.
>
> (Einstein to Born 7 September 1944; translation modified. See Born 1971, 149)

Here, as elsewhere, Einstein regards theory-construction not driven by the rationalist impulse for comprehensive understanding as no better than the science of the philistines, that is, of the positivists. In point of fact, a closer reading of the Herbert Spencer Lecture shows that even these hyper-rationalist declarations are tempered by the setting posed by the perennial problem of philosophy and of science, the "eternal antithesis between Reason and Experience". Phrased in this way, and confronted with Einstein's late appreciation of "the truly valuable in Kant" encapsulated in the cryptic catechism *nicht gegeben, sondern aufgegeben,*

Einstein cannot be thought unaware of the significance of its message in the Transcendental Dialectic: that Reason in science, as elsewhere, has only a purely *regulative* employment; it poses a mold for an understanding that requires unity. It is a necessary role because "intelligibility" matters, and for Einstein, intelligibility is the child of reason. The constructs of fundamental physical theory must manifest to the greatest extent possible the unity of systematic interconnection of *an ideal order of nature*, an ideal projected by the hypothetical employment of reason.

In Einstein, this employment took the form of mathematical speculation guided by "logical simplicity". Hegel, poster child of rationalism run amuck, ignored Kant's message that such a use of reason must be "tamed" or disciplined. In fact, Kant's most expansive discussion of mathematical knowledge in *The Critique of Pure Reason* occurs in a chapter entitled "The Discipline of Pure Reason", where he noted that "discipline" is to be understood in the punitive sense: the purpose is to *humiliate* reason's pretensions to knowledge. More might be said about why the regulative role of reason is regarded as a necessary part of the Kantian, and especially neo-Kantian, conceptions of science (see Buchdahl 1969; Neiman 1994). But this broadly neo-Kantian aspect of Einstein's methodology of theory-construction by mathematical speculation has largely been overlooked (a notable exception is Beller 2000). To Einstein's "tamed metaphysicist", without reason's Don Quixote-fixation on a "passion for understanding", "there would be neither mathematics nor natural science".[15]

8.7. Conclusion

Distinct components, rooted in his struggles to find the field equations of general relativity, constitute Einstein's objections to positivist or logical empiricist accounts of scientific theory:

First, "realism as a program": the motivational or aspirational aspect affirming the aim of fundamental physical theory to provide a model of physical reality. This is a realism without the semantic apparatus of truth and reference of scientific realism. It also lacks the "epistemic optimism" encouraging the scientific realist to commit to claims of "approximate truth" for current theories. Nonetheless, following Planck's example, Einstein would endorse claims that certain elements of the current "conceptual model [*Konstruction*]" may well be permanent fixtures of the future "world image" that gives meaning to the term "physical reality". Such elements are not entities or laws of particular theories but meta-level constraints on theories, values of physical constants, principles of connection (e.g., between entropy and probability), or symmetries that remove the bias of particular observers or frames of reference.

Second, the contesting currents of Einstein's "believing rationalist" and the method of mathematical speculation can best be understood through the prism of what he had come to regard as "the truly valuable in Kant".

In choosing to identify that message with the cryptic catechism *nicht gegeben, sondern aufgegeben*, Einstein was certainly aware of its import in the context of the regulative use of reason in the search for unity, a mandate for projecting a comprehensibility of nature at the level of fundamental physics. Both components play roles in a common dialectical purpose of opposition to positivist conceptions of physical theory that Einstein perceived to be widely, and largely uncritically, canonically promulgated by logical empiricism and tacitly assumed by many quantum physicists under the aegis of Bohr's complementarity.

Notes

1. See the detailed discussion in Ryckman (2005, Ch. 3).
2. This section based upon Ryckman (2017, Ch. 7)
3. Holton acknowledged non-positivist aspects in the philosophy of the early Einstein but considered them subordinate. For example, the young Einstein clearly identified with Boltzmann rather than Mach in the atomism controversy ca. 1900. See Ryckman (2017, Ch. 3).
4. Einstein to Lanczos, 21 March 1942, (EA 15–294) as cited and translated in Lanczos (1998, 2–1526, note 9).
5. This is an accurate observation given Einstein's reverence for Spinoza; see Ryckman (2017, 265).
6. See Einstein (1929b, 114). The full passage reads: "The characteristics which especially distinguish the General Theory of Relativity and even more the new third stage of the theory, the Unitary Field Theory, from other physical theories, are the degree of formal speculation, the slender empirical basis, the boldness in theoretical construction, and finally the fundamental reliance on the uniformity of natural laws and their accessibility to the speculative intellect". This is one of several remarks on the methodological novelty posed by unified field theory, made in the context of the theory based on the concept of "distant parallelism".
7. This section based in part upon Ryckman (2017, Ch. 10).
8. "A propos de la philosophie de Kant, je crois que chaque philosophe a son Kant propre et je ne puis répondre à ce que vous venez de dire, parce que les quelques indications que vous avez données ne me suffisent pas pour savoir comment vous interprétez Kant. Je ne crois pas, pour ma part, que ma théorie concorde sur tous les points avec la pensée de Kant telle qu'elle m'apparaît. Ce qui me paraît le plus important dans la philosophie de Kant, c'est qu'on y parle de concepts a priori pour édifier las science. Or on peut opposer deux points de vue: l'apriorisme de Kant, dans lequel certains concepts préexistent dans notre conscience et le conventionnalisme de Poincaré. Ces deux points de vue s'accordent sur ce point que la science a besoin, pour être édifiée, de concepts arbitraires; quant à savoir si ces concepts sont donnés a priori, ou sont des conventions arbitraires, je ne puis rien dire." *Bulletin de la Société Française de Philosophie*, XVII (1922), p. 101.
9. See, for example, the letter to Moritz Schlick dated 14 December 1915, in Schulmann et al. (1999), doc. 165; or see the there undated letter to Max Born from the summer of 1918, doc. 575.
10. This section based upon Ryckman (2017, Ch. 10).
11. See the insightful review by Harry T. Costello ("H.T.C.") of Ernst Zimmer's *The Revolution in Physics*. See Costello (1936).
12. This section based upon Ryckman (2017, Ch. 8).

13. This section based upon Ryckman (2017, Ch. 8).
14. In a 1928 review of Meyerson's book, Einstein (1928) notes the aptness of Meyerson's "very ingenious" comparison of the deductive-constructive character of relativity theory with the systems of Descartes and Hegel because "the human mind wants not only to propose relationships, it wants to *comprehend*".
15. This section based upon Ryckman (2017, Ch. 8).

References

Barnett, L. (1950), 'The Meaning of Einstein's New Theory', *Life Magazine*, January 9: 22–25.

Beller, M. (2000), 'Kant's Impact on Einstein's Thought', in D. Howard (ed.), *Einstein: The Formative Years 1879–1909*. Basel-Boston-Berlin: Birkhäuser, pp. 83–106.

Born, M. (ed.) (1971), *The Born-Einstein Letters*. Translated by Irene Born. London: Macmillan Press.

Buchdahl, G. (1969), *Metaphysics and the Philosophy of Science: The Classical Origins: Descartes to Kant*. Oxford: Basil Blackwell.

Carnap, R. (1927), 'Eigentliche und uneigentliche Begriffe', *Symposion*, 1 (4): 355–374.

Costello, H. T. (1936), 'Review of Zimmer's the Revolution in Physics', *The Journal of Philosophy*, 33: 527–528.

Einstein, A. (1918/2002), 'Motive des Forschens', in M. Jannsen et al. (eds.), *The Collected Papers of Albert Einstein, Volume 7: The Berlin Years: Writings, 1918–1921*. Princeton and Oxford: Princeton University Press, pp. 55–59.

——— (1921/2002), 'Geometry and Experience', in M. Jannsen et al. (ed.), *The Collected Papers of Albert Einstein, Volume 7: The Berlin Years: Writings, 1918–1921*. English translation supplement. Princeton and Oxford: Princeton University Press, pp. 208–222.

——— (1923), 'The Theory of the Affine Field', *Nature*, 112: 448–449.

——— (1928), 'A propos de La Déduction relativiste de M. É. Meyerson', *Revue philosophique de la France et de l'etranger*, 105: 161–166.

——— (1929a), 'Über den gegenwärtigen Stand der Feld-Theorie', in E. Honegger (ed.), *Festschrift Prof. Dr. A. Stodola zum 70. Geburtstag*. Zurich und Leipzig: Orell Füssli Verlag, pp. 126–132.

——— (1929b), 'The New Field Theory II: The Structure of Space-Time', *The Times*, February 5. London. Reprinted in *The Observatory*, 659 (April): 114–118.

——— (1934), 'On the Method of Theoretical Physics', *Philosophy of Science*, 1 (2): 163–169.

——— (1944), 'Remarks on Bertrand Russell's Theory of Knowledge', in P. A. Schilpp (ed.), *The Philosophy of Bertrand Russell*. Evanston, IL: Northwestern University Press, pp. 278–291.

——— (1949), 'Replies to Criticisms', in P. A. Schlipp (ed.), *Albert Einstein: Philosopher-Scientists*. New York: Tudor Publishing Company, pp. 665–688.

——— (1950/1954), 'On the Generalized Theory of Gravitation', in *Ideas and Opinions*. New York: Crown Publishers, pp. 341–355.

——— (1953a), 'Einleitende Bemerkungen über Grundbegriffe', in *Louis de Broglie: Physicien et Penseur*. Paris: Éditions Albin Michel, pp. 4–14.

—— (1953b), 'Elementare Überlegungen zur Interpretation der Grundlagen der Quanten-Mechanik', in *Scientific Papers Presented to Max Born*. Edinburgh: Oliver & Boyd, pp. 33–40.

Fine, A. (1996), *The Shaky Game*. 2nd ed. Chicago: University of Chicago Press.

Frank, Ph. (1947), *Einstein: His Life and Times*. New York: Alfred A. Knopf.

—— (1949), 'Einstein's Philosophy of Science', *Reviews of Modern Physics*, 21 (3): 349–355.

—— (2021), *The Humanistic Background of Science*. Edited by G. Reisch and A. T. Tuboly. New York: SUNY Press.

Holton, G. (1968/1988), 'Mach, Einstein, and the Search for Reality', in *Thematic Origins of Scientific Thought: Kepler to Einstein*. Cambridge, MA: Harvard University Press, pp. 237–277.

Kuhn, Th. (1977), 'Objectivity, Value Judgment, and Theory Choice', in *The Essential Tension: Selected Studies in Scientific Tradition and Change*. Chicago and London: University of Chicago Press, pp. 320–329.

Lanczos, C. (1931), 'Die neue Feldtheorie Einsteins', *Ergebnisse der exakten Naturwissenschaften*, 10: 97–132.

—— (1998), *Collected Published Papers with Commentaries*. Vol. 4. Raleigh, NC: North Carolina State University Press.

Neiman, S. (1994), *The Unity of Reason: Rereading Kant*. New York and Oxford: Oxford University Press.

Planck, M. (1933), *Where Is Science Going?* Prologue by Albert Einstein. Translated and Edited by J. Murphy. New York: W.W. Norton and Co.

—— (1934), 'Positivismus und reale Aussenwelt', in *Wege zur physikalischen Erkenntnis: Reden und Vorträge*. Zweite Auflage. Leipzig: S. Hirzel Verlag, pp. 208–232.

Quine, W. V. O. (1951), 'Two Dogmas of Empiricism', *The Philosophical Review*, 60 (1): 20–43.

Ryckman, Th. (2005), *The Reign of Relativity: Philosophy in Physics 1915–1925*. New York: Oxford University Press.

—— (2017), *Einstein*. New York: Routledge.

Schlick, M. (1931/1979), 'Causality in Contemporary Physics', in H. Mulder and B. F. B. van de Velde-Schlick (eds.), *Moritz Schlick: Philosophical Papers, Volume 2 (1925–1936)*. Dordrecht: D. Reidel, pp. 176–209.

—— (1932/1979), 'Positivism and Realism', in H. Mulder and B. F. B. van de Velde-Schlick (eds.), *Moritz Schlick: Philosophical Papers, Volume 2 (1925–1936)*. Dordrecht: D. Reidel, pp. 259–284.

Schulmann, R. et al. (eds.) (1999), *The Collected Papers of Albert Einstein, Volume 8A: The Berlin Years: Correspondence, 1914–1918*. Princeton and Oxford: Princeton University Press.

Stadler, F. and Uebel, Th. (eds.) (2012), *Wissenschaftliche Weltauffassung. Der Wiener Kreis*. Reprint of the first edition on behalf of the Institute Vienna Circle on the occasion of its 20th anniversary. Wien and New York: Springer.

Uebel, Th. (2013), '"Logical Positivism": "Logical Empiricism": What's in a Name?', *Perspectives on Science*, 21 (1): 58–99.

Wuest, A. (2015), *Philipp Frank: Philosophy of Science, Pragmatism, and Social Engagement*. Ph.D. Dissertation. Department of Philosophy, University of Western Ontario.

9 'Geometrization of Physics' Vs. 'Physicalization of Geometry'. The Untranslated Appendix to Reichenbach's *Philosophie der Raum-Zeit-Lehre*

Marco Giovanelli

9.1. Introduction

For today's readers of Hans Reichenbach's *The Philosophy of Space and Time* (Reichenbach 1958), it might come as a surprise that the book is missing the translation of an Appendix entitled "Weyl's Extension of Riemann's Concept of Space and the Geometrical Interpretation of Electromagnetism". A reference to a no-longer-existing §46 on page 17 is the only clue of its existence. The Appendix covered some 50 pages of the German original, the *Philosophie der Raum-Zeit-Lehre* (Reichenbach 1928) – not a few considering that Reichenbach dedicated only half a score pages more to general relativity. The editors of the English translation, Maria Reichenbach and John Freund, had prepared a typescript of the translation of the Appendix (HR, 041–2101), including the transcription of the mathematical apparatus, which was considerably heavier than that in the rest of the book.[1] However, they must have decided not to include it in the published version eventually. By the end of the 1950s, Weyl's geometrical interpretation of electromagnetism was at most of antiquarian interest. The mathematical effort required from the readers to familiarize themselves with the subject might have appeared not worth the modest philosophical gain. With the important exception of a pathbreaking paper by Alberto Coffa (1979), the case of the missing Appendix does not seem to have attracted attention in the theoretical and historical literature since then (see Giovanelli 2016).

The context in which the Appendix was written is briefly recounted by Reichenbach himself in an unpublished autobiographical sketch (HR, 044–06–25). Reichenbach started to work on the *Philosophie der Raum-Zeit-Lehre* in March 1925. The drafting of the manuscript was interrupted several times, but some of its core parts were finished at the turn of 1926, when Reichenbach was negotiating a chair for the philosophy of physics in Berlin (Hecht and Hoffmann 1982). As Reichenbach writes in the autobiographical note: "In March-April 1926, Weyl's theory was dealt with, and the peculiar solution of §49 was found. At that time, the entire Appendix was written. (correspondence with Einstein). Talk at the physics conference in Stuttgart" (HR, 044–06–25). This short reconstruction is confirmed by independent textual evidence (see Giovanelli 2016). In March 1926, after making some critical remarks on Einstein's newly published metric-affine theory (Einstein 1925b), Reichenbach sent Einstein a ten-page 'note' that would turn out to be an early draft of §49 of the Appendix. The typescript of the note is still extant (HR, 025–05–10). Einstein's objections and Reichenbach's replies reveal that, contrary to Coffa's (1979) claim, Reichenbach's defense of geometrical conventionalism against Weyl's geometrical realism (see Ryckman 1995) was not the motivation behind Reichenbach's note.

Reichenbach was concerned with the more general problem of the meaning of a 'geometrization' of a physical field. At that time, it was a widespread opinion that, after general relativity had successfully geometrized the gravitational field, the next obvious step was to 'geometrize' the electromagnetic field (see, e.g., Eddington 1921; Weyl 1918a, 1921b). However, according to Reichenbach, this research program – the so called unified field theory project (see Goenner 2004; Sauer 2014; Vizgin 1994) – was based on a fundamental misunderstanding of the nature of Einstein's geometrical interpretation of the gravitational field. To prove his point, Reichenbach put forward his own attempt of a geometrical interpretation of the electromagnetic field. Thus, he hoped to demonstrate that such geometrization was not much more than mathematical trickery that in itself did not constitute a gain in physical knowledge. Einstein, in spite of having found some significant technical mistakes in Reichenbach's theory, agreed at least superficially with Reichenbach's 'philosophical' message (Lehmkuhl 2014). Against Einstein's advice, Reichenbach presented this material in public at the Stuttgart meeting of the German Physical Society (Reichenbach 1926). Later, he included it in the manuscript of the book he was working on as part of a longer Appendix. After some struggle in finding a publisher, the manuscript of the *Philosophie der Raum-Zeit-Lehre* was finished in October 1927 and published the following year by De Gruyter (HR, 044–06–25).

The correspondence between Reichenbach and Einstein has been already discussed elsewhere (Giovanelli 2016). In the present chapter, I aim to offer an introduction to the Appendix itself. Besides the note that Reichenbach sent to Einstein, which became §49, the Appendix contains over 40 pages of additional material, which are worth further investigation. As the present chapter will try to demonstrate, the philosophical message of the Appendix should be considered an integral part of the line of argument of *Philosophie der Raum-Zeit-Lehre* and, in particular, of its last chapter dedicated to general relativity. As Reichenbach pointed out, according to general relativity, the universal effect of gravitation on all kinds of measuring instruments defines a *single geometry*, an, in general, non-flat Riemannian geometry. "In this respect, we may say that gravitation is *geometrized*" (Reichenbach 1928, 294; original emphasis; tr. 256). We do not speak of deformation of our measuring instruments "produced by the gravitational field", but we regard "the measuring instruments as 'free from deforming forces' in spite of the gravitational effects" (Reichenbach 1928, 294; tr. 256). However, such a *geometrical explanation* is, according to Reichenbach, not an explanation at all but merely a codification of a matter of fact. Reichenbach insisted that, also in general relativity, it was still necessary to provide a *dynamical explanation* of the observed behavior of rods and clocks, although a dynamical explanation of a new kind. Even if "we do not introduce a force to explain the *deviation* of a measuring instrument from some normal geometry", we must still invoke a force as a *cause* for the fact that "*there is a general correspondence [einheitliches Zusammenstimmen] of all measuring instruments*" (Reichenbach 1928, 294; original emphasis; tr. 256) that all agree on a non-flat Riemannian geometry depending on the matter distribution.

Indeed, according to Einstein's theory, general relativity teaches us that we may consider the "effect of gravitational fields on measuring instruments to be of the same type as all known effects of forces" (Reichenbach 1928, 294; tr. 257). What is characteristic of gravitation with respect to other fields is the *universal coupling* of gravitation and matter. Measuring instruments made of whatever fields and particles can be used to explore the gravitational field, and the result of such measurements is independent of the device. As a consequence, it becomes impossible to separate the measuring instruments that measure the background geometry (rods and clocks, light rays, uncharged test particles) from those that measure the dynamical field (charged test particles). The geometrical measuring instruments have become indicators of the gravitational field. However, this does not imply that it is "*the theory of gravitation that becomes geometry*"; on the contrary, it implies that "*it is geometry that becomes an expression of the gravitational field*" (Reichenbach 1928, 294; original emphasis; tr. 256). For the reader of the English translation, Reichenbach's line of argument makes a short appearance at the end of the sections dedicated to general relativity and is then interrupted abruptly. However, in the original German, Reichenbach's line of argument is picked up again and developed for further 50 pages in the Appendix. Thus, the latter is nothing but the second half of an argumentative arc whose first half had been erected in the last chapter of the book.

This chapter demonstrates how in the Appendix, starting from the affine connection instead of the metric, Reichenbach formulated a theory that seems to 'geometrize' both the gravitational and the electromagnetic fields. However, unlike general relativity, Reichenbach's theory does not add any new physical knowledge that was not already entailed by previous theories. Thus, the geometrization of a field is not in itself a physical achievement. A comparison with a geometrization of Newtonian gravity suggested by Kurt Friedrichs (1928) at around the same time provides the simple reason Reichenbach's attempt was bound to fail.[2] Nevertheless, Reichenbach's theory is revealing of the philosophical message that *Philosophie der Raum-Zeit-Lehre* was meant to convey. Undoubtedly, general relativity has dressed the distinctive feature of gravitation (its universal coupling with all other physical entities) in a shiny geometrical 'cloak' (a Riemannian geometry with variable curvature). However, in Reichenbach's preferred analogy, one should not mistake "the cloak [*Gewand*] for the body that it covers" (Reichenbach 1928, 354; tr. [493]). Contrary to widespread belief, general relativity was not the dawn of the new era of 'geometrization of physics' that was supposed to dominate 20th-century research for a unified field theory; general relativity was the culmination of a historical process of 'physicalization of geometry' that had begun in the 19th century (see also Reichenbach 1929a, §4).

9.2. The Appendix to the *Philosophie der Raum-Zeit-Lehre.* Overview and Structure

General relativity rests formally on Riemannian geometry. The latter is based on the 'hypothesis' that the squared distance ds between two neighboring points, x_v and $x_v + dx_v$, is a homogeneous, second-order function of the four coordinate differentials. In Einstein's notation (where summation over repeated indices is implied), it reads as follows:

$$ds^2 = g_{\mu v} dx_\mu dx_v. \qquad [9.1]$$

The coefficients $g_{\mu v} = g_{v\mu}$ are at the same time (a) the components of the metric ('measurement') field – a set of ten numbers that serve to convert coordinate distances dx_v between two close-by spacetime points into real distances $ds = \pm 1$ – and (b) components of the gravitational field. The numerical value of ds has a physical meaning if it can be considered the result of a measurement, which inevitably demands a unit in which to measure. In relativity theory, rods and clocks at relative rest in a free-falling frame perform orthogonal measurements: clocks supply $ds = -1$, and rods $ds = +1$. "Why are these measuring instruments adequate for this purpose?" (Reichenbach 1928, 331; tr. [463]). The fundamental property that makes them suitable for measuring ds is outlined by Reichenbach in §4 of the *Philosophie der Raum-Zeit-Lehre* (Reichenbach 1928, 26–27; tr. 17). One can choose, say, n spacings between the atoms of a rock-salt crystal as a unit of length $ds = 1$. As it turns out, two identical crystals that have the same length when lying next to each other are always found to be equally long after having been transported along different paths to a distant place. The same holds for unit clocks. One can arbitrarily choose n_1 wave crests of cadmium atom emitting the red line or n_2 of a sodium atom emitting the yellow line as a unit of time $ds = -1$. However, the ratio n_1/n_2 happens to be a natural constant.

General relativity, as a testable theory, stands or falls with this empirical fact (see Giovanelli 2014). However, it would be possible to think of a world in which rods and clocks do not have this peculiar 'Riemannian' behavior. In this world, it would still be possible to formulate a definition of congruence, that is, a definition of the equality of *ds*'s. However, such a definition would not be unique. The units of length would have to be given for every space point, and we could not simply rely on the Paris standard meter. In a Riemannian world, if we know the length of a room, we also know the number of unit rods that we can place along one of its walls (Reichenbach 1928, 333; tr. [464]). In a non-Riemannian world, such a number would depend upon the path by which the rods were actually brought into the room (Reichenbach 1928, 333; tr. [464]). "Such conditions may seem very strange, but they are certainly possible, and if they were real, we would surely have adapted ourselves to them" (Reichenbach 1928, 333; tr. [464]). Obviously, setting randomly different units of measure at every point would be of little use for the people in such a world. Instead, they would search for a "geometrical method which would characterize the law of *change in length* during transport; that is, they would search for 'the law of displacement' [*Verschiebungsgesetz*]" (Reichenbach 1928, 333; tr. [464]), which describes how much lengths change when transported between infinitely close points.

Thus, the geometrical problem of formulating such a 'law of displacement' arises. According to Reichenbach, this problem was addressed and solved by Weyl (1918a, 1918b). Weyl's "solution certainly constitutes a mathematical achievement of extraordinary significance regardless of its physical applicability" (Reichenbach 1928, 333; tr. [464]). One can think of dx_v as the components of a vector A^r. Weyl realized that there are two separate operations of comparison of vectors A^r. Using somewhat idiosyncratic language, Reichenbach calls the *metric* the operation of *distant-geometrical* comparison of lengths of vectors. At every point, once we know the numerical values of the components of a vector A^μ in a certain coordinate system, the metric $g_{\mu v}$ allows us to calculate its length, a single number $l^2 = g_{\mu v} A^\mu A^v$. If a different unit of measure were chosen (inches instead of cm), then we would obtain a different number $l' = \lambda l$. However, the ratio λ is regarded an absolute constant (see Weyl 1919, 102). Weyl realized that, on the contrary, in the general case, it is not possible to establish whether two vectors A^r and A'^r at different places have the same direction by simply inspecting their components. Reichenbach typically calls *displacement* the operation of *near-geometrical* comparison of the direction of vectors that takes into account the intermediary steps needed to 'displace' or transfer a vector from one place to another. In Reichenbach's characterization, Weyl discovered a type of space more general than the Riemannian space, in which the near-geometrical operation of displacement, rather than the distant-geometrical metrical comparison, represents the most fundamental operation.

In particular, Weyl envisaged a geometrical setting in which "[t]he comparison of lengths by means of a metric is . . . replaced by a comparison of lengths through displacement" (Reichenbach 1928, 336; tr. [469]). The ratio of units is allowed to change from point to point; thus, $l' = \lambda(x_v)l$ is an arbitrary function of the coordinates. Weyl found that in such a geometry, the change of length l of a vector transported to a nearby point is expressed by the formula $d(\log l) = \kappa_\sigma dx_\sigma$, where κ_σ is a vector field. The mathematical 'discovery' of the independence of the operation of displacement and of the metric had important implications in spacetime physics. On the one hand, it turns out to be advantageous to present general relativity as a theory based not solely on the metric (Eq. [9.1]) but also on two separate geometrical structures and to impose as a compatibility condition that the length of equal parallel vectors is the same. On the other hand, by weakening such a compatibility condition, one can open new mathematical degrees of freedom that can be used to incorporate the electromagnetic field into the geometry of spacetime. In particular, Weyl (1918a) identified k_σ with the electromagnetic four-potential (see, e.g., Scholz 1994). Other strategies of 'geometrizing' the electromagnetic field were evaluated when Reichenbach was writing the Appendix: one can keep Riemannian geometry but increase the number of dimensions (Einstein 1927a, 1927b; see Kaluza 1921) or abandon the restriction that ds is a quadratic form (Reichenbächer 1925). However, Reichenbach did not consider these theories and only discussed Weyl's *Ansatz* and its further developments.

The goal of the Appendix to the *Philosophie der Raum-Zeit-Lehre* is to outline a generalization of Weyl's mathematical approach based on the work of Eddington (1923) and Schouten (1922b). In particular, Reichenbach's terminology and notation are taken from the German translation (Eddington 1925) of Eddington's textbook on relativity (Eddington 1923). However, Reichenbach follows Schouten (1922b) in adopting a more general definition of the operation of displacement. This mathematical treatment, in Reichenbach's parlance, amounts to providing (1) a *conceptual definition* of the operation of displacement, without asking about its physical realization or interpretation. The second step (2) is to investigate the physical applicability of such a mathematical apparatus. This means dealing with the empirical question of whether there are objects in nature that behave according to the operation of displacement. This second step amounts to the provision of a *coordinative definition* of the operation of displacement. After a coordinative definition has been chosen, Reichenbach (3) presents an example of *physical application* of such a mathematical apparatus. Reichenbach demonstrates that, by resorting to a sufficiently general definition of the operation of displacement with an appropriate physical interpretation, one can provide a geometrical interpretation of the electromagnetic field that is 'just as good' as the geometrical interpretation of the gravitational field provided by Einstein. (4) Finally, Reichenbach reflects on the *philosophical conse-quences* of his theory. By demonstrating that his geometrization of the electromagnetic field, although impeccable as geometrical interpretation, does not lead to any new physical results, Reichenbach concludes that geometrization in itself is not the reason general relativity is a successful physical theory. Accordingly, after a brief introduction of §46, the Appendix is roughly structured in this way:

1. The conceptual definition of the operation of displacement (§47)
2. The coordinative definition of the operation of displacement (§48)
3. An example of a physical application of this mathematical apparatus (§49)
4. The philosophical consequences (§50)

9.3. The Conceptual Definition of the Operation of Displacement

9.3.1. *Displacement Space and Metrical Space*

Let us assume that a coordinate system is spread over an spacetime region, so that each point is identified by a set of four numbers x_v (where $v = 1,2,3,4$). This coordinatization is of course completely arbitrary. A vector A^τ placed at some point P with coordinates x_v can be thought of as an arrow or as the sum of its components A^μ, the four numbers (A^1, A^2, A^3, A^4) that we associate with some point P. Given a vector with components A^μ in the coordinate x_v, the components A'^μ of the vector in the coordinates x'_v is $A'^\mu = (\partial x'^\mu / \partial x^\rho) A^\mu$. A collection of four numbers A^μ that change according to this rule is defined as the contravariant components of a vector A^τ. For example, the displacement vector d_v (the separation between two neighboring points) leading from x_v to $x_v + dx_v$ is the prototype of a contravariant vector. In Euclidean geometry, it is always possible to introduce a Cartesian coordinate system in which two vectors are equal and parallel when they have the same components:

$$A'^\tau - A^\tau = 0 \qquad [9.2]$$

One can move A^τ at P with coordinates x_v to a neighboring point P' with the coordinates $x_v + dx_v$ from P and place a vector A''^τ there. dx_v might be called a 'displacement'. If the vector does not change its components, then it is the 'same' vector at a different point P'.

However, this simple relation does not hold if we introduce curvilinear coordinates, for example, polar coordinates. The components of the vector change according to $A^\mu = (\partial x'^\mu / \partial x^\rho) A^\rho$. Since the partial derivatives vary from point to point, we cannot compare two vectors, not even at neighboring points. Two vectors might have different components because they are not equal and parallel or because the components have changed in a different way at different points. Thus, if one moves the vector to a neighboring point dx_v, one does not know whether the vector has remained the 'same' by simply looking at its components. In other words, we have lost the 'connection' (*Zusammenhang*) from a point to another. Since affine geometry is the study of parallel lines, Weyl (1918a, 1918b) used to speak of the necessity of establishing a 'affine connection' (*affiner Zusammenhang*). However, it is relation of 'sameness' rather than parallelism that is relevant in this context. According to a nomenclature that was also widespread at that time, Reichenbach typically referred to the operation of 'displacement' (*Verschiebung*) as the small coordinate difference dx_v along which the vector is transferred. Since the 'displacement' also indicates the vector dx_v, the expression 'transfer' (*Übertragung*) was preferred by other authors (see, e.g., Schouten 1922b, 1923).

Displacement To reinstate the 'connection', one needs to establish a rule for comparing vectors at infinitesimally separated points.

$$dA^\tau = A'^\tau - A^\tau.$$

Given a vector A^τ at x_v in any coordinate system, we need to determine the components of the vector A'^τ at $x_v + dx_v$ that is to be considered the 'same vector' as the given vector A^τ, that is, $dA^\tau = 0$. We expect that the vector at $x_v + dx_v'$ will depend linearly on the vector at x_v; further, the change in the components between the two points will be proportional to the coordinate shifts dx_v. Thus, we expect a rule that takes the form:

$$dA^\tau = \Gamma^\tau_{\mu v} A^\mu dx_v. \qquad [9.3]$$

The quantity $\Gamma^\tau_{\mu v}$ has three indices, that is, entails τ possible combinations of $\mu \times v$ coefficients, which can vary arbitrarily from worldpoint to worldpoint. In other words, the $\Gamma^\tau_{\mu v}$ are arbitrary continuous functions of the x_v. Notice that Reichenbach, following Schouten (1922a), did not impose the symmetry of the μ, v, so that in general $\Gamma^\tau_{\mu v} \neq \Gamma^\tau_{v\mu}$. Thus, in four-dimensions, one has $4\times16 = 64$ coefficients, which reduce to $4\times10 = 40$, if $\Gamma^\iota_{\mu v} = \Gamma^\iota_{v\mu}$.

Starting with a vector A^τ at a point P with coordinates x_v, one may displace this vector along dx_v to the worldpoint P' with coordinates $x_v + dx_v$; using Eq. [9.3], one can compute the components vector A'^τ at P' that is equal and parallel to A^τ at P in any coordinate system. $\Gamma^\tau_{\mu v}$ is not a tensor. $\Gamma^\tau_{\mu v} = 0$ in a Cartesian coordinate system, so that the components of a vector do not change under parallel transport. However, $\Gamma^\tau_{\mu v} \neq 0$ in a different coordinate system, so that the components of a vector do change under parallel transport. Continuing this process step after step from A'^τ to A''^τ, to A'''^τ and so on, we obtain a broken-line curve. As the size of each displacement goes to zero, this broken line becomes a continuous curve $x_v(s)$:

$$\frac{dA^\tau}{ds} = \Gamma^\tau_{\mu\nu} A^\mu dx_\nu. \qquad [9.4]$$

Let P and Q be two worldpoints connected by a curve. If a vector is given at P, then this vector may be moved parallel to itself along the curve from P to Q; for given initial values of A^τ, Eq. [9.4] gives the unknown components of the vector A'^τ, which is being subjected to a continuous parallel displacement, in which step is labeled by the parameter s. Thus, Eq. [9.4] picks up the *straightest among all possible curves* between P and Q, that is, the lines whose direction along itself is parallel transported. At this stage, neither the notion of distance nor of angles between vectors have been defined. Nevertheless, the operation of displacement is sufficient for comparing the lengths of parallel vectors, that is, lengths along the *same* straightest line. We can compare the lengths of any two sections of this curve, that is, determine the ratio of the 'number of steps' s (the so-called affine parameter) involved in each of them.

The displacement is not assumed to be symmetric, $\Gamma^\tau_{\mu\nu} \neq \Gamma^\tau_{\nu\mu}$. This implies that, if four neighboring infinitesimal vectors are parallel in pairs and equally long in the sense of the displacement, they will not form a quadrilateral. Thus, by transporting a vector parallel to itself starting from P, one will not arrive at the same point P. In the general displacement space, there are generally no infinitesimal parallelograms $(dx)^\mu + (\delta x)^\mu - \Gamma^\mu_{\nu\alpha}(\delta x)^\alpha (dx)^\nu$.[3] If we require the connection to be symmetric, the vectors will form a quadrilateral, and we will arrive at the same point P (Reichenbach 1928, 348–349; tr. [485–486]). However, the vector thus obtained will not be, in general, the same vector as the one we have started from; that is, it will not be equal and parallel to it. The parallel transport of a vector A^τ from A to B and from B to A is reversible (by transferring back along the same curve, you get back the initial vector at P); however, in general, parallel transport along a curve depends on the curve, not only on the initial and final points. Thus, the components of A'^τ at B depend on the path chosen:

$$A'^\tau - A^\tau = \int_s \Gamma^\tau_{\mu\nu} A^\mu dx_\nu$$

where the integral s depends on the path. Given a vector at one point P, using $\Gamma^\tau_{\mu\nu}$, one can determine which is the 'same' vector at the neighboring point P'. To determine which is the 'same' vector at a point Q that is a finite distance from P, we will have to 'transport' the initial vector along a succession of infinitesimal steps to reach the final point Q. However, the vector thus obtained will generally depend on the path chosen between P and Q. Thus, it is meaningless to speak of the 'same vector' at different distant points. The difference between A^τ and A'^τ might vanish or not depending on the path of transportation. As a consequence, a vector A^τ transported parallel around any closed curve might not return to the same vector. It is generally said that parallel transport is, in the general case, non-integrable.

Thus, integrability occurs only in a particular class of spaces, in which it is allowable to speak of the same vector at two different distant points P and P'. Such spaces are characterized by the fact that $\Gamma^{\tau}_{\mu\nu}$ can be made to vanish everywhere by a suitable choice of coordinates, that is, by introducing linear coordinates (such as Cartesian coordinates). Given the 64 coefficients of the connection $\Gamma^{\tau}_{\mu\nu}$ at every point, it would be difficult to decide by sheer inspection whether this is the case. One needs, therefore, to introduce a criterion of integrability. From the connection alone, one can construct the following tensor:

$$R^{\tau}_{\mu\nu\sigma}(\Gamma) = \frac{\partial \Gamma^{\tau}_{\mu\nu}}{\partial x^{\sigma}} - \frac{\partial \Gamma^{\tau}_{\mu\sigma}}{\partial x^{\nu}} + \Gamma^{\tau}_{\alpha\nu}\Gamma^{\alpha}_{\mu\sigma} - \Gamma^{\tau}_{\alpha\sigma}\Gamma^{\alpha}_{\mu\sigma}. \qquad [9.5]$$

Weyl called this four-index symbol 'direction curvature', and it is always antisymmetrical in $\nu\sigma$ but possesses no other symmetry properties. Thus, in general, it has $4 \times 4 \times 6 = 96$ components. If the tensor $R^{\tau}_{\mu\nu\sigma}(\Gamma)$ vanishes, one can introduce 'linear' coordinate systems, which are characterized by the fact that in them, the *same* vectors have the same components at different points of the systems. If it does not, it is impossible to introduce such a 'linear' coordinate system. Thus, from the operation of displacement alone, one can construct an analogon of the Riemann tensor $R^{\tau}_{\mu\nu\sigma}(g)$ without any reference to the metric $g_{\mu\nu}$.

Metric If only the operation of displacement is defined, it does not make sense to say that a vector has a magnitude and direction, since non-parallel vectors are not comparable. When vectors lie along different straightest lines, we need to specify an additional operation to compare their magnitudes and directions. The notions of length and angles are defined by means of the dot product of two vectors. By the summation convention, the dot product $A^{\mu}B_{\mu}$ stands for the sum of the four quantities $A_1 B^1, A_2 B^2, A_3 B^3, A_4 B^4$. The squared length of a vector is defined as the dot product of the vector with itself, $l^2 = A_{\mu}A^{\mu}$. The angle θ of two unit vectors is given by $A^{\mu}B_{\mu} = \cos\theta$. These expressions use two different kinds of vector components of the same vector, one with a subscript and one with a superscript. The components A^{μ} change inversely to changes in scale of coordinates, $A'^{\mu} = (\partial x'^{\mu}/\partial x^{\rho})A^{\rho}$; consequently, they are called 'contravariant' components. A_{μ} change in the same way as the changes in the scale of coordinates $A'_{\mu} = \left(\partial x^{\rho} / \partial x'^{\mu}\right)A_{\rho}$; consequently, they are called 'covariant' components of a vector. Since one change compensates the other, the length of a vector $l^2 = A_{\mu}A^{\mu}$ is the same in the new coordinate system. One can write the same vector A^{τ} in terms of its covariant and contravariant components $A^{\tau} = e_{\mu}A^{\mu} = e^{\mu}A_{\mu}$. By setting $e^{\mu} \cdot e^{\nu} = g^{\mu\nu}$ and $e_{\mu} \cdot e_{\nu} = g_{\mu\nu}$, the dot product of a vector with itself can be written:

$$l^2 = g_{\mu\nu}A^\mu A^\nu = g^{\mu\nu}A_\mu A_\nu = A^\mu A_\mu. \qquad [9.6]$$

$g_{\mu\nu}$ is the so-called metric (i.e., measurement) tensor. Its primary role is to indicate how to compute an invariant length l of a vector A^τ from its components A^μ, which are in general different in different coordinate systems; its secondary role is to allow the conversion between contravariant and covariant components $A^\nu = g_{\mu\nu}A^\mu$ and $A^\nu = g^{\mu\nu}A_\mu$. The contravariant vector A^τ could be the displacement vector dx^ν.[4] Then Eq. [9.6] is nothing but Eq. [9.1], which extracts the distance between two neighboring points from their coordinates. If A^τ is dx^ν/ds, (where ds is the timelike interval which is an element of the four-dimensional trajectory of a moving point), then l is the length of the four-velocity vector u^ν, d^2x^ν/ds^2 is the four-acceleration vector, and so on. What is worth noting is that in Reichenbach's parlance, the 'metric' $g_{\mu\nu}$ is so defined that it allows the comparison of the length of two vectors l and l' not only at the same point in different directions but also at distant points independently of the path of transportation:

$$l' - l = \sqrt{g_{\mu\nu}A^\mu A^\nu} - \sqrt{g'_{\mu\nu}A'^\mu A'^\nu}.$$

In other words, if two vectors are equal at P (that is, $l'-l = 0$), they will be equal at P', whatever the path they are transported along. In Reichenbach's parlance, for a manifold to be a metrical space, it is not sufficient that the dot product be defined at every point (i.e., it is possible to compare the lengths of vectors at the same point in different directions); in addition, the dot product should not change under parallel transport.

Compatibility between the Metric and the Displacement The two operations defined by Reichenbach, the displacement and the metric, relate to different subjects. The metric does not say anything about whether the two vectors at different points have the same direction, whereas, by contrast, the displacement does not supply a number for vector lengths and therefore cannot be used for the comparison of unequal lengths. However, even though the purely affine notion of vectors is not enough to define the length of a vector in general, it does allow for the comparison of lengths of parallel vectors, that is, relative to other lengths along the same straightest line. In this case, the two operations, the displacement and the metric, refer to a common subject. Therefore, they might contradict each other. Two vectors at different points that are of unequal lengths according to the metric $l'-l \neq 0$ might be of equal lengths in the sense of the displacement $A'^\tau - A^\tau = 0$. Thus, every time a comparison of lengths is at stake, we would always have to specify which of the two operations of comparison we are referring to. In Reichenbach's view, although this situation is "logically permissible, it is geometrically unsatisfactory" (1928, 339; tr. [473]). It seems reasonable to require that the two operations be so defined such that the assertions they make in common are not contradictory. Vectors that are of equal lengths, $A'^\tau - A^\tau = 0$, according to the displacement should be of equal lengths, $l'-l = 0$, according to the metric (see Reichenbach 1929b).

Reichenbach intended to demonstrate how it was possible to construct a class of '*balanced spaces*', that is, spaces in which some degree of compatibility between the metric and the displacement is assured. Although none of these possible geometries might turn out to be "applicable to reality, the problem still has purely geometrical interest" (Reichenbach 1928, 339; tr. [473]). In any case, the class of balanced spaces represents "an even more general geometrical frame than Riemannian geometry for the description of reality" (Reichenbach 1928, 339; tr. [473]). How to construct a balanced space? According to Reichenbach, there are two possible ways to accomplish this goal. (1) One may "limit the scope of the metric"; that is, one may weaken the full compatibility condition between the metric and affine connection, "so that it will no longer refer to statements that result from the use of the displacement" (Reichenbach 1928, 339; tr. [474]). Only the ratio of $g_{\mu\nu}$ is preserved by parallel transport, and the length of vectors becomes path dependent: the square of the length of equal and parallel vectors l^2 is only proportional. (2) One may "limit the displacement so that statements common to the displacement and the metric no longer contradict one another" (Reichenbach 1928, 339; tr. [474]); that is, one might impose the full compatibility condition between the affine connection and the metric. The absolute values of $g_{\mu\nu}$ are preserved under parallel transport, and the length of vectors is path independent: the square of the lengths of equal and parallel vectors l^2 are the same. Reichenbach calls the type of space obtained by the first method a *displacement space*, because in it, the displacement is the dominant principle to which the metric will have to be adapted. The type of space that results from the second approach is called a *metrical space*, because in this case, the metric dominates, and the displacement is subordinate to it.

Since the metric and the displacement are two independent geometrical operations, to define a 'balanced space', Reichenbach needs to introduce a formal measure of their reciprocal compatibility. Following Eddington, Reichenbach introduces a mathematical object that determines how much the length l of a vector changes $d(l^2)$ under parallel transport:

$$d\left(l^2\right) = (\underbrace{\frac{\partial g_{\mu\nu}}{\partial x_\sigma} + \Gamma_{\mu\sigma,\nu} + \Gamma_{\nu\sigma,\mu}}_{})A^\mu A^\nu dx_\sigma . \qquad [9.7]$$

$$\hookrightarrow K_{\mu\nu,\sigma} = \nabla g_{\mu\nu}$$

The tensor $K_{\mu\nu,\sigma}{}^5$ measures the degree compatibility of the metric and the connection, that is, the degree of covariant constancy of the metric. A space in which $K_{\mu\nu,\sigma}$ is defined is a 'balanced space'.

9.3.2. 'Balanced' Spaces

Based on the tensor $K_{\mu\nu,\sigma}$, Reichenbach introduces a formal classification of balanced spaces (see section 9.8):

General Displacement Space The contradiction between the metric and the displacement is avoided by relaxing the metric-compatibility condition:

$$K_{\mu\nu,\sigma} = g_{\mu\nu} \cdot \kappa_\sigma \quad d\left(l^2\right) = l^2 \kappa_\sigma dx_\sigma.$$

The dot product of vectors is defined at one point (one can compare the length of two vectors at the same point in different directions). Still, it is not preserved under parallel transport but only rather proportional.[6] This means that the angle between two parallel transported vectors, but not their lengths, is preserved. To compare the lengths of vectors attached to different points, such a length has to be transported from one point to another, and, in general, the result would depend on the path on transportation. In addition to the knowledge of $g_{\mu\nu}$, the determination of the length l of a vector requires knowledge of four more quantities κ_σ. For two infinitesimally close points, a change in length dl of a vector satisfies the relation $dl/l = \kappa_1 dx + \kappa_2 dy + \kappa_3 dx + \kappa_4 dl = \kappa_\sigma dx_\mu$. If one further imposes the condition $\Gamma^\tau_{\mu\nu} = \Gamma^\tau_{\nu\mu}$, one arrives, as a special case, at the geometry originally introduced by Weyl (1918a), the Weyl space. In this geometry, the affine connection takes the form:

Christoffel Symbols←

$$\Gamma^\tau_{\mu\nu} = -\left\{\begin{matrix}\mu\nu\\\tau\end{matrix}\right\} + \frac{1}{2}g^\tau_\mu \kappa_\nu + \frac{1}{2}g^\tau_\nu \kappa_\mu - \frac{1}{2}g_{\mu\nu}\kappa^\tau. \qquad [9.8]$$

→ additional terms

As one can see, the metric $g_{\mu\nu}$ does not determine the components of $\Gamma^\tau_{\mu\nu}$ alone but only together with four-vector κ_σ. The straightest lines are defined as usual by the condition that vectors transported along them should always remain parallel to them. However, such lines cannot be interpreted as the shortest lines since the concept of length along different curves is not meaningful. In this setting, starting from the Weyl connection, one can construct the analogon of the Riemann tensor $B^\tau_{\mu\nu\sigma}$ from the connection as in Eq. [9.5]. Weyl demonstrated that $B^\tau_{\mu\nu\sigma}$ splits into two parts:

$$B^\tau_{\mu\nu\sigma} = R^\tau_{\mu\nu\sigma} - \frac{1}{2} g^\tau_\mu f_{\nu\sigma},$$

where $R^\tau_{\mu\nu\sigma}$ corresponds to the Riemann tensor, and $f_{\mu\nu}$ is an antisymmetric tensor of rank 2. Weyl realized that in this geometry, there are two kinds of curvature, a direction curvature (*Richtungskrümmung*) $R^\tau_{\mu\nu\sigma}$ and a length curvature (*Streckenkrümmung*) $f_{\mu\nu}$. The tensor $R^\tau_{\mu\nu\sigma}$ vanishes when the parallel displacement of a vector subjected to a change of direction is integrable. The tensor $f_{\mu\nu}$ vanishes when and only when the transfer of lengths is integrable.

General Metrical Space The alternative way of constructing a 'balanced space' is to impose a more restrictive condition on the displacement,

$$d\left(l^2\right) = 0 \quad K_{\mu\nu,\sigma} = 0.$$

This implies that the dot product of two vectors is preserved under parallel transport; not only the angle between two parallel transported vectors but also their lengths remain unchanged. As a consequence, one can compare not only the length of vectors at one point in different directions but also at distant points. Thus, the absolute values of $g_{\mu\nu}$ are defined, not only their ratios. Due to the existence of a metric, the shortest lines are defined. However, in general, they are not identical with the straightest lines defined by the displacement. To make these two special lines coincide, one needs to impose the additional restriction $\Gamma^\tau_{\mu\nu} = \Gamma^\tau_{\nu\mu}$. With this imposition, one obtains Riemann connection:[7]

$$\Gamma^\tau_{\mu\nu} = -\left\{\begin{matrix}\mu\nu \\ \tau\end{matrix}\right\} = \frac{1}{2} g^{\tau\sigma}\left(\frac{\partial g_{\mu\sigma}}{\partial x_\nu} + \frac{\partial g_{\nu\sigma}}{\partial x_\mu} \quad \frac{\partial g_{\mu\nu}}{\partial x_\sigma}\right). \quad [9.9]$$

This condition guarantees that the affine 'straights' are at the same time the lines of extremal 'length'. The components of $\Gamma^\tau_{\mu\nu}$ have the same numerical values of the so-called Christoffel symbols of the second kind, as they are calculated from the metric $g_{\mu\nu}$ and its first derivatives. They measure the variability of $g_{\mu\nu}$ with respect to the coordinates. Thus, the metric $g_{\mu\nu}$ and its derivatives uniquely determine the components of $\Gamma^\tau_{\mu\nu}$. If one starts with a symmetric metric $g_{\mu\nu}$, the Christoffel symbols are indeed the only possible choice; thus, the full compatibility of the metric and the connection is assured from the outset. By contrast, if one defines the operation of displacement independently from the metric, the Riemannian connection (Eq. [9.9]) appears only as a special case that is achieved by introducing a series of arbitrary restrictions. There are, of course, different Riemannian connections of different curvatures $R^\tau_{\mu\nu\sigma}(\Gamma)$. If one imposes the further condition that the direction curvature vanishes, one obtains the Euclidean space:

$$R^\tau_{\mu\nu\sigma}(\Gamma) = 0$$

In the Euclidean space, not only the length but also the direction of vectors is comparable at a distance. Thus, it is meaningful to speak of the same vector at different points. Indeed, in Cartesian coordinates, two vectors with the same components are equal and parallel.

9.4. The Coordinative Definition of the Operation of Displacement

Reichenbach's classification of geometries (see section 9.8)[8] opens mathematical possibilities that, in principle, could be used in physics. As one would expect from Reichenbach, he claims that to give physical meaning to the metric and the displacement, one must provide a 'coordinative definition' of both operations. "Only after a coordinative definition has been chosen, can we define the judgments as 'true' or 'false'" (Reichenbach 1928, 357; original emphasis; tr. [498]). "The choice of the indicator is, of course, *arbitrary*, since no rule can tell us what entities we should use for the realization of the process of the metric or displacement" (Reichenbach 1928, 357; tr. [498]). However, a choice is necessary. In Reichenbach's view, unless the geometrical operations introduced into the foundations of the theory can be directly identified with real objects, the theory cannot be compared with experience. The success of relativity theory lies in the fact that spacetime measurements carried out with real physical systems (rods and clocks, light rays, free-falling particles, etc.) are more correctly described in that theory than in previous theories.

In three-dimensional space, one can use rigid rods to measure lengths of three-dimensional vectors *l*; in this way, the metric acquires a physical meaning. Rigid rods (that do not change their lengths when transported) define a comparison of *length* but no comparison of *direction*; thus, they are not suitable indicators of the operation of parallel displacement. For this purpose, one might use the axis of a gyroscope, whose angular momentum vector maintains its direction. By displacing the gyroscope step by step parallel to itself from *P* to *P'*, one defines the straightest lines between those two points. We can introduce a similar interpretation of the two operations in four dimensions. The metric $g_{\mu\nu}$ is measured not only by rigid rods but also by ideal clocks, which give physical meaning to the length of the four-dimensional vector *l*. Thus, $g_{\mu\nu}$ represents the *chronogeometrical structure* of spacetime. A similar coordinative definition should be provided for the displacement $\Gamma^\tau_{\mu\nu}$. A gyroscope alone is not sufficient, because we now have to maintain the direction of a four-dimensional vector. Reichenbach suggested that one can tentatively adopt the velocity four-vector u^τ as the physical realization of the operation of displacement (Eddington 1923). When the particle is not accelerating (that is, it moves inertially), the direction of the velocity vector does not vary. Thus, the motion of force-free particles can be used to define physically the straightest line between two spacetime points. In this sense, $\Gamma^\tau_{\mu\nu}$ is sometimes said to represent the *inertial structure* of spacetime.

In the last chapter of the *Philosophie der Raum-Zeit-Lehre*, Reichenbach presents general relativity as a theory based on one primary geometrical structure, the metric $g_{\mu\nu}$. In the Appendix, the theory of general relativity appears as a theory based on two different geometrical structures, the affine connection $\Gamma^\tau_{\mu\nu}$ and the metric $g_{\mu\nu}$, and their compatibility condition $K_{\mu\nu,\sigma}$. Recognizing the autonomous role of the affine connection has an immediate advantage. The latter allows one to pick out the straightest lines directly, without the detour via the metric $g_{\mu\nu}$ and the shortest line. Indeed, a test particle in any given point on its trajectory does not 'know' about the integral length of the timelike curve between the point where the particle is and some other point where the particle is directed to. Thus, it is more suitable to claim that test particles follow the 'straightest' or auto-parallel line, that is, the line having the differential property of preserving the direction of the velocity vector u^τ unaltered while the vector is displaced by dx_ν. Given the velocity vector u^τ at one point, $\Gamma^\tau_{\mu\nu}$ allows one to determine the components of the equal and parallel vector u'^τ at an infinitesimally later point on its path. When an uncharged particle moves freely, its velocity vector is carried along by parallel displacement; that is, the velocity does not change. Thus, the particle does not accelerate and move along an auto-parallel curve $x_\nu(s)$. To appreciate the power of this formalism, it is useful to consider as a warm-up exercise the 'geometrized' formulation of Newtonian gravitation theory suggested by Kurt Friedrichs (1928) at around the same time Reichenbach was working on the Appendix.

Let us introduce a flat Newtonian spacetime with independent, mutually orthogonal spatial and temporal metrics $h^{\mu\nu}$ ($h^{11} = h^{22} = h^{33} = -1$ and = 0), $g_{\mu\nu}$ ($g_{44} = 1$ and the rest = 0). In a prototypical field theory (Maxwell electrodynamics, Newton theory of gravitation, etc.), the field equations (Maxwell's equations, Poisson equation, etc.) relate the field variables to the source variables (four-current, matter density), and the latter are related to the possible trajectories of our particle by equations of motion (Friedman 1983, 193ff.). The latter take the general form:

$$\underbrace{\mu \frac{du^\tau}{ds}}_{\text{four-acceleration}} \overset{\text{flat affine connection}}{\underbrace{-\bar\Gamma^\tau_{\mu\nu} u^\mu u^\nu}} = [\text{field strengths}] \times \overset{\text{force term}}{[\text{coupling factor}]}. \quad [9.10]$$

In a world with the electro-magnetic force $f_{\mu\nu}$ but without gravity, objects that carry no electrical charge ρ move on the straightest lines of the flat connection $\bar{\Gamma}^{\tau}_{\mu\nu}$; that is, the four-velocity vector is displaced parallel to itself, and there is no change in velocity $du^{\tau}/ds = 0$. A connection is 'flat' if a coordinate system can be introduced in which $\bar{\Gamma}^{\tau}_{\mu\nu} = 0$ everywhere, that is, $R^{\tau}_{\mu\nu\sigma}(\Gamma) = 0$. Charged objects (objects subject to the forces) are pulled off their inertial path by the electromagnetic field depending on their charge $\rho f^{\tau}_{v}u^{v}$. Thus, the velocity vector of charged particles is not transported parallel to itself along their path. The ratio of the inertial mass μ and the coupling factor ρ is different from particle to particle. Objects with the same mass and smaller electric charge and objects with the same charge and greater mass are less susceptible to electromagnetic fields. Thus, one is able to approximate the straightest paths by using objects of progressively greater masses or smaller charges. As a consequence, motions are divided into two classes: inertial, force-free motions represented by straightest worldlines determined by $\bar{\Gamma}^{\tau}_{\mu\nu}$ and motions caused by the action of forces. One can apply the same approach to the gravitational field. In this case, the field variable is the gravitational scalar potential Φ, and the coupling factor is the gravitational charge μ_{g}, that is, the gravitational mass. Thus, the force term in the equation of motion would include $-\mu_{g}\partial\Phi/\partial x_{v}$.

However, there is a key difference between gravitation and electromagnetism. The ratio of the gravitational 'charge' density μ_{g} to the inertial mass μ is the same for *all* particles. Thus, the coupling factor on the right-hand side of Eq. [9.10] and the inertial mass on the left-hand side cancel out. All bodies move in the same way in a gravitational field. As a consequence, one would not be able to approximate a force-free motion using objects of progressively greater inertial mass or smaller gravitational charge. In these circumstances, it becomes natural to change the standard of non-acceleration and incorporate the force term in Eq. [9.10] into the suitably defined connection $\Gamma^{\tau}_{\mu\nu}$. The latter can be written as the sum of the flat connection $\bar{\Gamma}^{\tau}_{\mu\nu}$ and mixed symmetric tensor of the third rank:

$$\Gamma^{\tau}_{\mu\nu} = \gamma^{\tau}_{\mu\nu} + \varphi^{\tau}_{\mu\nu}.$$

$$\gamma^{\tau}_{\mu\nu} = \bar{\Gamma}^{\tau}_{\mu\nu} \qquad\qquad \varphi^{\tau}_{\mu\nu} = -g_{\mu\nu}h^{\rho\sigma}\frac{\partial\Phi}{\partial x_{\sigma}} \qquad [9.11]$$

The sum of an affine connection and a tensor of the this type is again a symmetric connection. Thus $\Gamma^{\tau}_{\mu\nu}$ is a symmetric connection. Since the gravitational charge-to-mass ratio is equal for all particles, it can be set at = 1 and eliminated from Eq. [9.10]. The field variable Φ that appears in the force term in Eq. [9.10] can be then absorbed into the definition of the connection $\Gamma^{\tau}_{\mu\nu}$ Eq. [9.11], which becomes, in general, non-flat and dependent on Φ. In this way, one can get rid of the force term and transform Eq. [9.10] into a geodesic equation of the form

$$\frac{du^\tau}{ds} \Gamma^\tau_{\mu\nu} u^\mu u^\nu = 0. \qquad\qquad [9.12]$$

$\hookrightarrow$ non-flat affine connection

Because the equation of motions of gravitational test charges can be written in the form of Eq. [9.12], one can say that the gravitational force has been 'geometrized'. The planet does not follow its curved path because it is acted upon by the gravitational scalar field but because the affine connection $\Gamma^\tau_{\mu\nu}$ "leaves it, so to speak, no alternative path" (Reichenbach 1928, 295; tr. 257). The theory still admits a standard of non-acceleration, the motion on the straightest lines of $\Gamma^\tau_{\mu\nu}$. However, the latter is not fixed once and for all, since the numerical values of its components depend on the gravitational potential Φ. In Friedrichs' (1928) theory, $\Gamma^\tau_{\mu\nu}$ represent the inertial field, and $g_{\mu\nu}$ and $h^{\mu\nu}$ represent the metric. The inertial field $\Gamma^\tau_{\mu\nu}$ is dynamical and curved and determined by the distribution of matter, whereas the metrics $g_{\mu\nu}$ and $h^{\mu\nu}$ have fixed constant values. Thus, the metric does not uniquely fix the connection $\Gamma^\tau_{\mu\nu}$ but allows just enough leeway to incorporate the scalar potential Φ (see Havas 1964).

In this way, Friedrichs (1928) is able to demonstrate that, with some good will, it is possible to geometrize even Newton's scalar theory of gravity without changing its content. Indeed, in Newton's non-relativistic theory, the equivalence principle already suggests that gravitational phenomena are best incorporated into a non-flat affine connection $\Gamma^\tau_{\mu\nu}$ subject to certain dynamical field equations involving $R^\tau_{\mu\nu\sigma}(\Gamma)$. In hindsight, general relativity can be seen as the attempt to construct a theory of the same type but compatible with the metric structure of Minkowski spacetime. Instead of starting from the metric alone, as Einstein historically did, one could have obtained general relativity by starting with two separate structures, the operation of displacement $\Gamma^\tau_{\mu\nu}$, the metric $g_{\mu\nu}$, and their compatibility condition $K_{\mu\nu,\sigma}$ (Stachel 2007). At this point, one would have had two options:

(a) One can decide to keep the flat Minkowski spacetime, in which $g_{\mu\nu} = \bar{g}_{\mu\nu}$, with a suitable choice of coordinates and drop the unique compatibility condition between $g_{\mu\nu}$ and the non-flat affine connection $\Gamma^\tau_{\mu\nu}$. This means using what Reichenbach called an *unbalanced space*. Free-falling particles, that is, particles moving under the influence of the gravitational field alone, are indicators of $\Gamma^\tau_{\mu\nu}$, which is generally non-flat, that is, $R^\tau_{\mu\nu\sigma}(\Gamma) \neq 0$. However, free-falling rods and clocks do not reliably measure temporal and spatial intervals; that is, they are distorted by the gravitational field. Thus, one can maintain a flat Minkowski metric $R^\tau_{\mu\nu\sigma}(g) = 0$.[9]

(b) One can require the full compatibility between the connection $\Gamma^\tau_{\mu\nu}$ and the metric $g_{\mu\nu}$, that is, use a fully *balanced space* $K_{\mu\nu,\sigma} = 0$, and drop the requirement that $R^\tau_{\mu\nu\sigma}(g) = 0$. Freely falling rods and clocks reliably measure temporal and spatial intervals and determine the $g_{\mu\nu}$ of a geometry, which is generally non-flat $R^\tau_{\mu\nu\sigma}(g) \neq 0$. In a given coordinate system, the components of the $R^\tau_{\mu\nu\sigma}(\Gamma)$ as measured by free-falling particles agree with those of $R^\tau_{\mu\nu\sigma}(g)$ as measured by rods and clocks. Since $\Gamma^\tau_{\mu\nu}$ has become dynamic, so $g_{\mu\nu}$ must be. In analogy with an electric field that is the gradient of electric potentials, $g_{\mu\nu}$ can be said to provide the 'gravitational potential field' and the affine connection 'gravitational gradient field' (Reichenbach 1928, 271; tr. 236f).

Reichenbach's famous 'relativity of geometry' can be reformulated as the choice between (a) and (b) (1928, 271f; tr. 236f). It is always possible to save Minkowski spacetime by introducing the universal distorting effect of gravitation on all measuring instruments but at the expense of using an unbalanced space, in which the affine connection (the gravitational field) determining the motion of free-falling particles and the metric (the inertial field) determining the rods and clocks and light rays contradict each other. A 'distorted, but measurable' geometry measured by neutral test particles under the influence of the gravitational field would differ from the 'true but hidden' Minkowski geometry measured by ideal rods and clocks (Stachel 2007). However, the effect of this separation has no physically observable consequences. In Maxwell's electrodynamics, it is always possible to choose non-charged rods and clocks that are not accelerated by the electromagnetic field. However, in Einstein's theory, rods and clocks free-falling in a gravitational field are indistinguishable from rods and clocks at rest in an inertial frame.

Thus, the choice (a) is not matter of 'truth', but it is a matter of 'simplicity' (see Reichenbach 1924, §2). Free-falling particles, on the one hand, and free-falling rods and clocks, on the other hand, determine the *same geometry*. In particular, free-falling clocks traveling along a geodesic path also measure the affine parameter s in Eq. [9.12] along the path. As a result, the straightest lines defined by the parallel displacement of u^μ coincide with the lines of extremal lengths as measured by a clock. Admitting the full compatibility of $g_{\mu\nu}$ and $\Gamma^\tau_{\mu\nu}$ implies that, in a certain coordinate system, the components of $\Gamma^\tau_{\mu\nu}$ are numerically equal to the Christoffel symbols, as in Eq. [9.9]. The Riemann tensor would have two distinct interpretations, Weyl's 'direction curvature' $R^\tau_{\mu\nu\sigma}(\Gamma)$ and Riemann's curvature tensor $R^\tau_{\mu\nu\sigma}(g)$ interpreted as a generalization of the 'Gaussian' curvature. As Reichenbach pointed out, at first sight, it is natural to consider the choice of (b) a 'geometrization' of gravitation. As in Friedrichs' (1928) theory, in Einstein's theory, the affine connections and $R^\tau_{\mu\nu\sigma}(\Gamma)$ can be determined experimentally by observing the motion of free-falling particles. However, in Einstein's theory, the affine connections $\Gamma^\tau_{\mu\nu}$ can be expressed in terms of the metric $g_{\mu\nu}$ alone, and $R^\tau_{\mu\nu\sigma}(g)$ can be measured using rods and clocks. In this sense, free-falling particles measure the same numerical values of the components of $R^{\tau}_{\mu\nu\sigma}$ in a given coordinate system; that is, they agree on the same geometry. The gravitational field has become indistinguishable from the geometry of spacetime.

However, as we have seen, Reichenbach wanted to resist this conclusion. In Reichenbach's view, there is no need to use rods and clocks for the representation of the gravitational field. In principle, it is sufficient to recognize the gravitational field by the motion of free-falling particles, that is, by considering the operation of displacement. The use of free-falling particles as indicators of the gravitational field also suggests a geometrical interpretation, since the motion of free-falling particles can be represented as motion along geodesics. These are, at the same time, the timelike worldlines of extremal lengths as measured by a clock free-falling with the particle. Yet, Reichenbach claimed "this assertion goes beyond what is given by the motion of the mass points alone since it puts this motion into a relation with the geometrical behavior of the measuring instruments (measurement of length by ds^2)" (1928, 353; tr. [492]). We are not compelled to think of the relation between straightest and longest worldline at all times, and one can consider the inertial structure encoded in $\Gamma^{\tau}_{\mu\nu}$ without considering its relations with the chronometrical structure encoded in $g_{\mu\nu}$. The geometrical interpretation of gravitation is a 'visualization' (*Veranschaulichung*) of the full compatibility $K_{\mu\nu,\sigma} = 0$ between $g_{\mu\nu}$ and $\Gamma^{\tau}_{\mu\nu}$, but it is not a necessary representation of the gravitational field:

> The field of the force of gravitation affects the behavior of measuring instruments. Besides serving in their customary capacity of determining the geometry of space and time, they also serve, therefore, as indicators of the gravitational field. The geometrical interpretation of gravitation is consequently an expression of a real situation, namely, of the actual effect of gravitation on measuring rods and clocks. . . . The geometrical interpretation of gravitation is merely the *visual cloak* in which the factual assertion is dressed. It would be a mistake to *confuse the cloak with the body* it covers; *rather, we may infer the shape of the body from the shape of the cloak it wears.* After all, only the body is the object of interest in physics. . . . The new insight of Einstein consists merely in recognizing the fact that the well-known complex of relations concerning the motions of mass points is supplemented by their relations to the behavior of measuring instruments [rods and clocks].
>
> (Reichenbach 1928, 353–354; my emphasis; tr. [491–492])

The opposition between the body and the cloak that covers it is somewhat a more poetic formulation of the same message that Reichenbach had announced in the last chapter of the *Philosophie der Raum-Zeit-Lehre*. According to Reichenbach, one must attribute to the gravitational field a physical reality that is comparable to that of any other physical field. The peculiarity of gravitation with respect to other fields is that it is "the cause of *geometry itself*, not as the cause of the *disturbance* of geometrical relations" (Reichenbach 1928, 294; original emphasis; tr. 256). Non-gravitational fields (as the electromagnetic field) are *differential forces*. One 'defines' the geometrical measuring instruments (rods and clocks, light rays, force-free particles) as those that (up to a certain degree of approximation) can be shielded from action of the field (non-charged objects) and the dynamical ones (charged test particles), which react to the field depending on a coupling factor (charge). However, gravitation is a *universal force*: it cannot be neutralized or shielded. Thus, one cannot sort out the geometrical measuring instruments from the dynamical ones. Thus, it is more appropriate to *decide* to set universal forces equal to zero. The geometrical measuring instruments become at once indicators of the gravitational field. Nevertheless, the effect of the gravitational field on these instruments does not transform the gravitational field into geometry but rather deprives geometry of its independent status. The goal of the Appendix was to develop this intuition into a full-fledged argument. According to Reichenbach, one needs to separate the geometrical cloak (which can be chosen with some arbitrariness) from the physical body (the fundamental fact of universal coupling of gravitation with all other physical entities).

However, this was not conventional wisdom. "The great success, which Einstein had attained with his geometrical interpretation of gravitation, led Weyl to believe that similar success might be obtained from a geometrical interpretation of electricity" (Reichenbach 1928, 352; tr. [491]). Just after general relativity was accepted by the physics community, the search for a suitable geometrical cloak that could cover the naked body of the electromagnetic field began. To this end, one needed something analogous to the equivalence principle, a physical fact that relates the electrical field to the behavior of measuring instruments. "However, the fundamental fact which would correspond to the principle of equivalence is lacking" (Reichenbach 1928, 354; tr. [493]). Thus, physicists had to proceed more speculatively. At this point, the separation of operation of displacement from the metric acquired a central role. Weyl did not separate $\Gamma^{\tau}_{\mu\nu}$ and $g_{\mu\nu}$ merely for mathematical reasons; his goal was their physical application (Reichenbach 1928, 354; tr. [491]). Since $g_{\mu\nu}$ were already appropriated by the gravitational field, Weyl (1918a) constructed a more encompassing geometrical setting that contained some unassigned geometrical elements that he could ascribe to the electromagnetic field (see, e.g., Scholz 2008).

Weyl's strategy can be described as an attempt to keep a non-flat spacetime, as in general relativity, but weaken the compatibility condition between the metric $g_{\mu\nu}$ and the connection $\Gamma^{\tau}_{\nu\mu}$ by resorting to a 'displacement space' in which $K_{\mu\nu,\sigma} = \kappa_\sigma$. Weyl's space is a special case of

Reichenbach's general displacement space, since Weyl imposed the condition that $\Gamma^{\tau}_{\mu\nu}$ is symmetric. As we have mentioned, in Weyl geometry, there are two kinds of curvature, a direction curvature (*Richtungkrümmung*) $R^{\tau}_{\mu\nu\sigma}$ and a length curvature (*Streckenkrümmung*) $f_{\mu\nu}$. Weyl demonstrated that the length curvature is expressed by the curl of κ_{σ}. As is well known, in the four-dimensional representation of Maxwell electrodynamics, the electromagnetic tensor field is the curl of the electromagnetic four-potential vector. Thus it was natural to interpret κ_{σ} as the electromagnetic four-potential vector and its curl $f_{\mu\nu}$ as the electromagnetic tensor. The absence of the electromagnetic field $f_{\mu\nu} = 0$ is represented by a space in which the displacement of lengths is integrable. The absence of the gravitational field is represented by a space in which the transfer of directions is integrable, that is, where $R^{\tau}_{\mu\nu\sigma}(\Gamma) = 0$. Thus, only in a flat Minkowski spacetime, there is neither electromagnetism nor gravitation.

This geometrical representation of the electromagnetic field is still a rather formal analogy. Thus, it remains to be decided whether Weyl's geometrical apparatus describes the behavior of actual physical systems. In Reichenbach's parlance, it is necessary to introduce a coordinative definition of the operation of displacement. Weyl's geometry is a balanced space in which the comparison of lengths is defined, although not at a distance. Thus, it is natural to assume that the length of vectors can be measured with rods and clocks. Weyl uses rods and clocks as indicators of the gravitational field and, at the same time, indicators of the electromagnetic field. As long as a gravitational field exists alone, $f_{\mu\nu} = 0$, the geometry is Riemannian; the behavior of the measuring rods can be integrated; that is, they define a comparison of length independent of the path. As soon as an electrical field is added, however, $f_{\mu\nu} \neq 0$, and the integrability fails. The behavior of the measuring instruments is describable only in terms of the operation of displacement. Thus, "the Weylian space now constitutes the natural cloak for the field, which is composed of electricity and gravitation" (Reichenbach 1928, 354; tr. [494]).

Unfortunately, however, the theory does not agree with the physical facts. Even if the electromagnetic field is introduced, the behavior of rods and clocks is still integrable. This is confirmed by a large amount of experimental knowledge about spectral lines of atoms that are typically employed as clocks. Those spectral lines are always sharp, well-defined spectral lines. If atomic clocks changed their periods as a function of their spacetime paths, one would expect that atoms with different pasts would radiate different spectral lines (Reichenbach 1928, 355; tr. [494]).[10] Weyl might have also used the velocity vectors of freely moving mass points as a realization of the operation of displacement. If one assumes that charged particles move on geodesics of the Weyl connection (Eq. [9.8]), one runs into a further difficulty. The affine connection in Weyl's theory (Eq. [9.8]) and, therefore, the right-hand side of the geodesic equation depends on both $g_{\mu\nu}$ and κ_{σ}. As a consequence, uncharged particles will be affected by the electromagnetic four-vector potential.[11] This is, however, not the case. Thus, neither rods and clocks nor charged particles behave as predicted by Weyl's theory.

Thus, Weyl's displacement space is not suited to describe the behavior of rods and clocks and charged mass points in a combined electrical and gravitational field. "This means that we have found a cloak in which we can dress the new theory, but we do not have the body that this new cloak would fit" (Reichenbach 1928, 353; tr. [493]). What alternatives do we have at our disposal? According to Reichenbach, physicists had tried to "forgo . . . such a realization of the process of displacement" (Reichenbach 1928, 371; tr. [519]). Weyl (1921a) reformulated his theory by keeping the 'balanced' Weyl space, in which the dot product of vectors is defined but not preserved under parallel displacement; however, he rejected rods and clocks as indicators of the operation of displacement $\Gamma^{\tau}_{\mu\nu}$.[12] At around the same time, Eddington (1921) moved beyond Weyl and adopted an unbalanced space in which Riemannian geometry is maintained as true geometry of spacetime and a symmetric $\Gamma^{\tau}_{\mu\nu}$ is introduced without reference to the metric. Lengths, even at the same point in different directions, are not comparable. From a symmetric $\Gamma^{\tau}_{\mu\nu}$ alone, one can construct a Ricci tensor $R_{\mu\nu} = R^{\Gamma}_{\mu\nu\Gamma}$ that is generally non-symmetric:

$$R_{\mu\nu} = -\frac{\partial \Gamma^{\alpha}_{\mu\nu}}{\partial x_{\alpha}} + \Gamma^{\alpha}_{\mu\beta}\Gamma^{\beta}_{\nu\alpha} + \frac{\partial \Gamma^{\alpha}_{\mu\alpha}}{\partial x_{\nu}} - \Gamma^{\alpha}_{\mu\nu}\Gamma^{\beta}_{\alpha\beta}.$$

This tensor has an antisymmetric part as well as a symmetric part:

$$R_{\mu\nu} = G_{\mu\nu} + F_{\mu\nu}$$

It is natural to identify the antisymmetrical tensor $F_{\mu\nu}$ with the electromagnetic field $f_{\mu\nu}$ and the symmetrical tensor $G_{\mu\nu}$ with the metrical/gravitational field, after some rescaling, by setting $G_{\mu\nu} = \lambda g_{\mu\nu}$. This identification is, of course, based on a merely formal analogy. However, Einstein (1923a, 1923b, 1925a) became convinced that Eddington's purely affine approach might constitute a good starting point to 'guess' the right field equations (see Sauer 2014). Moreover, he hoped that by integrating the field equations, one could obtain solutions corresponding to the positive and negative electron. If these results were achieved, then Eddington's/Einstein's choice of the symmetric $\Gamma^{\tau}_{\mu\nu}$ as the fundamental geometrical structure of spacetime would be justified, so to speak, *post facto*. The latter would be the case, even if the operation of displacement has no physical meaning in itself. Only the theory as a whole (geometry plus physics) can be compared with experience (Einstein 1921, 1924, 1926).

Reichenbach, like many others (e.g., Pauli 1926), was pessimistic about the feasibility of this formal approach. According to Reichenbach, in a

good theory, one must provide a physical interpretation of the operation of displacement *ex ante* before one establishes the field equations,
just as in general relativity, the metric was interpreted in terms of rods
and clocks behavior from the outset. Reichenbach, so to speak, wanted
to combine the best of both worlds. Contrary to Eddington, he considered the adoption of a balanced space, such as Weyl geometry, important;
however, at the same time, contrary to Weyl, he did not want to forgo a
coordinative definition of the operation of displacement. Reichenbach's
reasoning was roughly the following. If rods and clocks are indicators of
the metric $g_{\mu\nu}$, then the only 'balanced space' we would still have at our
disposal is the general metrical space. However, this does not imply that
we have to adopt a Riemannian geometry. Indeed, one has still the option
of dropping the symmetry of the connection $\Gamma^{\tau}_{\mu\nu} \neq \Gamma^{\tau}_{\nu\mu}$ and obtaining a
non-Riemannian geometry with additional degrees of freedom. In this
way, Reichenbach believed it to be possible to define an operation of displacement that contains the effect of the electrical field but that, on the
other hand, does not contradict the metric. "The geometrical interpretation of electricity would then be expressed by the special kind of displacement of direction, but no longer by an effect upon the comparison of
length" (Reichenbach 1928, 357; tr. [498]).

9.5. An Example of a Geometrical Interpretation of Electricity

To introduce an indicator for the process of displacement $\Gamma^{\tau}_{\mu\nu}$, one must
select a physical phenomenon in which the gravitational and electric
fields together produce a 'geometrical effect'. The gravitational and electromagnetic fields together determine the motion of particles. Thus, a
natural, but still arbitrary, choice is to use the motion of charged and
uncharged particles as indicators of the displacement. $g_{\mu\nu}$ are measured
with rods and clocks, and the velocity four-vector u^{ν} of mass points
becomes the physical realization of the displacement $\Gamma^{\tau}_{\mu\nu}$. The general
relativistic equation of the motion of charged particles is, therefore, the
starting point of Reichenbach's investigation:

9.13

$$\mu \frac{du^{\tau}}{ds} = -\left\{\begin{matrix}\mu\nu\\\tau\end{matrix}\right\} u^{\mu} u^{\nu} - \rho f^{\tau}_{\nu} u^{\nu}.$$

charge →

→ $g^{\mu\nu} f^{\tau}_{\nu} = f_{\mu\nu}$ electromagnetic field strengths

Eq. [9.13] is a force equation of the type of Eq. [9.12]. Reichenbach's
goal was to rewrite Eq. [9.13] in the form of a geodesic equation like
Eq. [9.12], in which $\Gamma^{\tau}_{\mu\nu}$ will not be equal to the Christoffel symbols.

Not dissimilarly to Friedrichs' (1928) theory, the idea was to get rid of the force term by absorbing it in the definition of $\Gamma^\tau_{\mu\nu}$. In this way, both charged and uncharged particles would not experience acceleration under the influence of the combined gravitational/electromagnetic field. Their velocity-vector u^ν would be carried along by parallel displacement according to a suitably defined non-Riemannian connection $\Gamma^\tau_{\mu\nu}$. The four-velocity is not the velocity through space, which can of course take on different magnitude, but a velocity through spacetime which is fixed (up to a constant). Thus, the length l of the four-vector velocity u^ν is given by

$$l^2 = g_{\mu\nu} u^\mu u^\nu = 1 \quad \text{(by a suitable choice of units).} ^{13} \quad [9.14]$$

Thus, the length l of this velocity vector must remain unchanged under parallel transport, $d(l^2) = 0$. As we have seen, if one does not require the connection to be symmetric, then one can work in a metrical space that is not identical to the Riemannian space. In this space, the straightest lines are not identical to the shortest ones (see Misner, Thorne and Wheeler 1973, 248–251). Reichenbach exploited this fact to define an operation of displacement that expresses the effect of both the gravitational and electromagnetic fields. Charged mass points move (that is, their four-vector velocity is parallel-transported) along the straightest lines, and uncharged particles move on the straightest lines that are at the same time the shortest ones (or rather, the timelike worldlines of extremal length). Since a more philological account of Reichenbach's theory and a comparison with the manuscript he had sent to Einstein have been provided elsewhere (Giovanelli 2016), I will introduce in the following a slightly modified and simplified account of Reichenbach's geometrization of the electromagnetic field.

First Version of the Theory Following the model of Eddington's (1921) theory, Reichenbach first introduces the fundamental tensor $G_{\mu\nu}$:

$$G_{\mu\nu} = \overset{\displaystyle \overset{\longrightarrow \text{ electromagnetic field}}{\big\lceil}}{g_{\mu\nu}} + \underset{\underset{\text{gravitational field} \longleftarrow}{\big\rfloor}}{f_{\mu\nu}}. \qquad [9.15]$$

Just like Eddington (1921), Reichenbach demonstrates how this tensor can be decomposed into a symmetric part $g_{\mu\nu}$ and an anti-symmetric part $f_{\mu\nu}$ that, as one might expect, are identified with the gravitational/metrical field and the electromagnetic field, respectively. The metric can be defined as $ds^2 = G_{\mu\nu} dx_\mu dx_\nu = g_{\mu\nu} dx_\mu dx_\nu$. In the absence of the electromagnetic field ($f_{\mu\nu} = 0$) (the more $G_{\mu\nu}$ is nearly approximate to $g_{\mu\nu}$), ds is measured by using rods and clocks. Thus, rods and clocks are not indicators of $f_{\mu\nu}$, which is measured by the motion of charged test particles. The two fields are governed by Einstein and Maxwell's equations, which are léft unchanged. Reichenbach's aim was only to modify the equations of motions of particles in both fields without considering the relation of such fields with their sources.

The second step is to introduce the displacement $\Gamma^{\tau}_{\mu\nu}$ that depends on the fundamental fields $g_{\mu\nu}$ and $f_{\mu\nu}$:

$$\Gamma^{\tau}_{\mu\nu} = \gamma^{\tau}_{\mu\nu} + \varphi^{\tau}_{\mu\nu} \qquad [9.16]$$

$$\gamma^{\tau}_{\mu\nu} = -\begin{Bmatrix} \mu\nu \\ \tau \end{Bmatrix} \quad \varphi^{\tau}_{\mu\nu} = -kf^{\tau}_{\nu}u_{\mu} \qquad [9.17]$$

where $k = \rho/\mu$. Without pointing it out explicitly, Reichenbach relies on the fact that a non-symmetric displacement is always the sum of symmetric displacement and a skew symmetric tensor with two lower indices (see Schouten 1924, 851). $\gamma^{\tau}_{\mu\nu}$ is a symmetric connection that is set equal to the negative of the Christoffel symbols of the second kind, which, in turn, are functions of $g_{\mu\nu}$ and their first-order partial derivatives. $f^{\tau}_{\nu}u_{\mu}$ is the product an mixed anti-symmetrical tensor of rank two and a covariant vector (a tensor of first rank). The direct product of two tensors (multiplying components from the two tensors together, pair by pair) increases the rank of the tensor by the sum of the ranks of each tensor, keeping the character of the indices. Thus, $\varphi^{\tau}_{\nu\mu}$ is a mixed tensor of third rank with lower indices.

In this formulation, the reason for the definitions of Eq. [9.17] is immediately apparent. Using Eq. [9.17], Reichenbach can rewrite Eq. 9.13 so that the force term is absorbed into a suitably defined $\Gamma^{\tau}_{\mu\nu}$:

$$\frac{du^{\tau}}{ds} = \Gamma^{\tau}_{\mu\nu}u^{\mu}u^{\nu}.\qquad [9.18]$$

$$\longrightarrow \gamma^{\tau}_{\mu\nu} + \varphi^{\tau}_{\mu\nu}$$

According to Eq. [9.16], this equation is equivalent to the following one:

$$\frac{du^{\tau}}{ds} = \gamma^{\tau}_{\mu\nu}u^{\mu}u^{\nu} + \varphi^{\tau}_{\mu\nu}u^{\mu}u^{\nu}.\qquad [9.19]$$

The three-index symbol $\gamma^{\tau}_{\mu\nu}$ is defined as the Christoffel symbol of the second kind; thus, the first summand of Eq. [9.19] is simply the first summand of the left-hand side of the general relativistic force equation Eq. [9.13]. Substituting the definition $\varphi^{\tau}_{\mu\nu}$ in the second summand of Eq. [9.19], one obtains $-kf^{\tau}_{\nu}u_{\mu}u^{\mu}u^{\nu}$. Since, according to the definition of the four-velocity Eq. [9.14], the dot product $u_{\mu}u^{\mu} = 1$, we have $-kf^{\tau}_{\nu}u^{\nu}$, where $k = \rho/\mu$. Thus, after multiplying both sides of the equation by μ, the final result is nothing but Eq. [9.13], from which Reichenbach had started.

By defining the displacement space $\Gamma^{\tau}_{\mu\nu}$ using Eq. [9.16] and Eq. [9.17], Reichenbach was able to dress the physical fact expressed by Eq. [9.13] in the geometrical cloak of Eq. [9.18]. Just like Eq. [9.13] in general relativity, Eq. [9.18] in Reichenbach's theory describes the motion of charged and uncharged test particles under the influence of the combined gravitational and electromagnetic fields. However, now the difference in the behavior of charged and uncharged particles can be expressed in terms of geometrical differences. The velocity vectors of charged particles of mass are parallel transported along the straightest lines defined by $\Gamma^\tau_{\mu\nu}$ and uncharged particles on shortest lines. When the charge ρ of these particles is zero and the tensorial component of $\varphi^\tau_{\mu\nu}$ vanishes, $\Gamma^\tau_{\mu\nu} = \gamma^\tau_{\mu\nu}$; that is, it reduces to the Christoffel symbols. Thus, the straightest lines coincide with the shortest ones, as in Einstein's theory of gravitation.

Reichenbach conceded that the theory has a manifest problem.[14] Eq. [9.13] is supposed to be valid for particles of *arbitrary mass* and *arbitrary charge*, that is, of arbitrary k. Thus, particles of the same charge-to-mass ratio "will engender [their] own displacement geometry" (Reichenbach 1928, 362; tr. [506]) and run along their 'own' straightest lines defined by it (Reichenbach 1928, 363; tr. [508]). Indeed, the values of the components of $\Gamma^\tau_{\mu\nu}$ depend on the values of ρ/μ. Thus, for every different value of k, the numerical values of the components of $\Gamma^\tau_{\mu\nu}$ would be different; thus, $R^\tau_{\mu\nu\sigma}(\Gamma)$ would also be different. However, Reichenbach recognized that this approach is "questionable" (1928, 363; tr. [508]), since the existence of a field should not depend on the properties of the test particles. Indeed, in Friedrichs' (1928) geometrization of Newtonian gravity, the coupling factor and mass are dropped because they are equal; only the field variable Φ appears in the tensorial part of the connection.

"To avoid this peculiarity of our formulation" (Reichenbach 1928, 367; tr. [506]), Reichenbach suggested an alternative version of the theory. The definition of the fundamental tensor Eq. [9.15] remains unchanged, but the definition of the connection Eq. [9.16] is modified by setting $k = 1$ in the tensorial part $\varphi^\tau_{\mu\nu}$ of Eq. [9.17]. Instead of Eq. [9.17], Reichenbach introduces the following definition of the displacement:

$$\Gamma^\tau_{\mu\nu} = \gamma^\tau_{\mu\nu} + \varphi^\tau_{\mu\nu}.$$

$$\gamma^\tau_{\mu\nu} = -\left\{{\mu\nu \atop \tau}\right\} \quad \varphi^\tau_{\mu\nu} = -f^\tau_\nu u_\mu. \qquad [9.20]$$

The tensorial part of the displacement is now so defined that it depends only on the field variable and the electromagnetic field $f_{\mu\nu}$ and not on any property of the test particles, that is, the coupling factor ρ and the inertial mass (since k is set $= 1$). The equations of motion now apply only to *unit mass* particles of a certain *unit charge* (Reichenbach 1928, 363f.; tr. [508ff.]). Under the influence of the electromagnetic field, a class of charged particles with an arbitrarily chosen charge-to-mass ratio move on the straightest lines, and uncharged particles always move on the shortest lines. The norm $k = 1$ can be chosen arbitrarily. However, it is advantageous to choose the natural unit represented by the fixed ratio e/m between the charge and mass of the electron. Since there are two types of electrons, positive (the nucleus of the hydrogen atom) and negative, with different charge-to-mass ratio, the natural choice of the geometry is not unique. There are two 'natural' geometries, that is, two connections $\Gamma^r_{\mu\nu}$ with different curvatures depending on the choice of k.

9.6. The Epistemological Meaning of Reichenbach's Geometrical Interpretation of Electricity

Somewhat surprisingly, Reichenbach regarded both versions of his theory as successful examples of a 'geometrization' of the electromagnetic field. "In the preceding section", he writes, "we have carried through a complete geometrical interpretation of electricity" (1928, 365; tr. [510]). What is the physical significance of this theory? Reichenbach concedes that the theory does not add anything to the physical content of Einstein's and Maxwell's theories. Maxwell's and Einstein's field equations remain unchanged, and Eq. [9.18] is nothing but a geometrical reformulation of Eq. [9.13]. With a rather cheap trick, the force term in Eq. [9.13] has been absorbed in the definition of the connection and thus disappears from Eq. [9.18]. *The force equation of the type Eq. [9.10] is transformed into a geodesic equation of the type Eq. [9.12]. In this sense, the electromagnetic field has been 'geometrized' like the gravitational field in Einstein's theory.* Elaborating on a distinction introduced by Eddington (1925), Reichenbach points out that most readers would conclude that his geometrical interpretation of electricity is merely a *graphical representation* of the combined electromagnetic/gravitational field and not a proper *geometrical interpretation* (1928, §15). Yet, according to Reichenbach, this conclusion is due to a misconception of the nature of a geometrical interpretation.

Weyl's theory in its first form had the ambition to be a proper *geometrical interpretation* of the electromagnetism, just like general relativity was a geometrical interpretation of gravitational field; like the latter, it was a theory concerning the behavior of rods and clocks. The scale factor κ_σ was supposed to determine a change in the rate of ticking of clocks, which could be observed empirically. However, the theory was contradicted by experience. Real rods and clocks do not behave as predicted by the theory. Thus, Weyl forwent the coordinative definition of the transport of lengths in terms of rods and clocks. According to Eddington, Weyl's theory, in this second form, should be regarded a mere *graphical representation* (like a pressure-temperature diagram), which does not aim to describe the actual structure of spacetime but simply summarizes empirically well-confirmed laws in a unitary framework (see Ryckman 2005, §8.3). The reason the scale four-vector κ_σ is identified with the electromagnetic four-potential is that they both behave formally in the same way (see Section 9.4). In Reichenbach's view, his *"geometrical inter-pretation of electricity is not a* graphical representation, *but a genuine geometrical interpretation"* (1928, 365; tr. [512]). Like the first version of Weyl's theory, both the metric and the displacement are coordinated with the behavior of existing physical systems (rods and clocks and motion of charged and uncharged particles). However, the theory is not in con-flict with experience like the first version of Weyl's theory. In this sense, Reichenbach considers his geometrical interpretation of the electromag-netic field no 'worse' than the geometrical interpretation of gravitation provided by general relativity (Reichenbach 1928, 366; tr. [512]).

Reichenbach concedes that an obvious objection can be raised against this conclusion. At first sight, the geometrical interpretation of electric-ity lacks a basic physical fact analogous to the equivalence principle, that is, the equality of inertial and gravitational mass. According to this principle, the trajectory of an uncharged particle in a gravitational field depends only on its initial position and velocity. As Reichenbach has explained a few pages earlier, if the gravitational and inertial masses were not equal, freely falling particles would not travel on the same geodesics of the same non-flat spacetime geometry; *"different geometries would result in various materials of the mass points"* (Reichenbach 1928, 293; my emphasis; tr. 256). To keep all particles with different gravitational charge-to-mass ratio, m_g/m, one would need to introduce displacement geometries of different curvatures depending on structure and material of test particles. At first sight, this is precisely the difference between the electromagnetic field and the gravitational field. Indeed, charged particles of different charge-to-mass ratio e/m, starting from the same initial con-ditions, cannot travel on the same paths of the same connection with the same curvature. Thus, "different substances would supply us in this case with different geometries" (Reichenbach 1928, 367; tr. [513]). In other terms, the electromagnetic field is not a universal force like gravitation. However, Reichenbach believed that the geometrical setting he had intro-duced in the Appendix "was chosen wide enough to express, within a *single geometry*, the corresponding difference in the behavior of charged and uncharged unit mass points" (1928, 367; my emphasis; tr. [513]).

Reichenbach admits that in the first version of his theory, $\varphi_{\mu\nu}^{\tau} = -kf_v^{\tau}u_{\mu}$, each charged particle moves on the geodesics of its own connection depending on its charge-to-mass ratio e/m; however, this is not the case in the second version of Reichenbach's theory, $\varphi_{\mu\nu}^{\tau} = -f_v^{\tau}u_{\mu}$, where the components of the affine connection do not depend on the property of the particles. Thus, Reichenbach concludes, "[e]xpression [9.20] which prescribes only the unit ratio of charge and mass is therefore more advan-tageous than [9.17]" (1928, 367; tr. [513]). It is true that, in this way, one obtains "different natural geometries for the positive and the nega-tive charge" (Reichenbach 1928, 367; tr. [513]). However, Reichenbach insists that it is "of extraordinary significance that this procedure yields only two natural geometries" (1928, 367; tr. [513]). It is a physical fact that in electromagnetism, there is a difference between positive and negative charge, which produces a geometrical asymmetry; this asymmetry is absent in the case of gravitation where there is only one gravitational charge and thus only one natural geometry. In this sense, Reichenbach argues that this bifurcation may "*be taken as an analogy to Einstein's principle of equivalence*" (1928, 367; my emphasis; tr. [514]). According to Reichenbach, after all, "the equality of gravitational and inertial mass originally represents only a proportionality" (1928, 367; tr. [513]). Indeed, also in the case of general relativity, we set the gravitational charge-to-mass ratio $m_g/m = 1$, just as in Reichenbach's theory, one normalizes the electric charge-to-mass ratio of the electron as $e/m = 1$.

   This conclusion is, as we shall see, puzzling, to say the least. However, it is on its basis that Reichenbach tries to convince his readers that the geometrical cloak he has tailored fits the physical body of the combined electromagnetic/gravitational field well. This conviction is, unfortunately, the keystone of Reichenbach's argument. Einstein had found in Riemannian geometry a good cloak that fits the body, the gravitational field. This cloak was also particularly suitable to reveal something new about the form of body that it covered, that is, about properties of the gravitational field that were previously unknown (say, the gravitational, redshift). Reichenbach's non-Riemannian geometry is an equally good cloak. However, that cloak does not reveal anything new about the underlying body, the combined gravitational/electromagnetic field. Thus, covering a field with a geometrical cloak might be the manifestation of great mathematical ingenuity, but it is no guarantee of physical insight:

> Is it not true, though, that the geometrical interpretation of gravitation has brought about an advance of physics? It has *brought* about it, yes, but it is not *identical* with this progress. It has led, in its effects, to a physical discovery, but it in itself *is* not this discovery. . . . We must therefore recognize that the geometrical interpretation of gravitation has attained its important position in the historical development of science, because it has led to new physical insights. The geometrical interpretation itself is merely a formulation, a visualization of these new insights. What we have attained with our geometrical interpretation of electricity is an analogous formulation of physical insights regarding electricity, but these insights are not physically new. As long as the geometrical interpretation of electricity does not act as a heuristic principle, its sole value will lie in the visualization it provides.
>
> (Reichenbach 1928, 368; tr. [516])

Einstein's geometrical interpretation of gravitation has led to a new set of field equations and equations of motion. The latter replaced the old

Newtonian theory of gravitation and led to new confirmed predictions. Reichenbach's theory has also provided a complete geometrical interpretation of the two fields; however, Reichenbach continued to say that such "geometrical interpretation of electricity constitutes no more advance in physical knowledge" (1928, 368; tr. [514]). Indeed, the theory reproduces the empirical content of Maxwell's or Einstein's theory, the field equations are the same, and equations of motion have received only a cosmetic redesign. "Rewriting the theory in this fashion would tell us nothing about the reality that we did not know before" (Reichenbach 1928, 368; tr. [514]). The recognition of this difference was, according to Reichenbach, an important epistemological achievement. Relativists, especially under the influence of Weyl, have always "defended the idea that the geometrical interpretation of electricity constitutes something which is physically essential" (Reichenbach 1928, 368; tr. [515]). Reichenbach could demonstrate that this is not the case. He had provided a good geometrical reinterpretation of already well-known physical laws that does not bring anything physically new. "This epistemological insight might very well be helpful to the physicist, by showing him the limitations of his method and making it easier for him to free himself from the enchantment of a unified field theory" (Reichenbach 1928, 368; tr. [519]).

9.7. Conclusion

At the beginning of 1928, Einstein published a brief review of the *Philosophie der Raum-Zeit-Lehre* (Einstein 1928b) in the *Deutsche Literaturzeitung*. Einstein clearly recognized the importance of the Appendix, part of which he had read as a draft two years earlier (Giovanelli 2016). Einstein explicitly shared Reichenbach's skepticism toward the rhetorics of the 'geometrization of physics' that were widespread in technical and philosophical literature: "In this chapter, [the Appendix], *just as in the preceding* – in my opinion quite rightly – it is argued that the claim that general relativity is an attempt to *reduce physics to geometry* is unfounded" (Einstein 1928b, 20; my emphasis). Thus, Einstein explicitly identified the question of 'geometrization' as the fundamental issue of the last chapter on general relativity and of the Appendix in Reichenbach's book. At that time, this matter was close to Einstein's heart (Lehmkuhl 2014), as is testified by Einstein's contemporary review (1928a) of a book by French philosopher Emile Meyerson (1925). There, Einstein criticizes Meyerson for having given too much weight to the idea that general relativity has reduced physics to geometry (Giovanelli 2018).

However, Einstein's endorsement of Reichenbach's criticism of the geometrization parlance should be taken with a grain of salt. As Einstein had tried to explain to him in private correspondence, Reichenbach was right for the wrong reasons (Giovanelli 2016). Indeed, Reichenbach laid down a surprisingly bad argument to make a good point. The reason we speak of a 'geometrical' interpretation of gravitation is the universality of free fall. Because of the weak equivalence principle, which establishes the equality of gravitational charge and inertial mass, it is possible to ensure that all particles of whatever gravitational charge-to-mass ratio m_g/m, given the same initial conditions, follow the same trajectory – a geodesic of a generally non-flat Riemannian spacetime – under the influence of the sole gravitational field (Reichenbach 1928, 293; tr. 256). In this sense, one can say that geometry replaces the concept of gravitational force, and the trajectories of free-falling particles are determined not by a force equation but by a geodesic equation (Reichenbach 1928, 293; tr. 256). If the weak equivalence principle did not hold, the geometrization of gravitation would fail. Test particles with different gravitational charge-to-mass ratio m_g/m would all move on geodesics only if one introduced connections with different curvatures of each type of particles. However, the properties of a real field cannot depend on the properties of the test particles. Thus, one should conclude that without the weak equivalence principle, the geometrical interpretation of the field breaks down.

A comparison with Friedrichs' (1928) geometrization of Newtonian gravity is instructive (see section 9.4). Such a geometrization works because the inertial mass and the coupling factor cancel out, and only the scalar potential Φ enters in the tensorial part $\varphi^\tau_{\mu\nu} = -g_{\mu\nu}h^{\rho\sigma}\partial\Phi/\partial x_\sigma$ of the connection in Eq. [9.11]. It does not seem to be possible to apply this stratagem to other types of interaction, where the ratio of the inertial mass to the coupling factor is not the same for all bodies (Friedman 1983, 197). Reichenbach's theory proves exactly this point, and it is puzzling that Reichenbach tries to defend the very opposite claim. In the first version of his theory, indeed, all particles of any electrical charge-to-mass ratio e/m travel on geodesics but of connections of *different* curvature. In fact, the components of Reichenbach's connection depend on the charge-to-mass ratio, which is encoded in the tensorial part of the connection $\varphi^\tau_{\mu\nu} = -kf^\tau_v u_\mu$, where $k = e/m$. In the second version of the theory $k = 1$, and therefore the ratio e/m does not appear in the tensorial part of the connection $\varphi^\tau_{\mu\nu} = -f^\tau_v u_\mu$. However, the cure is worse than the disease.

If the charge-to-mass ratio e/m cannot be absorbed in the definition of the connection, it must appear explicitly in the equations of motion. This implies that only one type of particle travels on geodesics, that is, particles with a certain fixed and arbitrarily chosen electrical charge-to-mass ratio e/m, say, electrons. Reichenbach claims that this is an analogon of the equivalence principle, because in both cases, the charge-to-mass ratio is set equal to 1. However, this claim is, to say the least, preposterous. The basis of the weak equivalence principle is the experimental fact that *all* particles have the same gravitational charge-to-mass ratio, which therefore can be set as = 1. The equivalence principle does not claim that particles have in general different charge-to-mass ratio, and we can arbitrarily set the charge-to-mass ratio of *one* class of particles = 1.

Thus, Reichenbach's theory proves the opposite of what Reichenbach wanted to prove. In a four-dimensional setting, it is *impossible* to make *all* charged and uncharged particles move on the geodesics of the *same* connection with the *same* curvature. Reichenbach had found a geometrical cloak, but the cloak does not fit the body, the combined electromagnetic/gravitational field, however one tries to stretch it. What Reichenbach's theory demonstrates is that the electromagnetic field *cannot* be geometrized, at least if one takes Reichenbach's own definition of geometrization. Unfortunately, Reichenbach did not seem to have ever questioned the content of the Appendix (Reichenbach 1929b). I was not able to find a convincing explanation for Reichenbach's resounding blunder. Reichenbach's Appendix, in spite of the admirable knowledge of the differential geometry of his time, is ultimately somewhat underwhelming for the philosopher of physics. If Reichenbach had read Friedrichs' (1928) paper (see section 9.4), he might have found there a much better example of a geometrization of a physical theory that does not change its physical content. At the same time, he might have realized that for non-gravitational interactions, the trick of absorbing the force term into the definition of the connection inevitably fails. Without the weak equivalence principle, the geometrization of a physical field is a non-starter (cf., however, Droz-Vincent 1967).

Nevertheless, the Appendix is a significant document for the historian of the philosophy of science. The Appendix gives more weight to a central issue of the *Philosophie der Raum-Zeit-Lehre*, the critique of the 'geometrization' program, which, in spite of being explicitly emphasized by Einstein (1928b) in his review, has been completely neglected in Reichenbach scholarship – including Coffa's (1979) paper, the only one dedicated to the Appendix. Reichenbach's interpretation of general relativity appears in a very different light once he has presented the theory as based on two separate but compatible structures, the displacement and the metric, the inertial and the chronogeometrical structure. The peculiarity of general relativity, in Reichenbach's view, is the fact that the metric and the connection are fully compatible. Physically, this means that moving particles, and rods and clocks, when under the influence of the sole gravitational field, define a *single geometry*; they measure the same values of the components of the curvature $R^\tau_{\mu\nu\sigma}$. Nevertheless, Reichenbach invites his readers not to linger to admire this fancy geometrical cloak. What is essential in Einstein's theory is what the cloak covers, the universal coupling of gravitation with all other fields. In Reichenbach's view, only a yet-to-be-developed theory of matter (Reichenbach 1928, 233; tr. 201) can ensure that all coupling constants are indeed constant. If this were not the case, different material devices, made up of different non-gravitational fields and particles, would yield different geometrical results. "The theory of relativity did not convert a part of physics into geometry. On the contrary, even more physics is involved in geometry, than was suggested by the empirical theory of physical geometry" (Reichenbach 1928, 295; tr. 256).

9.8. Appendix: Reichenbach's Classification of Geometries

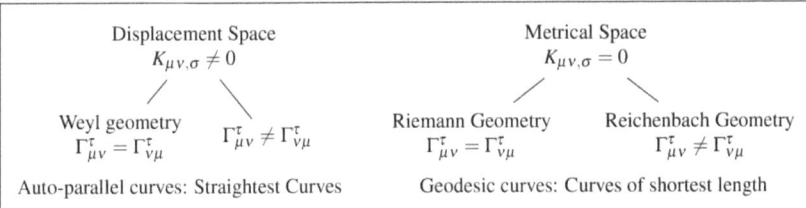

$$K_{\mu\nu,\sigma} = \nabla_\sigma g_{\mu\nu}$$
metricity tensor

$$J_{\mu\nu,\tau} = \tfrac{1}{2}\Gamma^\tau_{\mu\nu} - \Gamma^\tau_{\nu\mu}$$
asymmetry tensor

$$K_{\mu\nu,\sigma} \neq 0$$
Dot products and lengths of vectors not preserved

$$J_{\mu\nu,\tau} \neq 0$$
Infinitesimal parallelograms do not exist

Displacement Space
$$K_{\mu\nu,\sigma} \neq 0$$

Metrical Space
$$K_{\mu\nu,\sigma} = 0$$

Weyl geometry
$$\Gamma^\tau_{\mu\nu} = \Gamma^\tau_{\nu\mu}$$

$$\Gamma^\tau_{\mu\nu} \neq \Gamma^\tau_{\nu\mu}$$

Riemann Geometry
$$\Gamma^\tau_{\mu\nu} = \Gamma^\tau_{\nu\mu}$$

Reichenbach Geometry
$$\Gamma^\tau_{\mu\nu} \neq \Gamma^\tau_{\nu\mu}$$

Auto-parallel curves: Straightest Curves   Geodesic curves: Curves of shortest length

---

Reichenbach's Geometry

$$K_{\mu\nu,\sigma} = 0 \qquad\qquad J_{\mu\nu,\tau} \neq 0$$

Auto-parallel Curves $\quad \neq \quad$ Geodesic Curves

Abbreviations

CPAE   Albert Einstein (1987–). *The collected papers of Albert Einstein.* Ed. by John Stachel et al. 15 vols. Princeton: Princeton University Press, 1987– .

HR   *Archives of Scientific Philosophy* (1891–1953). *The Hans Reichenbach Papers.* 1891–1953.

Notes

1. Henceforth, the English translations of *Philosophie der Raum-Zeit-Lehre* are taken from Reichenbach (1958); translations of the Appendix are taken from HR, 041–2101. In the latter case, page numbers are enclosed in square brackets.
2. This geometrization is known as Newton-Cartan theory, since it was developed independently by Cartan (1923, 1924); see Malament (2012).
3. The difference between a symmetric connection and a non-symmetric one $\Gamma^\tau_{\mu\nu} - \Gamma^\tau_{\nu\mu}$ is a tensor of third rank $J_{\mu\nu,\tau} = \dfrac{1}{2}\Gamma^\tau_{\mu\nu} - \Gamma^\tau_{\nu\mu} = 0$. This tensor is called an asymmetry tensor or torsion. However, Reichenbach did not introduce this notation (see section 9.8).
4. As Reichenbach rightly noticed, "writing of a coordinate differential with a lower index is a mistake" (Reichenbach 1928, 348; tr. [485; fn.]), since coordinate differentials are the prototype of contravariant vectors. However, it is standard in physical literature.

5. The non-metricity tensor, in modern parlance.
6. For this reason, to avoid confusion, it is important to emphasize that in Reichenbach's parlance, the general displacement space is *not* the affine space. It is a semi-metric space.
7. Also called the Levi-Civita connection.
8. For a similar classification, see also Infeld (1928a, 1928b).
9. A theory of this type might look like the bimetric theory of Rosen (1940a, 1940b).
10. This is, of course, the celebrated objection against Weyl's theory Einstein's (1918). See Ryckman (2005) for more details, and also see Giovanelli (2014).
11. This is a less famous objection that Reichenbach raised in correspondence with Weyl.
12. The idea that there are two versions of Weyl's theory was suggested by Pauli (1921). Cf. also Weyl (1921b) and Reichenbach (1922, 367–368).
13. In geometrized units.
14. It is possible that Einstein pointed out this problem to him in 1926. See Giovanelli (2016).

References

Cartan, É. (1923), 'Sur les variétés à connexion affine et la théorie de la relativité généralisée. Premiére partie', *Annales scientifiques de l'École Normale Supérieure*, 40: 325–412.
——— (1924), 'Sur les variétés à connexion affine, et la théorie de la relativité généralisée. Premiére partie, Suite', *Annales scientifiques de l'École Normale Supérieure*, 41: 1–25.
Coffa, A. J. (1979), 'Elective Affinities: Weyl and Reichenbach', in W. C. Salmon (ed.), *Hans Reichenbach, Logical Empiricist*. Dordrecht and Boston: Reidel, pp. 267–304.
Droz-Vincent, P. (1967), 'Electromagnetism and Geodesics', *Il Nuovo Cimento*, 51: 555–556.
Eddington, A. S. (1921), 'A Generalization of Weyl's Theory of the Electromagnetic and Gravitation Fields', *Proceedings of the Royal Society London*, 99: 104–121.
——— (1923), *The Mathematical Theory of Relativity*. Cambridge: Cambridge University Press.
——— (1925), *Relativitätstheorie in mathematischer Behandlung*. Mit einem Anhang 'Eddingtons Theorie und Hamiltonsches Prinzip' von Albert Einstein. Translated by A. Ostrowski and H. Schmidt. Berlin: Springer. German translation of Eddington 1923.
Einstein, A. (1918), 'Review of Weyl, Raum – Zeit – Materie', *Die Naturwissenschaften*, 6: 373. Repr. in CPAE, Vol. 7, Doc. 10.
——— (1921), *Geometrie und Erfahrung*. Erweiterte Fassung des Festvortrages gehalten an der Preussischen Akademie der Wissenschaften zu Berlin am 27. January 1921. Berlin: Springer. Repr. in CPAE, Vol. 7, Doc. 52.
——— (1923a), 'The Theory of the Affine Field', *Nature*, 2812: 448–449. Repr. in CPAE, Vol. 14, Doc. 123.
——— (1923b), 'Zur allgemeinen Relativitätstheorie', *Sitzungsberichte der Preußischen Akademie der Wissenschaften, Physikalisch-mathematische Klasse*: 32–38, 76–77. Repr. in CPAE, Vol. 13, Doc. 425.

―――― (1924), 'Review of Elsbach, *Kant und Einstein* [Elsbach 1924]', *Deutsche Literaturzeitung*, 45: 1685–1692. Repr. in CPAE, Vol. 14, Doc. 321.

―――― (1925a), 'Eddingtons Theorie und Hamiltonsches Prinzip', in A. S. Eddington (ed.), *Relativitätstheorie in mathematischer Behandlung*. Mit einem Anhang 'Eddingtons Theorie und Hamiltonsches Prinzip' von Albert Einstein. Translated by A. Ostrowski and H. Schmidt. Berlin: Springer. Repr. in CPAE, Vol. 14, Doc. 282.

―――― (1925b), 'Einheitliche Feldtheorie von Gravitation und Elektrizität', *Sitzungsberichte der Preußischen Akademie der Wissenschaften, Physikalisch-mathematische Klasse*: 414–419. Repr. in CPAE, Vol. 15, Doc. 17.

―――― (1926), 'Space-Time', in J. L. Garvin (ed.), *Encyclopædia Britannica*. 13th ed. London and New York: Encyclopædia Britannica, Inc., pp. 608–609. Repr. in CPAE, Vol. 15, Doc. 148.

―――― (1927a), 'Zu Kaluzas Theorie des Zusammenhanges von Gravitation und Elektrizität', Erste und zweite Mitteilung. *Sitzungsberichte der Preußischen Akademie der Wissenschaften, Physikalisch-mathematische Klasse*: 23–25, 26–30.

―――― (1927b), 'Zu Kaluzas Theorie des Zusammenhanges von Gravitation und Elektrizität', Zweite Mitteilung. *Sitzungsberichte der Preußischen Akademie der Wissenschaften, Physikalisch-mathematische Klasse*: 26–30.

―――― (1928a), 'A propos de 'La Déduction Relativiste' de M. Émile Meyerson [Meyerson, 1925]', *Revue philosophique de la France et de l'étranger*, 105: 161–166.

―――― (1928b), 'Review of Reichenbach, *Philosophie der Raum-Zeit-Lehre* [Reichenbach 1928]', *Deutsche Literaturzeitung*, 49: 19–20.

Friedman, M. (1983), *Foundations of Space-Time Theories: Relativistic Physics and Philosophy of Science*. Princeton: Princeton University Press.

Friedrichs, K. (1928), 'Eine invariante Formulierung des Newtonschen Gravitationsgesetzes und des Grenzüberganges vom Einsteinschen zum Newtonschen Gesetz', *Mathematische Annalen*, 98: 566–575.

Giovanelli, M. (2014), 'But One Must Not Legalize the Mentioned Sin: Phenomenological vs. Dynamical Treatments of Rods and Clocks in Einstein's Thought', *Studies in History and Philosophy of Science: Part B: Studies in History and Philosophy of Modern Physics*, 48: 20–44.

―――― (2016), ''. . . But I Still Can't Get Rid of a Sense of Artificiality": The Einstein-Reichenbach Debate on the Geometrization of the Electromagnetic Field', *Studies in History and Philosophy of Science: Part B: Studies in History and Philosophy of Modern Physics*, 54: 35–51.

―――― (2018) ''Physics Is a Kind of Metaphysics": Émile Meyerson and Einstein's Late Rationalistic Realism', *European Journal for Philosophy of Science*, 8: 783–829.

Goenner, H. F. M. (2004), 'On the History of Unified Field Theories', *Living Reviews in Relativity*, 7.

Havas, P. (1964), 'Four-Dimensional Formulations of Newtonian Mechanics and Their Relation to the Special and the General Theory of Relativity', *Reviews of Modern Physics*, 36: 938–965.

Hecht, H. and Hoffmann, D. (1982), 'Die Berufung Hans Reichenbachs an die Berliner Universität', *Deutsche Zeitschrift für Philosophie*, 30: 651–662.

Infeld, L. (1928a), 'Zum Problem einer einheitlichen Feldtheorie von Elektrizität und Gravitation', *Zeitschrift für Physik*, 50: 137–152.

——— (1928b), 'Zur Feldtheorie von Elektrizität und Gravitation', *Physikalische Zeitschrift*, 29: 145–147.

Kaluza, T. (1921) 'Zum Unitätsproblem der Physik', *Sitzungsberichte der Preußischen Akademie der Wissenschaften*: 966–972.

Lehmkuhl, D. (2014), 'Why Einstein Did Not Believe That General Relativity Geometrizes Gravity', *Studies in History and Philosophy of Science: Part B: Studies in History and Philosophy of Modern Physics*, 46: 316–326.

Malament, D. B. (2012), *Topics in the Foundations of General Relativity and Newtonian Gravitation Theory*. Chicago and London: University of Chicago Press.

Meyerson, É. (1925), *La déduction relativiste*. Paris: Payot.

Misner, C. W., Thorne, K. S., and Wheeler, J. A. (1973), *Gravitation*. New York: W.H. Freeman and Company.

Pauli, W. (1921), *Relativitätstheorie*. Leipzig: Teubner.

——— (1926), 'Review of Eddington, *The Mathematical Theory of Relativity* [Eddington 1923]', *Die Naturwissenschaften*, 13: 273–274.

Reichenbach, H. (1922), 'Der gegenwärtige Stand der Relativitätsdiskussion. Eine kritische Untersuchung', *Logos*, 22: 316–378.

——— (1924), *Axiomatik der relativistischen Raum-Zeit-Lehre*. Vieweg: Braunschweig.

——— (1926), 'Die Weylsche Erweiterung des Riemannschen Raumes und die geometrische Deutung der Elektrizität', *Verhandlungen der Deutschen Physikalischen Gesellschaft*, 7: 25.

——— (1928), *Philosophie der Raum-Zeit-Lehre*. Berlin and Leipzig: Walter de Gruyter.

——— (1929a), 'Allgemeine Grundlagen der Physik', in H. Geiger and K. Scheel (eds.), *Ziele und Wege der physikalischen Erkenntnis*. Vol. 4: Handbuch der Physik. 2nd ed. Berlin: Springer, pp. 1–80.

——— (1929b), 'Zur Einordnung des neuen Einsteinschen Ansatzes über Gravitation und Elektrizität', *Zeitschrift für Physik*, 53: 683–689.

——— (1958), *The Philosophy of Space and Time*. Edited by Maria Reichenbach and Translated by Maria Reichenbach and John Freund. New York: Dover Publications.

Reichenbächer, E. (1925), 'Die mechanischen Gleichungen im elektromagnetischen Felde', *Zeitschrift für Physik*, 33: 916–932.

Rosen, N. (1940a), 'General Relativity and Flat Space', I. *Physical Review*, 57: 147–150.

——— (1940b), 'General Relativity and Flat Space', II. *Physical Review*, 57: 150–153.

Ryckman, T. (1995), 'Weyl, Reichenbach and the Epistemology of Geometry', *Studies in History and Philosophy of Science*, 25: 831–870.

——— (2005), *The Reign of Relativity: Philosophy in Physics 1915–1925*. Oxford and New York: Oxford University Press.

Sauer, T. (2014), 'Einstein's Unified Field Theory Program', in M. Janssen and C. Lehner (eds.), *The Cambridge Companion to Einstein*. Cambridge: Cambridge University Press, pp. 281–305.

Scholz, E. (1994), 'Hermann Weyl's Contributions to Geometry, 1917–1923', in S. Chikara, S. Mitsuo, and J. W. Dauben (eds.), *The Intersection of History and Mathematics*. Basel: Birkhäuser, pp. 203–229.

—— (2008), 'Weyl Geometry in Late 20th Century Physics', in D. E. Rowe (ed.), *Beyond Einstein: Proceedings Mainz Conference September 2008*. Birkhäuser: Basel, to appear.

Schouten, J. A. (1922a), 'Nachtrag zur Arbeit "Über die verschiedenen Arten der Übertragung in einer *n*-dimensionalen Mannigfaltigkeit, die einer Differentialgeometrie zugrundegelegt werden kann" [Schouten 1922b]', *Mathematische Zeitschrift*, 15: 168.

—— (1922b), 'Über die verschiedenen Arten der Übertragung in einer *n*-dimensionalen Mannigfaltigkeit, die einer Differentialgeometrie zugrundegelegt werden kann', *Mathematische Zeitschrift*, 13: 56–81.

—— (1923), 'Über die Einordnung der Affingeometrie in die Theorie der höheren Übertragungen', *Mathematische Zeitschrift*, 17: 183–188.

—— (1924), 'On a Non-Symmetrical Affine Field Theory', *Verhandelingen de Koninklijke akademie van Wetenschappen te Amsterdam*, 26: 850–857.

Stachel, J. (2007), 'The Story of Newstein or Is Gravity Just Another Pretty Force?', in J. Renn (ed.), *The Genesis of General Relativity*. 4 vols. Dordrecht: Springer, pp. 1962–2000.

Vizgin, V. P. (1994), *Unified Field Theories in the First Third of the 20th Century*. Translated by J. B. Barbour. Boston, Basel, and Stuttgart: Birkhäuser.

Weyl, H. (1918a), 'Gravitation und Elektrizität', *Sitzungsberichte der Preußischen Akademie der Wissenschaften*: 465–480. Repr. in Weyl (1968), Vol. 2, Doc. 31.

—— (1918b), 'Reine Infinitesimalgeometrie', *Mathematische Zeitschrift*, 2: 384–411.

—— (1919), 'Eine neue Erweiterung der Relativitätstheorie', *Annalen der Physik*. 4th ser., 59: 101–133.

—— (1921a), *Raum – Zeit – Materie. Vorlesungen über allgemeine Relativitätstheorie*. 4th ed. Berlin: Springer.

—— (1921b), 'Über die physikalischen Grundlagen der erweiterten Relativitätstheorie', *Physikalische Zeitschrift*, 22: 473–480.

10 Did Logical Positivism Influence the Early Interpretation of Quantum Mechanics?

Jan Faye and Rasmus Jaksland

10.1. Introduction

While quantum mechanics grew out of the experimental problems which physicists had to face in their attempts to understand atoms, one might expect that philosophical theories would have a profound influence on the interpretation of the emergent theory. Especially because physicists quickly realized that this new theory violated many of the fundamental principles of classical physics. And since the movement of logical positivism, or logical empiricism, was contemporary with the discovery of quantum mechanics, many philosophers have over the years supposed that the movement had a significant impact on physicists' early understanding of this new but revolutionary theory.

Many of the logical empiricists lost deep interest in quantum mechanics at the end of the 30s, most likely because they were more familiar with relativity theory. Among the few exceptions were Philipp Frank and Hans Reichenbach. The latter's most original contribution to the interpretation of quantum mechanics, entitled *Philosophic Foundations of Quantum Mechanics* was published in 1944. That is why Reichenbach had no bearing on the early discussions of quantum mechanics.

The next generation of philosophers who were in opposition to the logical empiricists, like Karl Popper, Mario Bunge and others, did a lot to spread their positivistic reading of the Copenhagen interpretation in myriad books supported by some positivists themselves. In particular, Patrick Heelan coined the phrase "facile positivism" to refer to Niels Bohr, but he speaks as though that were a well-established fact (1965, ix; quoted in Folse 1985, 23, n.24). Another important point was that the confusion of "The Copenhagen Interpretation" with Bohr's philosophy meant that anything people said in defense of "Copenhagen", much of which was explicitly positivistic, was automatically transferred to Bohr. Among philosophers of science in this generation, it was only Paul Feyerabend and Norwood Russell Hanson who didn't understand Bohr's philosophy of quantum mechanics as a genuflection for positivistic orthodoxy (see, e.g., van Strien 2020; Kuby 2021).

In the present chapter, we shall argue that logical positivism influenced neither the initial elaboration of the quantum mechanics formalism by Heisenberg and Schrödinger nor its early interpretation as championed by Bohr. Heisenberg might have formulated the matrix mechanics under the influence of Ernst Mach's positivism, but Bohr soon convinced him that the interpretation of quantum mechanics could not rely on positivist doctrines. And while Bohr was in contact with several of the leading figures of logical positivism, this contact was only established in the beginning of the 30s through the Danish philosopher Jørgen Jørgensen. Bohr's main view towards the new quantum theory was by then well developed and was motivated by physics more than empiricist epistemology, as we shall see.

This is not to say that there are no similarities between Bohr's view of quantum mechanics and the central tenets of logical positivism. Indeed, when Bohr finally got in touch with the logical empiricists, it was important for him to emphasize the similarities between his and their views. We can only guess, but it seems as if Bohr had hoped that because of these similarities philosophers were ready to embrace his ideas of complementarity even outside the domain of quantum mechanics. Little did he know that philosophy, like fashion, sees trends come and go.

10.2. Logical Positivism and Quantum Mechanics Formalism

The common view is that the Copenhagen interpretation due to Bohr and Heisenberg had established itself as the dominant interpretation among the physicists at the end of the 1920s. However, recent scholarship agrees very much with Max Jammer, who concluded in 1974:

> The Copenhagen interpretation is not a single, clear-cut, unambiguously defined set of ideas but rather a common denominator for a variety of related viewpoints. Nor is it necessarily linked with a specific philosophical or ideological position. It can be, and has been, professed by adherents to most diverging philosophical views, ranging from strict subjectivism and pure idealism through neo-Kantianism, critical realism, to positivism and dialectical materialism.
>
> (87)

Even between the two chief architects, Bohr and Heisenberg, one finds early discord about how to understand quantum phenomena, and later on, their views separated even more (for this development, see Camilleri 2009). That Schrödinger and Einstein stood in stark opposition to both Bohr and Heisenberg is no surprise because they both hoped that Schrödinger's wave mechanics could be interpreted differently from what Bohr proposed.

In summer 1925, Werner Heisenberg presented his matrix mechanics, which later that year he developed further in cooperation with Max Born and Pascual Jordan. About Heisenberg's attitude towards the construction of such a theory, Kristian Camilleri (2009, 5) writes that it "bears the influence of Einstein's theory, or, to be more precise, the strongly positivistic tendency which many of Heisenberg's contemporaries, Einstein included, considered integral to the theory". Camilleri points to three features in Heisenberg's thinking that could indicate such a positivistic leaning: (1) the principle of observability, (2) instrumentalism and 3) operationalism with respect to kinematic concepts. Camilleri (2009, 6) also remarks that "we can see the beginnings of the shift away from positivism in Heisenberg's thought in his discussion with Einstein and Bohr in 1926–7".

Heisenberg's revolutionary paper "Über quantentheoretische Umdeutung kinematischer und mechanischer Beziehungen" ("On the Quantum-Theoretical Reinterpretation of Kinematic and Mechanical Relations") excels by renouncing the classical orbit of an electron, which was characteristic of Bohr's atomic model of 1913. In this paper, Heisenberg stated in the Introduction that the aim is "to establish a basis for theoretical quantum mechanics, founded exclusively upon relationships between quantities which in principle are observable" (1925/1967, 261). The fact is that, according to the Bohr model, we can observe the transition of an electron between stationary states in the form of electromagnetic radiation, but we cannot observe the electron moving around in orbits, which implies that they might be excluded from the final theory. Around that time, it was generally assumed that unobservable quantities should be eliminated from physics, a view advocated by Pauli, Born and Jordan, and that Einstein, under the influence of Ernst Mach, had been successful in his attempt to meet this demand by first rejecting Newton's notion of absolute simultaneity and then Newton's absolute space and time. But Camilleri (2009, Ch. 2), like other scholars before him, argues convincingly that the principle of observability was not at the center of Heisenberg's consideration, and he seemed to have added it as an afterthought.

The second element in a positivist reading of Heisenberg is whether he considered the matrix mechanics only an instrument for prediction. Around 1926–1927, he sometimes expressed himself in this way, but a close analysis made by Camilleri (2009, 53 ff.) reveals that he might have regarded the mathematical structure of quantum mechanics as corresponding to a similar structure in nature. This matches Heisenberg's later adopted notions of Aristotle while talking about observation as actuality and quantum states as potentiality. Indeed, Bohr – if any of the two – had instrumentalist inclinations, as illustrated by his remark that the purpose of quantum mechanics "is not to disclose the real essence of phenomena but only to track down, so far as it is possible, relations between the manifold aspects of experience" (1954/1958, 71). The consequence is that "[t]he entire formalism is to be considered as a tool for deriving predictions of definite and statistical character" (Bohr 1948/1998, 144). However, as Henry Folse (1986) has convincingly argued, Bohr was at the same time an entity realist, which suggests that this instrumentalism was not motivated by, for instance, a principled antagonism towards metaphysics. What brought him to this conclusion was, in other words, not an allegiance to the positivist program. Rather, the finding that the wave function does not represent something that happens in real space motivated his attitude – therefore, the wave function could only have a symbolic meaning.

Finally, operationalism had brief favor in Heisenberg's thinking. He hoped to be able to give an operational definition of space and time, and his imaginary gamma-ray microscope was rather considered to give such analysis than to demonstrate the limits of measurement. However, Heisenberg's heated discussion with Bohr in early 1927 convinced him that such an analysis was not feasible. As Camilleri (2009, 87) mentions: "This realization signifies an important turning in Heisenberg's philosophy, in which he abandoned a verification theory of meaning, and finally accepted Bohr's doctrine of the indispensability of classical concepts". Although operationalism and verificationism of meaning are not exactly the same, Camilleri's observation bears witness to the fact that Heisenberg very early on moved away from his positivistic orientation.

Much less than Heisenberg was Erwin Schrödinger a standard-bearer for positivistic slogans in science. The motivation behind proposing his wave mechanics in 1926 was to comprehend the electron in stationary state of a hydrogen atom by describing it as a wave, because he was optimistic with respect to saving a causal space-time description in quantum physics. Even after Born had interpreted the wave function as a probability amplitude, and he himself had proved the formal equivalence between matrix mechanics and the wave mechanics, Schrödinger still kept thinking that the renunciation of a continuous space-time description of the electron, as Bohr and Heisenberg believed was inevitable, would be wrong.

Echoing Kant, Schrödinger wrote in 1928 against their understanding of the new quantum mechanics in which they had abandoned a unified casual and space-time description in order to avoid contradiction:

> This contradiction is so strongly felt that it has been doubted whether the phenomena in the atom can be described in the space-time form of thought at all. From the philosophical standpoint, I would consider such a definitive decision of this sort to be equivalent to complete surrender. For we cannot really change our forms of thought, and what we cannot understand within them, we cannot understand at all. There are such things – but I do not believe that the structure of the atom is one of them.
>
> (quoted by Camilleri 2009, 44)

Indeed, such a belief is far from the observational requirements we find among the logical empiricists. The problem for Schrödinger was that his own wave mechanics failed to provide him with such an understanding; in fact, it rather made things worse. The wave function of more than one electron does not represent the electron in real space-time but in a multi-dimensional abstract configuration space. Moreover, movement of a free election described by the wave function as a wave in real space would have a phase velocity greater than light and a group velocity smeared all over space-time.

These challenges to Schrödinger's program, however, do not change that his approach to quantum mechanics – failed or not – was to some extent the antithesis of the anti-metaphysical views of logical positivism. As we have seen, only Heisenberg came close to falling under the spell of positivistic ideas, but the influence was rather ideas adopted from Ernst Mach than the logical positivists. In other words, logical positivism did not influence the early formulations of the quantum formalism and is therefore not built into the theory itself. This, however, does not pre-clude an influence of logical positivism in the interpretation of quantum mechanics. We shall explore this theme in the remainder of this chapter.

10.3. The Motivations of Bohr's Interpretation

As already indicated, Bohr's interpretation of quantum mechanics has in particular been associated with logical positivism. There are several rea-sons one might think that logical positivism influenced Bohr's interpreta-tion of quantum mechanics. First, Bohr's emphasis on observation and the concepts of classical physics and his insistence that quantum mechan-ics introduced indeterminacy and not merely uncertainty have affinities to the logical positivists' views on verification, the observation language and anti-metaphysics. Second, Bohr expressed on several occasions that he agreed with logical positivism. Third, Bohr was in the audience when both Carnap and Neurath gave lectures in Copenhagen in the beginning of the 1930s, which might be taken to indicate a pre-existing interest in their work. In addition, Bohr was in correspondence with several leading logical positivists during the 1930s.

However, as we shall argue in the following, Bohr's interpretation of quantum mechanics is motivated by physics rather than philosophy. Still, the general epistemological lessons that Bohr derives from his interpreta-tion of quantum mechanics bear some resemblance to central theses of logical positivism. It is upon realizing this in the early 1930s – primarily through his acquaintance with Jørgen Jørgensen – that Bohr reaches out to the logical positivists. Bohr's expressions of agreement with the logical positivists seem to have been motivated by the hope that they could assist in the generalization of these epistemological lessons from quantum mechanics to other disciplines, a work Bohr had already begun himself.

It was, in other words, not so much logical positivism influencing Bohr but Bohr trying to influence logical positivism and philosophy in general. Bohr, however, ultimately regarded this as a failed attempt.

The starting point for Bohr in his approach to quantum mechanics, as he presented it in the Como lecture from 1927, was different from that of Heisenberg and Schrödinger. Where they respectively understood quantum mechanics in the light of the motivation behind the construction of the matrix formalism and the wave function formalism, Bohr focused on what experiments could say about these two conceptually different but mathematically equivalent versions with respect to the structure of atoms. The important point for Bohr was that we cannot separate the object under investigation from the measuring instrument as long as we describe their interaction in terms of quantum mechanics. However, at the same time, it does not make sense to say that the measuring instrument gives us knowledge of the object unless we distinguish between the apparatus that serves as the "agency of measurement" and the object being measured. Such a separation is "inherent in our very idea of observation" (Bohr 1928, 584), and in order to make an observation, we have to make use of terms like space, time, momentum and energy "since our interpretation of the experimental material rests essentially upon the classical concepts" (Bohr 1928, 580). In classical mechanics, the terms expressing these concepts allow us to give a causal space-time description of both the object and the measuring instrument.

However, in his Como lecture, Bohr maintained that an analysis of the conditions for observation and definition of the stationary states of an atom reveal the limited applicability of the classical concepts. As we shall explain in the following, he argued that in case one wants to define these stationary states, which are defined in terms of energy, then one is deprived of the possibility of observing the system due to the quantum postulate. Thereby, the concept of space and time loses its ordinary meaning with respect to such states because without observation, only the ascription of energy and momentum makes sense. But if one attempts to observe the stationary state of the atom, then the measuring device has to interact with the system. Such an interaction makes it impossible to define any precise stationary state. In this case, the electron's energy and momentum are indeterminate so that the concept of causation does not apply in any precise way. This reasoning brings Bohr to the conclusion:

> The very nature of the quantum theory thus forces us to regard the space-time co-ordination and the claim of causality, the union of which characterises the classical theories, complementary but exclusive features of the description, symbolising the idealisation of observation and definition respectively.
>
> (1928, 580)

After he had reached his view about the complementary nature of a causal and a space-time description in quantum mechanics, he applied his analysis on the wave-particle duality and the observation of kinematic and dynamic variables in accordance with Heisenberg's uncertainty principle. The observation of wave aspects and particle aspects appears from the two ways an electron manifests itself in different experimental arrangements. For instance, in the double slit experiment, the particle aspects and the wave aspects do not reside side by side but correspond to whether one slit is closed or both are open. With two slits open, the particle description of the photon does not apply. Only a wave description is meaningful relative to that particular experimental setup: there is one wave front that – due to the two holes – creates an interference pattern. On the other hand, if one slit is open at a time, no inference pattern appears, and the particle description applies. Hence, these two aspects are complementary, too. And when it comes to the precise measurement of the position of a free atomic system, it involves an experimental setup which excludes the possibility of the precise observation of its momentum. Also, here we face two complementary descriptions of the atomic object which exclude each other as predicted by Heisenberg' uncertainty relations.

Bohr's own model of the hydrogen atom had failed exactly because he had taken the electron to be orbiting the nucleus in classical states. As a consequence of his matrix mechanics, Heisenberg derived his uncertainty relations, which showed that it was impossible to gain mutually precise information about the movement of a free particle through space and time. During a quantum mechanical measurement, the object and the apparatus interact in an uncontrollable way such that if we gain exact knowledge of the object's position, our knowledge of its momentum becomes completely uncertain. It was not until 1935, when Bohr replied to the Einstein, Podolsky and Rosen paper, that he changed his terminology from talking about Heisenberg's uncertainty relation with respect to knowledge to Heisenberg's indeterminacy relation with respect to the application of classical concepts.[1]

Bohr was relentless in his belief that the use of classical concepts was necessary for understanding the outcome of an experiment. The dilemma that Heisenberg's uncertainty or indeterminacy relation raised for understanding quantum mechanics could not be solved by introducing new concepts to describe our observation. He was able to convince Heisenberg of this. Later, that is after 1935, Bohr maintained that the ascription of a precise momentum or position is contextual in the sense that the particular experimental arrangement defines the semantic condition under which it makes sense to ascribe a definite value of, say, momentum or position to the object under examination. Yet, these contextual descriptions, in keeping with Heisenberg's indeterminacy relations, are *complementary* to one another because they are mutually exclusive but exhaustive together.

Bohr's analysis started out by bringing to light the consequences of the fact that stationary states were not describable as classical orbits. The uncontrollable interaction between the agency of measurement and the quantum object prohibits that the stationary states could be physically defined and observed simultaneously. So well before Bohr came in touch with the logical positivists at the beginning of the 1930s, he had already developed his complementarity interpretation. If and how this encounter had any impact on Bohr's thinking is something we shall return to soon.

10.4. The Epistemological Lesson

Very early on, Bohr became aware that quantum mechanical experiments indicate that microscopic objects behave differently from macroscopic objects. This insight shaped the basis of his own atomic model from 1913. The experimental practice tells us that we cannot use classical physical theories to *predict* the outcomes of quantum experiments. Instead, we have to use quantum mechanics to make such *predictions*. Thus, the challenge for philosophy was to explain how this could be the case given that every macroscopic object is constituted by microscopic objects. However, in part, Bohr saw it differently, since he had given a physical explanation of why classical mechanics was impotent at the quantum level. The empirical discovery of the quantum of action ruled out the use of classical mechanics. However, since classical physics has formed the experimental practice in physics, we would still have to use concepts used by classical physics to interpret the results within quantum mechanics. A consistent use of these classical concepts only had to be restricted to specific experimental contexts.

Consequently, Bohr believed that quantum mechanics had an *epistemological* lesson to tell philosophy, namely that an objective description of nature cannot be separated from the conditions under which we explore natural phenomena. This feature, Bohr believed, relates not only to quantum mechanics but extends to every form of scientific knowledge. So it is not very surprising to see that when Bohr addressed an assembly of physicists and philosophers during the Unity of Science Conference in Copenhagen in 1936, he opened with a statement about the epistemological lesson:

> On several occasions I have pointed out that the lesson taught us by recent developments in physics regarding the necessity of a constant extension of the frame of concepts appropriate for the classification of new experiences leads us to a general epistemological attitude which might help us to avoid apparent conceptual difficulties in other fields of science as well.
>
> (Bohr 1937, 289)[2]

A little more can be said about what Bohr meant by this epistemological lesson.[3] The case of atomic physics has taught (or reminded) us of a point about human knowledge that in Bohr's view has been ignored (or forgotten). The lesson applies to all areas in the "description of nature", that is, to all natural sciences, but Bohr himself especially applies it in some detail in biology and psychology but also with quick references to art, anthropology, linguistics and religion. It arises because progress in atomic physics "disclosed unsuspected presuppositions for the unambiguous application of some of our most elementary concepts". The lesson teaches that concepts are used to describe nature in the context of a specific conceptual framework that determines how the descriptive concepts attach to nature. A change of descriptive framework can alter the presuppositions for the use of descriptive concepts. Since such presuppositions are often tacit, this change may go unnoticed, leading to misinterpretation of what the description is asserting. Furthermore, the new framework is a "rational generalization" of the preceding framework. (Perhaps this applies only in physics.) The lesson also teaches us that we cannot ignore "the inseparability of knowledge and our possibilities of inquiry" (Bohr 1960/1963, 12). Thus, the methods and goals of inquiry will partially affect the content of knowledge. The recognition of our position as observers in the world (spectators) affects the conditions for objective description of nature. They will change with the change of a conceptual framework in which that description is offered.

Why did the development of quantum mechanics contain such a lesson, according to Bohr? Bohr believed that the description of physics must always depart from the empirical observations we gain by making physical experiments. Classical concepts, like space-time position, momentum and energy, are (cognitively) necessary for describing these observations and experiments "so we can tell others what we have done and what we have learned" (Bohr 1949, 39). The classical concepts are also (cognitively) necessary for interpreting the quantum mechanical formalism by connecting its symbols with our observations and general physical experiences. But through the experimental practice of physics we realize that the applicability of the classical concepts for the description of quantum objects differs from their application in classical physics.

Thus, physicists have witnessed that the conceptual framework within which they formulate their theories can change over time. This was what happened with the discovery of the quantum of action. However, during such a transition from one conceptual framework to another, the scientific practice keeps constant since we have to describe our observation and experimentation by the same vocabulary as before. Although our theories may change, our common-sense experiences do not change. In physics, our common-sense experiences are expressed by ordinary language supplemented with the technical terms of classical concepts. Regardless of theory change and the development of knowledge, physicists must still describe their sensory experiences as they always have done in terms of space, time, movements and causal changes, because these terms are adapted to express our sensory experience.

The epistemological lesson has its roots in Bohr's principle of correspondence, which at the bottom demands that a consistent quantum mechanics should provide for high quantum numbers the same result as classical physics. Such a requirement can only be met in the case in which the language of possible observable predictions made by the two theories expresses the same physical content. For Bohr, but also for Heisenberg, the principle of correspondence stood as a methodological guideline in the invention of matrix mechanics. They both saw matrix mechanics, and thereby quantum mechanics, as a successful outcome of following the recommendation of the principle of correspondence, which Bohr had stated years before Heisenberg's discovery, at a time when he realized the explanatory problems with the Bohr-Sommerfeld model of the hydrogen atom.

The reader will probably recognize that Bohr's epistemological lesson concerning conceptual frameworks stands in opposition to Thomas S. Kuhn's suggestion that alternative scientific paradigms are incommensurable. According to Kuhn, there is a huge semantic difference between two succeeding paradigms making it impossible to translate the vocabulary of one paradigm into the other, although some of the basic terms are phonetically similar. He denied that the problem of incommensurability could be solved by an appeal to the scientific practice of observing and experimenting. Like Russell Hanson, Feyerabend, Popper and others, he argued that the observational language is theory laden, which means that observational terms get much of their meaning from the theory they supposedly sustain. Therefore, Kuhn believed that it was impossible to make a rational choice based on observation between competing paradigms. Bohr, in contrast, held that the construction of a new conceptual framework had to be a rational generalization of the old one in order for the new framework to meet our physical experiences and common practice.

10.5. Jørgen Jørgensen – The Danish Connection

At the beginning of the 30s, Bohr came in contact with some of the key figures of the logical positivism. Close to home, Bohr was able to get a firsthand glimpse of the philosophical ideas and the methodological approach that the movement embraced. One of Bohr's colleagues at the University of Copenhagen, Jørgen Jørgensen, who was professor of philosophy between 1926 and 1964, had become acquainted with a couple of the logical positivists in 1930 when he attended the Seventh International Congress of Philosophy in Oxford.[4] Soon after, Jørgensen was drawn into the discussions that took place among the members of the circle. Like many other positivists around that time, he had begun his philosophical career as a neo-Kantian, but eventually he became attached to the formal and anti-metaphysical attitude that the logical positivists had to philosophical and scientific problems. It was through Jørgensen that Bohr was introduced to Rudolf Carnap, Otto Neurath and others.

When Carnap in 1932 – on Jørgensen's invitation – gave a talk in Copenhagen on the character of philosophical problems in which he presented his ideas on syntax and/or semantics in transition, Bohr was in the audience. Bohr's presence should not be assigned too much significance, since he was an occasional guest at the lectures in "Selskabet for Filosofi og Psykologi" ("The Society of Philosophy and Psychology"), which hosted the event (Favrholdt 1992, 33). However, from Carnap's diary of November, Monday 14, we gather that Bohr found this talk interesting:

> 8h lecture "About the character of the philosophical problems", about 50 minutes, in the student dining house. Followed by tea and bread. Then lively discussion until 12. Niels Bohr speaks again and again, apparently very interested. Then a very long private conversation with him. He complains that one cannot understand how Einstein is now so conservative; he always comes up with counterexamples, but they are simply not correct. He (?) behaves exactly as Lenard used to do against him. It is simply a matter of stating that the observer cannot be separated. This is an insight that can no longer disappear from science. It is an epistemological question. The Wiener Kreis had said good things about the R Th [relativity theory], which Einstein himself had not seen so clearly. (Perhaps he thinks we would now like to express ourselves more clearly about quantum mechanics? Or he does not entirely agree with Schlicks or Frank's statements about it? It is not quite clear to understand because he expresses himself very carefully. He always stresses that he basically agrees with us). Until after ½ 1, 1 h at home.
>
> (2021; our translation)

From this note, we cannot see what Bohr and Carnap agreed upon, but since Carnap's lecture was dedicated to problems concerning syntax and semantics, we may guess that a part of this agreement was about the use of language in relation to expressing our common experience.

In spite of the acclaimed agreement between Bohr and Carnap, Jørgensen had also noticed some disagreement between them. A year later, on January 19, 1934, Carnap wrote a letter to Jørgensen in which he asked him about the possibility of approaching Bohr to write a recommendation for a Rockefeller-stipend, which Carnap unfortunately didn't get. In this letter, Carnap writes "I don't think he knows my work, but he seemed very friendly and interested at the time" (Jørgensen's correspondence with the logical positivists; The Danish Royal Library). Jørgensen sent Carnap's letter to Bohr accompanied by a letter of his own with the following remarks:

I do know that you do not agree with him on important points, but I did not think it would be right to fail to make you aware of his inquiry. And I would also like to add that the "direction" of philosophy, of which he is one of the leaders, seems to me to represent a much-needed reaction to the overpowering metaphysical-speculative verbiage philosophy that now flourishes in the German-speaking countries.

> (Jørgensen's letter of 22.1.1934 to Bohr; The Niels Bohr Archive, Copenhagen; our translation from Danish to English)

Although Bohr found himself unqualified to write Carnap a letter of recommendation, in his response to Jørgensen's request he expressed a sympathetic attitude towards Carnap's philosophy.

Thank you for your kind letter with the letter from Prof. Carnap, which I hereby return. However, I find it difficult to make a recommendation as he wished for his application for a Rockefeller stipend, because I do not think I have sufficient knowledge of his works and prerequisites to judge their value. I'm not even sure you're right when you note in your letter that I shouldn't agree with Carnap on significant points. On the contrary, as far as my knowledge of his endeavors goes, I have a very sympathetic impression of the honesty and sagacity, which characterize his position on ordinary philosophical problems. And the purpose of the more or less humorous remarks at the discussion after his lecture was merely to set off an exchange of views that could be fruitful to scientists like myself, whose language might at least seem superficially different from the dialectic cultivated by his school.

> (Bohr's letter of 23.1.1934 to Jørgensen; The Niels Bohr Archive, Copenhagen; our translation from Danish to English)

Bohr closed the letter by asking for both Jørgensen and Carnap's understanding of why he did not feel to be in a position to express his endorsement.

From Bohr's letter, we seem able to draw the conclusion that Bohr as late as 1934 knew very little of Carnap's work *despite* his general positive impression of Carnap. Carnap's remarks in his letter to Jørgensen corroborate this picture. What Bohr did know about Carnap's work was probably limited to what he had gathered during Carnap's visit to Copenhagen at the end of 1932. Moreover, we again see that Bohr believed that he was largely in agreement with what he knew about Carnap's work on general philosophical problems, and apparent differences were more terminological than substantial. Jørgensen might not have been so sure, but the conference on the unity of science in Copenhagen two years later could easily be interpreted as an attempt to clarify and overcome the remaining differences of opinion.

Also, Otto Neurath visited Copenhagen but not until the spring of 1934. On March 9, Carnap wrote Jørgensen a letter in which he says that he didn't receive the Rockefeller stipendium but that his main reason for writing Jørgensen was to inform him about Neurath, who was currently in Prague but would visit Gothenburg and would like to give a lecture, if possible, in Copenhagen. Carnap then suggested various topics for such a talk. Jørgensen responded to Carnap in a letter dated March 13, 1934, and wrote Neurath an invitation on March 23, 1934. On April 6, Neurath gave a speech on the topic "Psychologie und Sociologie auf physikalischer Grundlage". The next time Neurath came to Copenhagen was between October 18 and 24, 1934, when he gave six seminars concerning issues in epistemology ("The Minutes of the Society for Philosophy and Psychology," 1934). Niels Bohr took part in two of these sessions. A few weeks later, on November 14, 1934, Neurath wrote Carnap a letter mentioning his encounter with Bohr:

> Bohr. Idiosyncratic. An intense man. Came to two lectures and joined the discussion enthusiastically. . . . Basic line: he does not want to be considered a metaphysician. And he is able to express himself relatively non-metaphysically, when he is careful. Yet obviously there lies a certain tendency in the selection of problems, insofar as the question of life, etc. is discussed, as well as in the stress on uncertainty. In addition, his printed remarks are full of crass metaphysics. But he possesses certain basic attitudes which agree with mine, e.g., that in science one cannot clear up everything at once, but that the individual scientific-logical actions have to pay a price, as it were. An idea of compensation, which with him naturally tends to be connected with the uncertainty relation. Obviously tries to come into agreement with us. But since his circle confirms him in his habit to express himself somewhat unclearly, one would have to be able to work on him for a long time, which he would be prepared to do.
>
> (Faye 2010, 34)

Neurath's words inform us that he saw an obvious similarity between the ideas of the logical positivism and Bohr's thoughts on complementarity. Nevertheless, he was annoyed about how Bohr articulated his thinking. Neurath also hinted at his own analogy concerning revision of knowledge that knowledge is like a boat in open sea. It is impossible to change all the beams at once, but one can change one plank at a time. Here Bohr seemed to have agreed, since it was more or less in line with what he himself had argued in terms of his epistemological lesson.

The very same day Neurath left Copenhagen, Bohr sent him one of his books, possibly the German version of *Atomic Theory and the Description of Nature* (*Niels Bohr's Philosophical Writings Volume I*), along with a letter. In this letter he states his pleasure concerning the fact that their ideas were not so far apart from each other as one might otherwise think from their different ways of expression.[5]

Bohr and Neurath corresponded over the next couple of years. Therefore, we have a good impression of what Neurath's resentment was. On March 3, 1936, Neurath outlined in a long letter to Bohr various issues he thought should be the topics of discussion at the forthcoming Unified Science Conference in Copenhagen. Ulrich Röseberg (1995, 113) summarizes some objectives as follows:

> Neurath also asked how it would be possible to formulate terms that can be used in all scientific disciplines in the same way. For this reason Neurath proposed to Bohr that he should give up words like "free choice", "teleological argumentation", "irrationality", and criticized some of Bohr's specific formulations.

However, in spite of the fact that Bohr's interpretation of quantum mechanics had materialized long before he really became aware of logical positivism, this does not exclude their anti-metaphysical approach becoming useful at a certain moment in his constant struggle with Einstein's criticism of understanding quantum mechanics (see Faye [1991, Ch. 7] for further details). In 1935, Einstein published, with two other physicists, Boris Podolsky and Nathan Rosen (EPR), a paper, "Can Quantum-Mechanical Description of Physical Reality be Considered Complete?", containing a strong criticism of Bohr's interpretation of quantum mechanics. Their aim was to demonstrate that quantum mechanics was incomplete because it could not account for all elements of reality. The upshot of their thought experiment was that while the electron's momentum was measured, it must still have some pre-existing position, although the measuring process might disturb its definite value. EPR's claim of some hidden value would be grossly metaphysical according to the positivists, but as long as Heisenberg's uncertainty principle was considered a statement about knowledge, quantum mechanics seems to allow for such definite values that we could not empirically know.

At this point, Bohr seemed to have realized that Einstein and company failed because they saw Heisenberg's uncertain relation to be an expression of the limitation of knowledge obtainable by measurements. But for him, it was not a question about whether we could gain precise information of a particle's position in the case in which we measured its momentum. It was primarily not a question concerning uncertainty but indeterminacy. EPR got it wrong because the choice of a particular experiment constituted the conditions under which it makes sense to apply the concept of momentum, which at the same time exclude the conditions for applying the concept of position. The value of position is indeterminate, since the concept of position is not applicable. Obviously, if a concept is not applicable in a certain experimental situation, it goes without saying that we cannot gain knowledge about what this concept supposedly represents. This suggestion explains why Bohr afterwards always wrote "the Heisenberg indeterminacy relation".

That the positivists' anti-metaphysical attitude had an influence on Bohr's way of expressing himself becomes more plausible considering an exchange of letter between Bohr and Jørgensen. After Bohr had responded to the EPR-paper, he sent a copy to Jørgensen with an enclosed letter in which he wrote:

> Likewise, I have today written to Prof. Ph. Frank and told him that you had been kind enough to show me his critical-humorous article. I also sent him a copy of the answer to Einstein, hoping that – although he, as I jokingly wrote, would hardly here either be satisfied with my way of expression – it may, however, help to understand that we are all ultimately fighting for the same cause.
>
> (Bohr's letter of 10 July 1935 to Jørgensen; The Niels Bohr Archive; our translation from Danish to English)[6]

Three days later, Jørgensen responded very politely:

> Thank you very much for the article sent. It was very gracious of you to remember to send it to me and I have read it with great interest. Against your mode of expression here, I have not found the smallest thing to object to, nor can I imagine that Prof. Frank should be able to find something. And as far as the matter is concerned, I can of course only try to follow the professionals' discussion as best or badly I actually can. When, during conversations with you, I have raised objections to your considerations, it has always been intended as questions rather than objections, and the purpose of them has not been to convince you, but only to open opportunities for myself to understand your thinking as good as possible. A discussion at the Paris Congress or the post-conference in the Netherlands would certainly have contributed significantly to the understanding that we are all fighting for the same cause, and so far, I deeply regret that they cannot come – but there is nothing to do at this time.
>
> (Jørgensen's letter of 13 July 1935 to Bohr; The Niels Bohr Archive; our translation from Danish to English)

In the last part of his letter, Jørgensen addressed another part of Bohr's letter to him in which Bohr had mentioned that he had corresponded with Neurath about his possible participation in the First International Congress for the Unity of Science in Paris, although he had not been able to comply because of other duties. This shows the eagerness with which some of the logical positivists wanted to become as familiar with quantum mechanics as they were with relativity theory.

Just as importantly, Jørgensen's letter to Bohr also reveals that he, like some of the other logical positivists, had been rather critical of Bohr's ways of expressing himself while discussing the interpretation of quantum which they found belonged to a pre-positivist era of empirical philosophy. At the bottom, both parties shared an empiricist outlook, but Bohr did not share their logical approach to philosophical question.

However, it is important to issue a warning. Bohr came across the indeterminacy argument neither during his encounter with the EPR-paper nor under the influence of logical empiricism. It had been part of his thinking all along. In 1929, he made it explicitly clear that the limitation of measurement derives from the restricted application of concepts. "It is therefore, an inevitable consequence of the limited applicability of the classical concepts that the results attainable by any measurement of quantities are subject to an inherent limitation" (Bohr 1934, 95). Where Bohr and the positivists' thinking seems to meet is in the requirement that the criteria for the applicability of the classical concepts in quantum mechanics can only be specified with respect to the experimental situation of observation. Nevertheless, they had different opinions about why these criteria of application were grounded in experimental practice. The positivists saw it as a manifestation of their empirical criterion of significance, whereas Bohr took it as a consequence of the existence of the quantum of action and what this meant for the possible use of common language to describe quantum phenomena, a language that was originally adapted to our common-sense experience.

10.6. The Unity of Science Conference in Copenhagen

In the summer of 1936, many members of the positivist circle gathered in Copenhagen to participate in the Second International Congress for the Unity of Science, which Jørgensen, with Neurath, had arranged and which took place in Bohr's honoree residence at Carlsberg. The goal was, apparently, to reconcile Bohr's view on quantum mechanics and the logical empiricists' epistemology by eliminating possible misunderstandings (see Röseberg 1995, 113).

However, three significant members, Moritz Schlick, Hans Reichenbach and Rudolf Carnap, were absent for various reasons. Apart from Bohr's opening talk on quantum mechanics, two other main talks dealt with quantum mechanics. One was given by Philipp Frank (1936/1949) and the other, by Schlick (1936/1979), was read to the audience. Both authors touch upon the interpretation of quantum mechanics in the light of logical empiricism, and both saw Bohr's interpretation as a realization of the positivist program.

As a physicist, Philipp Frank had corresponded with Bohr before the conference, and we can see from this correspondence that it was partly Bohr's own fault that the logical empiricists had the impression that Bohr was on their side. In an extended letter to Bohr, written January 9, 1936,

Frank reflects upon the EPR-paper and Bohr's reply, and he expresses his view that Bohr had taken a positivist approach to physical reality, whereas Einstein subscribes to a metaphysical view. In response of January 14, Bohr wrote back, saying: "I am pleased to hear from your kind letter that you have given such care to the papers of Einstein and myself concerning the question of reality. I also think that you have caught the sense of my efforts very well".[7]

Among the participants at the unity of science conference was the German physicist Martin Strauss, whom Bohr had invited to spend time at his institute. Here he stayed as a refugee from September 1935 until December the following year, when he left to work with Frank in Prague (Jacobsen 2012, 129 ff.).[8] He also presented a small paper, "Komplementarität und Kausalität im Lichte der logischen Syntax". The title refers directly to Carnap's published work two years earlier. After the conference, Martin Strauss wrote Reichenbach a long letter about the conference in which he reported, "Also *Frank* who in Konigsberg still misunderstood quantum mechanics in a completely positivistic manner converted to complementarity" (see Röseberg 1995, 114). Apparently, Strauss realized differences between positivist doctrines like verificationism and Bohr's view on complementarity.

Undoubtedly, it was the positivist's empirical outlook and Bohr's more pragmatist concerns that here brought them together in the rejection of the meaningfulness of referring to things-in-themselves hidden behind the observable phenomena. However, neither Bohr nor the positivists themselves were always able to recognize the subtle epistemic differences between their respective views. They did not always share the same premises, though their conclusion seems to coincide. Jørgensen, who was closer to Bohr than any other of the logical empiricists, acknowledged this in a small paper published in the year after the unity of science conference.

> What I should like to emphasize in this place is the point that Niels Bohr and those agreeing with him in no way consider quantum-theoretical "indeterminism" or "acausality" a *consequence* of a positivistic view (epistemological understood) adopted beforehand but that in their opinion it is based on circumstances which presumably necessitate the assumption of "indeterminism" quite regardless of definite epistemological standpoint.
>
> (1937, 98)

Instead, Jørgensen saw the relationship differently, claiming "that the results of atomic physics may serve to support a positivistic epistemology". In other words, Jørgensen recognized that the oddities of quantum mechanics were driven by physics and not philosophy. In his view, Bohr's interpretation of quantum mechanics had developed quite independently of the positivistic considerations about the justification of knowledge.

10.7. Concluding Remarks

The organizers of the unity of science congress in Copenhagen had dedicated the conference to the problem of causality pertaining to physics and biology. One of the aims appears to have been to reconcile the logical empiricists and the quantum physicists; now the "scientific" philosophers had a chance to discuss with Bohr and some of the other leading quantum experts in person. However, the benefit of the congress seems to have varied. The logical positivists like Frank concluded that Bohr's interpretation of quantum mechanics was not in conflict with the epistemology of logical empiricism. However, Strauss reported to Reichenbach after the conference that Heisenberg, who participated in the conference, "swears a lot at the positivists . . . philosophy is simply = the great personality who has something to say, like Heidegger (who nevertheless is not great enough)" (Röseberg 1995, 114). Apparently, Heisenberg was not impressed by the contributions of the "scientific" philosophers.

Also, Bohr seems to have hoped for more. Most likely he wanted to see that the logical empiricists had borrowed complementarity as a general philosophical principle for providing the unity of knowledge. Many years later, he expressed his disappointment about the outcome of conference. In his paper about his discussions with Einstein, Bohr remarked that the talk at the Congress in 1936 had attempted to avoid the impression of mysticism, which one might have gotten by reading his paper "Light and Life" on the elaboration of complementarity outside physics.

> I therefore tried to clear up such misunderstandings and to explain that the only question was an endeavour to clarify the conditions, in each field of knowledge, for the analysis and synthesis of experience. Yet, I am afraid that I had in this respect only little success in convincing my listeners, for whom the dissent among the physicists themselves was naturally a cause of skepticism as to the necessity of going so far in renouncing customary demands as regards the explanation of natural phenomena.
>
> (Bohr 1949, 376)

Even the day before he died, Bohr expressed his annoyance with philosophers: "I felt that philosophers were very odd people who really were lost, because they have not the instinct that it is important to learn something and that we must be prepared really to learn something of very great importance".[9]

None of the logical empiricists adopted Bohr's philosophy of complementarity as a unifying principle of knowledge, and Jørgensen, who wrote very positively about Bohr's works in the 30s and 40s, became much more hesitant in the 50s, when he considered himself a critical realist. Nor had the logical empiricists any influence on the elaboration of quantum mechanics or on the subsequent interpretation developed mainly by Bohr. After the encounter in Copenhagen, the interests among philosophers seem to change, with Reichenbach as the only exception. Although they approached the understanding of human knowledge from the different angles represented by physics and logic, both Bohr and the logical empiricists saw themselves in the 30s as soldiers fighting for the same cause. By the end of the decade, another type of soldier terminated this endeavor.

Notes

1. Faye (1991, 184–189) has a discussion of Bohr's possible argument for such a change of terminology.
2. Those occasions, Bohr tells us in the second footnote, appear in "*Atomic Theory and the Description of Nature*, four essays and an introductory survey; Cambridge, 1934; quoted in the text as A_I, A_{II}, A_{III}, A_{IV}, and A_E; further, 'Light and Life' *Nature* 131, 421; 457, 1933; and 'Can Quantum Mechanical Description of Physical Reality be Considered Complete?' *Physical Review LXVIII*, 696, 1935; quoted as B and C respectively." These four papers and the Introduction were written in 1925, 1927 (the Como-paper) 1929, 1929 and 1929.
3. The following exposition of the epistemological lesson owes much to Henry Folse (1985).
4. About Jørgensen and the Vienna Circle, see Koch (2010), and about Bohr, Jørgensen and logical positivism, see Faye (2010).
5. Bohr to Neurath, October 24, 1934, quoted in Röseberg (1995).
6. The mentioned paper by Frank is most likely "Das Ende der mechanistischen Physik" (1935). This is supported by Bohr's letter to Frank on the 5th of July: "Vor einigen Wochen was Professor Jørgensen so freundlich, mir den Artikel zu zeigen, den Sie über die Verhältnisse zwischen Atomphysik und den allgemeinen Problemen der Biologie und Psykologie geschrieben haben. Obwohl mich Ihr Humor gefreut hat und ich im vielem mit Ihren ganz einverstandig bin, muss ich gestehen, dass es mir scheint, dass über den Tendenz meiner Bestrebungen nicht ganz im Bilde sind. Ich habe jedoch nicht die Hoffung aufgeben, mich einmal so deutlich auszudrücken, dass wir uns besser verständigen können."
7. The correspondence between Bohr and Frank (Niels Bohr Archive, BSC 19.3)
8. Probably Rosenfeld and Strauss were both interested in philosophy and left-wing politics and continued to correspond over many years.
9. Interview with Professor Niels Bohr on November 17, 1962, conducted by Th. S. Kuhn, Aage Petersen and Erik Rüdinger. *Archive for the History of Quantum Physics.*

References

Bohr, N. (1928), 'The Quantum Postulate and the Recent Development of Atomic Theory', *Nature*, 121: 580–590.
——— (1934), 'The Quantum of Action and the Description of Nature', in *Atomic Theory and the Description of Nature*. London: Cambridge University Press, pp. 92–101.

———— (1937), 'Causality and Complementarity', *Philosophy of Science*, 4: 289–298.

———— (1948/1998), 'On the Notion of Causality and Complementarity', in J. Faye and H. J. Folse (eds.), *Causality and Complementarity: Bohr's Philosophical Writings IV*. Woodbridge, UK: Oxbow Press, pp. 141–148.

———— (1949), 'Discussion with Einstein on Epistemological Problems in Atomic Physics', in P. A. Schilpp (ed.), *Albert Einstein: Philosopher-Scientist*. Evanston, IL: The Library of Living Philosophers, pp. 199–242.

———— (1954/1958), 'Unity of Knowledge', in *Atomic Physics and Human Knowledge*. New York: John Wiley & Sons, pp. 67–82.

———— (1960/1963), 'The Unity of Human Knowledge', in *Essays 1958–1962 on Atomic Physics and Human Knowledge*. New York: John Wiley & Sons, pp. 8–16.

Camilleri, K. (2009), *Heisenberg and the Interpretation of Quantum Mechanics*. Cambridge: Cambridge University Press.

Carnap, R. (2021), *Tagebücher 1908–1935*. Herausgegeben von Christian Damböck, unter Mitarbeit von Brigitta Arden, Brigitte Parakenings, Roman Jordan und Lois M. Rendl. Hamburg: Meiner Verlag, forthcoming.

Favrholdt, D. (1992), *Niels Bohr's Philosophical Background*. Copenhagen: Munksgaard.

Faye, J. (1991), *Niels Bohr: His Heritage and Legacy*. Dordrecht: Kluwer.

———— (2010), 'Niels Bohr and the Vienna Circle', in J. Manninen and F. Stadler (eds.), *The Vienna Circle in the Nordic Countries*. Dordrecht: Springer, pp. 33–45.

Folse, H. J. (1985), *The Philosophy of Niels Bohr: The Framework of Complementarity*. Amsterdam: North-Holland.

———— (1986), 'Niels Bohr, Complementarity, and Realism', *PSA: Proceedings of the Biennial Meeting of the Philosophy of Science Association*, 1: 96–104.

Frank, Ph. (1936/1949), 'Philosophical Misinterpretations of the Quantum Theory', in *Modern Science and Its Philosophy*. Cambridge, MA: Harvard University Press, pp. 158–171.

Heelan, P. (1965), *Quantum Mechanics and Objectivity: A Study of the Physical Philosophy of Werner Heisenberg*. The Hague: Martin Nijhoff.

Heisenberg, W. (1925/1967), 'On the Quantum-Theoretical Reinterpretation of Kinematic and Mechanical Relations', in B. L. van der Waerden (ed.), *Sources of Quantum Mechanics*. Amsterdam: North Holland, pp. 261–274. (Originally as 'Über quantentheoretische Umdeutung kinematischer und mechanischer Beziehungen', *Zeitschrift für Physik*, 32: 841–860).

Jacobsen, A. S. (2012), *Léon Rosenfeld*. Singapore: World Scientific Publishing.

Jammer, M. (1974), *The Philosophy of Quantum Mechanics*. New York: John Wiley & Sons.

Jørgensen, J. (1937), 'Causality and Quantum Mechanics', *Theoria*, 1: 115–117.

Koch, C. H. (2010), 'Jørgensen and Logical Positivism', in J. Manninen and F. Stadler (eds.), *The Vienna Circle in the Nordic Countries*. Dordrecht: Springer, pp. 153–166.

Kuby, D. (2021), 'Feyerabend's Reevaluation of Scientific Practice: Quantum Mechanics, Realism and Niels Bohr', in Karim Bschir and Jamie Shaw (eds.), *Interpreting Feyerabend: Critical Essays*. Cambridge: Cambridge University Press, forthcoming.

Röseberg, U. (1995), 'Did They Just Misunderstood Each Other? Logical Empiricism and Bohr's Complementarity Argument', in K. Gavroglu, J. Stachel, and M. W. Wartofsky (eds.), *Physics, Philosophy and the Scientific Community: Essays in the Philosophy and History of the Natural Sciences and Mathematics in Honor of Robert S. Cohen*. Dordrecht: Kluwer, pp. 105–123.

Schlick, M. (1936/1979), 'Quantum Theory and the Knowability of Nature', in H. Mulder and B. F. B. van de Velde-Schlick (eds.), *Moritz Schlick: Philosophical Papers, Volume II (1925–1936)*. Dordrecht: Kluwer, pp. 482–490.

van Strien, M. (2020), 'Pluralism and Anarchism in Quantum Physics: Paul Feyerabend's Writings on Quantum Physics in Relation to His General Philosophy of Science', *Studies in History and Philosophy of Science*, 80: 72–81.

11 Why Moritz Schlick's View on Causality Is Rooted in a Specific Understanding of Quantum Mechanics

Richard Dawid

11.1. Introduction

Moritz Schlick's ideas about causality and quantum mechanics dealt with a topic that was at the forefront of research in physics and philosophy of science at the time. In the present chapter, I want to discuss the plausibility of Schlick's ideas within the scientific context of the time and compare it with a view based on a modern understanding of quantum mechanics.

Schlick had already written about his understanding of causation in Schlick (1920/1979) ("Naturphilosophische Betrachtungen über das Kausalprinzip") before he set out to develop a new account of causation in "Causality in Contemporary Physics" (Schlick 1931/1979). The very substantial differences between the concepts of causality developed in the two papers are due to important philosophical as well as physical developments. Philosophically, Schlick (1931/1979) represents the shift from his earlier position of critical realism towards his endorsement of core tenets of logical empiricism. Physics-wise, his later paper was strongly inspired by the recent development and success of quantum mechanics. Unlike Carnap, who, though interested in scientific developments, preferred a philosophical level of discourse that remained independent from specific scientific statements, Schlick was a genuine expert on the physics of his time and tried to account for specific new developments in physics at a philosophical level. Schlick makes an attempt to account for philosophical as well as for scientific new developments and to build a coherent conception that can do justice to both of them. In his own understanding, he does find an interpretation that works for both sides and is strengthened by that fact.

The discussion of Schlick in the present chapter will focus on Schlick (1931/1979) and on his verificationist perspective on causation. Schlick's attempts to reconcile logical empiricism and a strict verificationism with a realist stance, which shows, for example, in Schlick (1932/1979), will not be of importance for the present analysis.[1]

About half a century after Schlick, Bas van Fraassen (1980), who is equally interested in analyzing philosophy of physics based on general empiricist conceptions, presented his constructive empiricism. The attempts of both authors to relate their respective understanding of empiricism to the topical understanding of quantum mechanics provide an instructive basis for comparing the compatibility of logical and constructive empiricism with quantum physics. While both positions have been developed primarily based on general philosophical reasoning, they seek to be coherent with the new developments in physics. van Fraassen's interest in quantum physics may suggest that he took his position's compatibility with the concepts of quantum mechanics to be of crucial importance. The question may be asked whether the development of philosophy of science from Schlick's interpretation of logical empiricism towards van Fraassen's constructive empiricism is related to the evolution of the understanding of quantum physics during that time span. In the present chapter, I will argue that connections between the physical and the general philosophical level of discussion can indeed be established. Schlick's 1931 position on causality is rooted in the prevalent understanding of quantum mechanics at the time. Likewise, the move away from a logical empiricist perspective on causation to van Fraassen's view reflects the shifts in the understanding of quantum mechanics that had occurred by then.

11.2. Schlick on Natural Laws and Causality

Early logical empiricism held that statements acquire meaning only through the conditions of their verification. This idea, which was intended to offer the basis for a clear-cut distinction between meaningful scientific statements and metaphysics without cognitive content, had an unpleasant consequence, however: it threatened to render all general scientific statements meaningless. Scientific theories are based on general natural laws which quantify over all events or objects of a given kind. Verification as understood by logical empiricism, however, must be based on individual observations and therefore can only pertain to individual facts.

To give an example, let us consider the natural law "All freely falling massive objects on earth accelerate with rate g". Verification can only be based on observations of the kind: the free-falling object A has accelerated with rate g at time t. Having measured accelerations with rate g consistently on many occasions will be considered a confirmation of the corresponding natural law by the scientific community. Inductive reasoning will lead from the set of confirming evidence to the prediction of future outcomes under comparable circumstances. The described set of measurements cannot constitute a verification of the law of acceleration, however. In the simplest logical empiricist view, verification must be logically conclusive, which would require deductive inference from the set of observations to the general statement. This, however, obviously cannot be provided.

Various ways of dealing with that problem have been suggested by exponents of logical empiricism in the early 1930s.[2] Rudolf Carnap, already starting in the late 1920s and culminating in his "Testability and Meaning" (1936), chose the path towards weakening the criterion of verification to an extent that would allow the notion of a gradual verification (confirmation) of natural laws.

Moritz Schlick rejected that strategy because he believed that any weakening of the criterion of verification would sacrifice the clear distinction between science and metaphysics. He therefore tried to develop a different strategy. He was ready to acknowledge that natural laws and generalizing sentences of similar kinds are indeed not verifiable. In fact, Schlick claimed that they do not constitute statements at all. He rather took them to constitute instructions for the generation of statements (*Anweisungen zur Bildung von Aussagen*, Schlick 1931/1979, 188). In other words, they were akin to commands or advices. According to Schlick, the scientist uses those laws in order to create statements on her predictions of individual future outcomes of experiments. Those predictions have the status of statements because they can be verified by empirical data. The natural laws are just guidelines on how to extract those statements based on inductive reasoning from other statements which have already been verified.

Schlick's understanding of natural laws plays directly into the more specific discussion on the status of the causality principle. The causality principle asserts that all events are fully determined by causal laws. It was widely considered an important and viable characterization of classical physics. However, a closer look at the principle's meaning seems to lead to an unfortunate dilemma: either one takes the principle to be tautological or one has to introduce arbitrary restrictions to the definition of causal law which turn the causality principle from a statement of universal importance into a specific claim about the applicability of an arbitrarily chosen conception.

Schlick presents the problem in several steps in Schlick (1931/1979). Causality denotes that a certain event in a given framework necessarily implies a specific consecutive event. The only way of understanding what this means, according to Schlick, leads through deploying the concept of a law. There has to be some kind of order (*Ordnung*) that determines the next event's outcome based on the first event. The problem arises, however, that any imaginable sequence of events might be declared some kind of order. There is no clear-cut and fundamental distinction between order and non-order. Any sequence can in principle be called an ordered sequence. The causality principle thus turns into a tautology.

One attempt to define the elusive distinction between order and non-order in a meaningful way relies on prediction. Following that idea, order allows for successful prediction, while non-order does not. However, no fundamental principle seems at hand that prevents even the least orderly and most arbitrary statement about an imagined sequence of events from being predictively successful. A time traveler coming from the future could use her records for providing predictions of that kind. Being extremely lucky might lead towards the same predictions by chance. Using predictive success as a criterion for order thus does not prevent us from accepting any imaginable sequence of events as the consequence of some law.

In order to free the causality principle from the tautology threat, Schlick suggests, one has to introduce specific additional conditions that must be met by natural laws. The most natural suggestion is to demand a certain degree of simplicity. The awkward nature of simplicity debates makes it obvious from the start, however, that any specific and precise condition introduced on that basis must remain highly arbitrary and is likely to be threatened by counter-examples of laws from real science which do not meet the given condition. Entering that discussion or a discussion on any other more specific definition of natural law leads into arbitrary constructions of law concepts whose significance does not reach beyond its creator's set of predilections. Such constructions seem inadequate for providing a stable foundation for universally acceptable statements on the viability of the causality principle. Schlick considers that route unpromising.

Schlick proposes to tackle the problem in a different way by applying his definition of scientific laws as instructions for generating statements. Applying the same way of thinking to causation, the principle of causality itself, which requires quantification over all events, does not constitute a statement with cognitive content at all. The question of whether it is tautological thus does not arise. The causality principle may be taken to be a kind of meta-law, which means that it does not constitute immediate advice regarding the generation of individual statements but rather advice regarding the generation of laws. The claim that all events are fully determined by causal laws must be interpreted as the meta-theoretical advice to look for causal laws which can suggest successful predictions of individual and verifiable events. "The causal principle does not directly inform us of a fact, such as that the world is regular, but represents a demand or a prescription to seek regularity and to describe events by means of laws" (Schick 1931/1979, 195–196).

Schlick's construction solves the problem of the delimitation of order from non-order. Seen from his perspective, it is not necessary to offer a rigid distinction between (sufficiently simple) laws and (overly complex) non-laws. The scientist can follow the advice given by the causality principle by looking for simple structures first and then adding complexity until she finds a law that is compatible with the known phenomena. The advice thus makes sense without offering any rigid definition of order.

11.3. Implications of Quantum Mechanics According to Schlick

The status of the causality principle was intensely debated in the context of quantum physics in the 1920s and 1930s. Werner Heisenberg took quantum mechanics to amount to a refutation of the principle of causality. In his view, the uncertainty relation implied that no natural law could predict the outcome of future measurements based on the present state of the world. Thus, there was no strict causality relation between the present state of the world and future events.

Max Born, to the contrary, endorsed the tautology interpretation of the causality principle. Like Heisenberg, he took quantum mechanics to imply that it made no physical sense to try to define a natural law that could implement a specific strict causality relation between the present state of the world and future events. Contra Heisenberg, however, he stressed the formal existence of a causality relation expressing the connection between old and new data. That causality expression was represented simply by the actual sequence of events. It could not be determined in advance based on the theory and the available data. This, however, could be interpreted in terms of the causal structure's lack of simplicity rather than in terms of the absence of a causal structure. Though the causality principle thus survived, applying the concept of causation lacked any significance with regard to a characterization of the scientist's activity. In Born's view, the impossibility to translate the uncertainty relation into a clear refutation of the causality principle was a nice illustration of that principle's tautological character.

While Heisenberg's understanding seemed to be an accurate characterization of the intuitive understanding of the canonical interpretation of quantum mechanics, Born's philosophical analysis appeared to be directly implied by a traditional universal conception of causation.

Schlick (1931/1979, Section 8) took the unsatisfactory state of the debate on quantum physics and causation to provide strong support for his position on causality. From his perspective, quantum physics clearly demonstrated the inadequate nature of the classical concepts of causation. In fact, it seemed to offer a physical argument that was in stupendous agreement with the philosophical arguments put forward by Schlick himself.

Schlick's reasoning relied on a strictly empiricist interpretation of quantum physics. According to that understanding, the indeterminacy of quantum mechanics was strictly identified with unpredictability: quantum mechanics was structurally incapable of providing precise predictions of future events. Ontological conceptualizations of indeterminacy which related the term to an irreducible characteristic of the quantum state were rejected by Schlick as metaphysically contaminated. This take on quantum physics did not constitute a consensual position among quantum physicists at the time. It was apparently endorsed by some physicists, however. In particular, Schlick (1931/1979, 190) attributes a position along those lines to Arthur Eddington.

Schlick then goes on to interpret this understanding of indeterminacy within the framework of his notion of a natural law. A natural law in

Schlick's eyes is a piece of advice for the generation of statements on individual events. If quantum physical laws are indeterminate laws, they give the advice *not* to generate statements which predict the outcomes of experiments at variance with the indeterminacy. The causality principle is located one level further up. It constitutes advice on which kind of natural laws one should formulate. Specifically, it constitutes advice to look for laws which give advice on how to generate statements on the specific outcome of experiments based on previously collected data. Rejecting the causality principle due to the success of quantum mechanics then amounts to rejecting the advice to look for new laws which give advice on how to build such statements. In other words, it means considering it a bad idea to try making precise predictions on future events in the given context.

Understood in this way, the causality principle is neither a statement that has been refuted by quantum physics, as Heisenberg suggests, nor is it tautological. It does not constitute a statement at all. It is not irrelevant either, however, since it conveys advice one can follow or reject. Whether one endorses the causality principle thus makes a significant difference with regard to scientific strategy.

Schlick's understanding may seem fairly convincing at first sight. As we will see in the next section, however, the foundational debate on deterministic alternative interpretations of quantum mechanics significantly changes that perception.

11.4. Hidden Variables and Bell's Inequalities

From very early on, quantum physicists thought about the question of whether the indeterminacy interpretation of quantum physics was unavoidable. Louis de Broglie (1930) in particular tried to develop a hidden variables theory, that is, a formulation of quantum physics that retained the principle that future events could not be precisely predicted based on earlier experimental data but nevertheless contained an ontological component that conformed to the principle of causality based on a tractable set of natural laws. What he was looking for was determinacy at an ontological level combined with indeterminacy at the epistemic level.

De Broglie's ideas did not attract much interest during the first decades of quantum physics. The possibility to formulate a question of the given kind in a scientifically meaningful way already demonstrates, however, that Schlick's approach to the causality principle is uncomfortably narrow. For Schlick, a theory's scientific statements do not have meaning beyond the theory's predictive qualities. Ontological statements which cannot be expressed in terms of predictions are either meaningless or must be reduced to purely mathematical statements about the theory's structure. Such purely mathematical statements, however, must be entirely unrelated to the concepts of indeterminacy and causality, since the latter, according to Schlick's understanding, are empirical rather than mathematical concepts. From Schlick's perspective, the question posed by de Broglie may be asked at a purely mathematical level about some structural properties of quantum physics but cannot have a meaningful rational reconstruction in terms of causality or the concept of indeterminacy.

De Broglie suggests that the answer to the question of whether a, as he would put it, deterministic interpretation of quantum mechanics exists clarifies an essential characteristic of the theory with regard to its causal structure. Schlick, based on his position, must deny the question's significance in this sense.

Note that finding the question interesting along the line suggested by de Broglie does not necessarily imply taking hidden variable theories to be physically meaningful. Even if one takes an empiricist stance and considers such theories physically uninteresting due to their lack of empirical implications, one may well be interested in the question of whether quantum physics allows for hidden variables. The answer to this question just tells one something about the structure of quantum mechanics.[3] If one decides to discuss the question, however, it seems perfectly natural to discuss it terms of determinacy versus indeterminacy, which simply denotes the intuitive framework of the analysis. It therefore turns out that Schlick's strictly empiricist definition of determinacy obstructs a full and intuitively accessible analysis of quantum mechanics. For that reason, once it had been established that the question of hidden parameters did constitute an important philosophical question regarding quantum mechanics, Schlick's perspective did not have a chance to survive. It is important to note that the analysis of hidden parameters in quantum mechanics did not just work against Schlick's radically empiricist understanding of causation in quantum physics. By providing an example where Schlick's reduction of causation to prediction did not allow an adequate characterization of an important debate in physics, it actually worked against Schlick's theory of causation in general. Schlick's understanding of the causation principle as a form of advice to aim at empirical predictions (and, correspondingly, of the denial of the causation principle as advice not to do so in some contexts) just did not seem capable of accounting for some important use physicists made of the term in their work. Schlick's concept thus was in danger of decoupling from the scientific discourse.

Going even further, by threatening the viability of Schlick's theory of causation, quantum mechanics arguably weakened his entire conception of natural laws as advice for the creation of statements.

The history of the debate on hidden parameters is, of course, well known. John von Neumann early on believed one could prove the inconsistency of hidden parameter models. This, however, turned out to be incorrect. In the 1950s, David Bohm (1952) started developing specific realizations of hidden parameter models. In 1964, John Bell showed that the class of local hidden parameter models was empirically inequivalent with canonical quantum mechanics (see Bell 1966). Later on, experiments carried out on those grounds confirmed canonical quantum mechanics against its local deterministic rival.

The proof that local hidden parameter models are empirically distinguishable from canonical quantum mechanics and the eventual empirical refutation of those models on that basis further strengthens the case against Schlick's understanding of the causality principle. Schlick relates the meaning of the causation principle to advice regarding the search for predictions in a specific context. Thereby, he goes beyond relating causality to observable features. He actually singles out one specific observable feature that is addressed by the causality principle: the question of whether one should look for a certain class of empirical predictions. Since hidden parameter models do not aim at offering more precise predictions than canonical quantum physics, this question plays no role in their discussion. Bell's analysis of local hidden parameter models deals with another empirically relevant aspect of causation. He asks whether the implementation of a certain type of deterministic system (i.e. the vindication of a causality principle) is empirically distinguishable from canonical quantum mechanics. By demonstrating that this is indeed the case, he establishes the empirical significance of the question of determinism (and thus also the question of causation) that goes beyond what Schlick is ready to acknowledge. What Bell shows is that determinism based on local hidden parameters can be tested in the context of quantum physics not by predicting the precise outcome of the determinate process but by looking at other experimental signatures. The complex of causation and determinacy thus becomes a part of the analysis of empirically relevant science without becoming expressible in terms of the range of the theories' predictions regarding the exact outcome of quantum processes. This, however, is at variance with Schlick's straightforward reduction of causation to prediction.

11.5. Quantum Mechanics and Constructive Empiricism

Logical empiricism has long been replaced as the leading paradigm of empiricist thinking by van Fraassen's constructive empiricism, which was formulated in 1980 within the framework of the by-then-popular realism debate (van Fraassen 1980). The core question to be answered by constructive empiricism is whether the success of scientific statements is in some way related to their truth.

van Fraassen's position was motivated by the search for an empiricist position that could withstand the massive philosophical criticism (which he partly recounts in Section 1) that had led to the virtual abandonment of logical empiricism in the 1960s and 1970s. Constructive empiricism differs from its predecessor in a number of ways.

First, van Fraassen endorses a literal understanding of scientific statements. He rejects the rigid theory of meaning that was defended by the logical empiricists and strongly criticized on various grounds by their opponents. In his view, we know what we are talking about when we distinguish, to give an example, the claim that electrons exist from a statement about a bundle of empirical phenomena explicable based on the concept of the electron. Unlike the logical empiricists, van Fraassen can thus distinguish in a meaningful way between truth and empirical adequacy.

Having introduced the distinction, van Fraassen can participate in the modern debate on scientific realism. He thereby departs from the logical positivist position of Carnap and Hempel, who reject the philosophical relevance of the realism debate based on denying that a realist or anti-realist position can have any significance beyond the purely pragmatic question of language choice.[4] van Fraassen defends empiricism based on an analysis of scientific activity and reasoning. He introduces the concept of empirical adequacy that can be distinguished from truth based on a literal understanding of scientific statements. Nonlocal hidden parameters and the canonical interpretation of quantum mechanics, to pick up the scenario discussed previously, are empirically equivalent and thus may be both empirically adequate. Nevertheless, in van Fraassen's understanding, it is possible to assume that one of the two interpretations is true and the other is false. van Fraassen now does not define his position of constructive empiricism in terms of the classical question whether we have reason to believe that our present theories are true or empirically adequate. He rather asks whether scientists *aim* at truth or empirical adequacy. He then proposes the constructive empiricist tenet that it makes more sense to attribute the aim of empirical adequacy to the scientist.

Based on his literal understanding of scientific theories, van Fraassen can state that we can know what we are talking about when discussing some theory's features even if they are not empirically testable. This wider theory of meaning allows meaningful talk about models in science without any immediate translation of that talk into verifiable observational sentences. Therefore, it opens the gates for van Fraassen's model theoretical account of scientific theories. van Fraassen rejects natural laws as universal statements about the world just like Schlick. Scientific theories in his understanding constitute sets of models which are used to represent the phenomena. "A scientific theory is empirically adequate if it has a model such that all appearances [i.e. structures which can be described in experimental and measurement reports] are isomorphic to empirical substructures of that model" (van Fraassen 1980, 64).

Constructive empiricism remains close to earlier empiricist conceptions in a number of respects. van Fraassen's understanding of causality, in particular, does show resemblances to Schlick's account. In van Fraassen's understanding, causality is placed at the level of models. It characterizes the model rather than the actual world. In this vein, one might call the causality principle a construction principle of a model, which would be an understanding reminiscent of Schlick's account.

The new conceptual elements introduced have generally been acknowledged to allow more successful answers to a number of questions in general philosophy of science than logical empiricism. van Fraassen's literal understanding of scientific statements and his semantic model theoretic approach were taken to offer a more adequate conception of how science in fact proceeded.

The previous discussion shows, however, that one of the most important new elements introduced by constructive empiricism, the literal understanding of scientific statements, can also be read in terms of a reaction to the altered character of the foundational debate in quantum mechanics. We have seen how the debate on hidden parameters from the 1950s and 1960s onwards strongly suggested an understanding of causation that was based on abstract conceptual considerations and could not be reduced to the question of the predictive range of quantum physics. Schlick's understanding of causation failed to account for the character of that debate. van Fraassen, to the contrary, can account for it based on his literal understanding of physical statements. Schlick decoupled universal laws from empirical confirmation by denying them the status of statements. Since causation for Schlick was an empirical concept, that decoupling implied that he could not understand de Broglie's hidden parameters in terms of causation. van Fraassen chooses a different strategy: he understands universal scientific statements as pertaining to models rather than to the world. At the level of models, however, the assertion that some model is a causal model, a determinate model, or an indeterminate model does constitute a statement about the model's characteristics with cognitive content. Based on this construction, van Fraassen can take seriously the debates on ontology in the context of quantum mechanics without acknowledging a realist implication of those debates. He therefore can account for the full range of scientific analysis while remaining faithful to his empiricist stance.

11.6. Conclusion

The comparison between Schlick's understanding of the causality principle and van Fraassen's understanding of causation can indeed be related to a shift in the interpretation of quantum mechanics. While in Schlick's time, the uncertainty principle and the indeterminacy of quantum mechanics could be understood in an uncompromisingly empiricist way in terms of the theory's predictive range, the later in-depth analysis of hidden parameter models changed that situation. Bell's analysis and the subsequent empirical tests demonstrated that the attribution of stochastic properties to the quantum state itself, that is, the distinction between indeterminacy due to our immutable lack of knowledge and indeterminacy due to the stochastic quality of the quantum state itself was in fact important for acquiring a full understanding of quantum physics. Based on that new understanding, even adherents to the canonical interpretation of quantum physics could not accept Schlick's strictly empiricist interpretation of quantum physics anymore. Quantum mechanics thus had changed from being a potential supporter of Schlick's interpretation of the causality principle to providing a strong argument against it.

van Fraassen's constructive empiricism is a position that accounts for the stronger emphasis on a characterization of the quantum state in today's quantum physics. His emphasis on a literal understanding of concepts stands in full agreement with that approach. van Fraassen's understanding of causation within his model theoretical approach allows for universal statements without requiring the assumption that they are verifiable by empirical data.

At a general philosophical level, van Fraassen aimed at developing an empiricist position that could withstand the criticisms that had been put forward against logical empiricism. The evolution from early logical empiricism towards van Fraassen's constructive empiricism at this level clearly reflects the results of a long process of analysis and criticism of empiricist ideas. The comparison of constructive empiricism with Schlick's version of logical empiricism carried out in the present work suggests that a similar story can be told at the level of the philosophy of physics. Schlick's application of logical empiricist ideas was based on an understanding of quantum physics that was legitimate at the time. The modern foundational debate on quantum physics, however, does not fit well with Schlick's theory of causation in quantum physics. van Fraassen's constructive empiricist take on causation has been strongly influenced by more recent debates on quantum mechanics and thus avoids the problems faced by Schlick.

11.7. Acknowledgments

I am grateful to Matthias Neuber, Sebastian Lutz, and an anonymous reviewer for very helpful comments on draft versions of this chapter.

Notes

1. For wider discussions of Schick's works on causality, see Stöckler (1996), Stöltzner (2008), Fox (2009).
2. See also Frank (1932) and his earlier (1907/1949) article on the law of causality.
3. In terms of an analysis of mathematical structure, the debate on empirically equivalent alternative interpretations of quantum mechanics thus can be of interest even from Schlick's perspective, as was emphasized by Stöltzner (2008).

4. Schlick, it must be noted, did envision a reconciliation of logical empiricism with realism without sacrificing verificationism; see, for example, Schlick (1932/1979). (Ideas on a rapprochement of principles of logical empiricism and realism were also developed by other exponents of the Vienna Circle, such as H. Feigl and V. Kraft, E. Kaila, or E. Nagel. See, e.g. Matthias Neuber [2011] and his chapter in this volume.)

References

Bell, J. S. (1966), 'On the Problem of Hidden Variables in Quantum Mechanics', *Reviews of Modern Physics*, 38: 447.

Bohm, D. (1952), 'A Suggested Interpretation of the Quantum Theory in Terms of "Hidden Variables"', *Physical Review*, 85: 166–179 and 180–193.

Carnap, R. (1936–37), 'Testability and Meaning', *Philosophy of Science*, 3: 420–471 and 4: 1–40.

De Broglie, L. (1930), *Electrons et Photon, Rapport au Ve Conseil Physicque Solvay*. Paris: Gautiers-Villars.

Fox, T. (2009), 'Die letzte Gesetzlichkeit – Schlicks Kommentare zur Quantenphysik', in F. Stadler and H. J. Wendel (eds.), *Stationen. Dem Philosophen und Physiker Moritz Schlick zum 125. Geburtstag*. Wien: Springer, pp. 212–258.

Frank, Ph. (1907/1949), 'Experience and the Law of Causality', in *Modern Science and Its Philosophy*. Cambridge, MA: Harvard University Press, pp. 53–60.

——— (1932/1998), *The Law of Causality and Its Limits*. Translated by M. Neurath and R. S. Cohen. Dordrecht: Springer.

Neuber, M. (2011), 'Feigl's "Scientific Realism"', *Philosophy of Science*, 78 (1): 165–183.

Schlick, M. (1920/1979), 'Philosophical Reflections on the Causal Principle', in H. Mulder and B. F. B. van de Velde-Schlick (eds.), *Moritz Schlick: Philosophical Papers, Volume I (1909–1922)*. Dordrecht: D. Reidel, pp. 295–321.

——— (1931/1979), 'Causality in Contemporary Physics', in H. Mulder and Barbara F. B. van de Velde-Schlick (eds.), *Moritz Schlick: Philosophical Papers, Volume II (1925–1936)*. Dordrecht: D. Reidel, pp. 176–209.

——— (1932/1979), 'Positivism and Realism', in H. Mulder and Barbara F. B. van de Velde-Schlick (eds.), *Moritz Schlick: Philosophical Papers, Volume II (1925–1936)*. Dordrecht: D. Reidel, pp. 259–284.

Stöckler, M. (1996), 'Moritz Schlick über Kausalität, Gesetz und Ordnung in der Natur', in R. Hegselmann and H.-O. Peitgen (eds.), *Modelle sozialer Dynamiken: Ordnung, Chaos und Komplexität*. Wien: Hölder-Pichler-Tempsky, pp. 225–245.

Stöltzner, M. (2008), 'Can Meaning Criteria Account for Indeterminism? Moritz Schlick on Causality and Verificationism in Quantum Mechanics', in F. O. Engler and M. Iven (eds.), *Moritz Schlick. Leben, Werk und Wirkung*. Berlin: Parerga, pp. 215–245.

van Fraassen, B. C. (1980), *The Scientific Image*. Oxford: Clarendon Press.

12 The Legacy of Logical Empiricism

Clark Glymour

> For the most part it is not a good sign if scholars are too eagerly occupied with the foundation and history of their discipline instead of producing new statements concerning the subjects treated by them.
>
> Otto Neurath (1930/1983, 32)

12.1. Visions of Philosophy

Passing through the 2019 meeting of the International Congress on Logic, Philosophy of Science, Scientific Method and Technology, I wondered what the logical empiricists would have thought of the professional philosophy of science now pursued by thousands but stimulated almost a century before by two little coffeehouse groups, one in Vienna, one in Berlin. My guess is that they would have been perplexed, all the more so if they opened the books offered there in Prague by various publishers. The literal point of philosophy according to Carnap had vanished – only a general attitude remained; the very point of philosophy according to Reichenbach was scarcely seen. Reichenbach could have found philosophical work he would have recognized and probably would have valued scattered thinly through the 21st-century literature.

The logical empiricist movement had two central, conflicting ideas of what philosophy should be. One, represented by Rudolf Carnap, was this:

> *Philosophy is to be replaced by the logic of science* – that is to say, by the logical analysis of the concepts and sentences of the sciences, *for the logic of science is nothing other than the logical syntax of the language of science.*
>
> (1934/1937, xiii, original emphases)

Philosophy was to be logical reconstruction, where "logic" meant something like proof theory. That cramped vision was realized in Carl Hempel's theories of confirmation and explanation and by Carnap's own theory of probability.

Hans Reichenbach (1930) had a different idea, which he announced in 1930 in an editorial in the first issue of *Erkenntnis*: "Philosophy should produce *results*, not manifestos". Perhaps taking Reichenbach's idea as a rejection of his own work, Moritz Schlick reportedly resigned from editorship of the journal in response to Reichenbach's manifesto against manifestos (Stadler 2011, 140–141). The results Reichenbach had in mind were undoubtedly various, but his own work suggests a more modest version of a grand vision given much later by Michael Friedman, the doyen of historical studies of Kant and the logical empiricists:

> Science, if it is to continue to progress through revolutions . . . needs a source of new ideas, alternative programs, and expanded possibilities that is not itself scientific in some sense – that does not, as do the sciences themselves, operate within a generally agreed upon framework of taken for granted rules. For what is needed . . . is precisely the creation and stimulation of new frameworks or paradigms, together with what we might call meta-frameworks or meta-paradigms – new conceptions of what a coherent rational understanding of nature might amount to – capable of motivating and sustaining the revolutionary transition to a new first-level or scientific paradigm. Philosophy, throughout its close association with the sciences, has functioned in precisely this way.[1]

(2001, 23)

More or less in accord with this conception of the role of philosophy, Reichenbach proposed a method for the discovery of "unobserved things", a construction of the direction of time from the macroscopic statistics of causal relations, and an entirely empirical theory of the truth conditions for counterfactuals. None of these efforts succeeded, although two of them presaged revolutionary developments that took four more decades to emerge.

Reichenbach and Carnap began with a common neo-Kantian spirit, but they ended poles apart. For Carnap, philosophy of science was to be "reconstruction", logical representation of what others, real scientists, did. For Reichenbach, philosophy of science was to be a productive enterprise, contributing to the sciences, whether in the grand way Friedman envisioned or less grandly by filling big conceptual holes in the science of the day. In Reichenbach's vision, as appropriately trained outsiders, philosophers might make significant, perhaps even revolutionary, contributions to science by untangling urgent problems, unsowing confusions, articulating generalizations, inaugurating new methods, refuting dogmas – indeed, just what many philosophers attempted from the 17th through the 19th centuries. In that vision, Reichenbach stood in the great tradition of modern philosophy, from Descartes to Peirce. To do so, he had to overcome the limits to the conception of philosophy he inherited from his neo-Kantian education. Carnap never overcame his. Carnap would have seen in the gathering of philosophers in Prague in the summer of 2019 that his passive conception of philosophy had won the century, even if his commitment to logical syntax had not. Reichenbach would have had to sift other literatures to see the survival of his optimism and ambition for the subject.

12.2. Contributions to Physics

Friedman gave no example from the logical empiricists of the contributions he envisioned. That was not from oversight. Physics, principally the theories of relativity, influenced the logical empiricists and their progenitors – notably Schlick – but they gave nothing back, nothing, at least, of scientific value. To the exclusion of almost all other sciences, the principal logical empiricists, champions of "scientific philosophy", loved physics and physicists, from Helmholtz and Planck to Einstein. They provided nothing in return except popularization and cheerleading. That was partly by design, partly by failure. At the time, the most explicit acknowledgement of a philosophical influence on physics was Hermann Weyl's (1922, 319, n1) nod to Husserl in *Space, Time, Matter*. Although educated in mathematics and physics, the logical empiricists left no trace on the substance of the science, on its mathematical and conceptual reformulations, or on its methodology. Even Percy Bridgman's "Operationalism" seems to have been uninfluenced by philosophers of the day. And because most of them were uninterested in almost everything scientific except physics, the logical empiricists made few contributions to any other science. Philip Frank, a physicist among the logical empiricists, collaborated with Richard von Mises on a monograph on differential and integral equations in mathematical physics.[2]

For all of the latter-day criticisms and condemnations of the movement, its barriers against productive engagement with physics and with other sciences still stand, unrecognized, powerful but invisible and only occasionally penetrated, which is why eminent physicists in our day have said with only statistical truth that philosophy of science has nothing to offer physics.

12.3. Problems, Silences, and Irrelevancies

Consider some of urgent foundational issues in the physics of the first half of the 20th century that the logical empiricists might have addressed but did not or did address but poorly:

- *Confirmation and Theory Testing*: What was the logic of the three "classical" tests of the general theory of relativity: which parts of the theory did the respective tests bear on and why? What alternative explanations did they remove? What further tests would inform us about the theory, and why? Was there a general strategy of inference at work? Only Reichenbach brings up the subject, and his discussion is cursory and uninformed. The issue was debated in the physics journals, sometimes rather hotly. The first "philosophical" – in the sense of dispassionate and quasi-formal – analysis came from Harold Jeffreys, then a schoolboy, later a distinguished statistician. Jeffreys' ideas were taken up (without credit) by Eddington and became the basis of the modern "post-Newtonian" classification of the empirical implications of metrical theories of the gravitational field. In two essays in the 1930s, Carnap offered an account of how "theoretical" claims can be tested by observational claims, ideas which he did not pursue and which he did not apply to physical problems. Hempel provided his "purely syntactic" account of confirmation, which, aside from the problem of predicate selection pointed out by Nelson Goodman, provided no account of confirmation of hypotheses with "unobservable" quantities and no account of errors of measurement. On Hempel's theory, the smallest error or omission implied that the data could provide no support for a theory. Both Carnap's and Hempel's accounts of confirmation were light years away from the problems of physics.

- *Identifiability in the Theory of Relativity:* What did the Lorentz signature of the metric field indicate about determinable and indeterminable physical quantities of space-time? Schlick drew conclusions about conventionality from the choice of inertial reference frames but added no formal insights. Reichenbach attempted an answer that was seriously criticized by Weyl and by von Neumann. A. A. Robb did a better job.

- *Meanings of Theories.* What was the proper interpretation of what quantum theory says about the world? That debate was famously carried out between Bohr and Einstein. Reichenbach later took credit for suggesting that the world might be indeterministic, but any insight into the problem was left to mathematicians such as von Neumann and, later, Bell. The logical empiricist literature about "the meaning of theoretical terms" helped not at all, nor was it intended to.

- *Theoretical Equivalence.* Poincaré had shown that a model of hyperbolic plane geometry could be embedded in the Euclidean plane, but were the theories literally equivalent under different measuring standards? For a while, physics had two competing theories of quantum phenomena, Schrödinger's and Heisenberg's. Was one of them true and the other false? Were they synonymous? Were they merely different mathematical formulations of a common mathematical structure? Von Neumann showed the third was the case. I can find no mention of the problem among logical empiricist writers, and no wonder, because they trivialized such problems. Even if they were sufficiently *au courant* with the physics, the matter was for them a pseudo-problem: either the theories have the same empirical implications, or they do not. If the latter, then they are not synonymous; if the former, then they are synonymous. In Reichenbach's later formulation, two theories are synonymous if they are equally confirmed by the same evidence, a proposal that helped not at all, especially since Reichenbach had no coherent account of how theories with "unobservable" claims could be confirmed.

- *Theoretical Unification.* Could the quantum theory and relativity be unified? The physicists offered proposals, but not the philosophers. The philosophers did not explore the methods or virtues of unifying distinct theories, and until Ernest Nagel's effort in 1949, there was no attempt a systematic account of theory unification.

- *Computability and Knowability.* Computability had always been essential to science, first in the need to estimate quantities from noisy data, which prompted the rapid conversion in the 19th century from uncertain but bounded inference to least-squares estimation. Computation came to the fore in physics during World War II, notably to compute the implications of various physical models. The theory of computation had grown directly out of logic and was as philosophical as anything in 20th century science, but with the exception of Carnap's discussions with Gödel, and his unsuccessful attempts to understand completeness and categoricity,[3] the logical empiricists had nothing to say about computation and its complexity, nor did their students until Hilary Putnam's work in the 1960s. The development of the theory of computation in the 1930s was left entirely to mathematicians – Turing, Post, Kleene, Church. The philosophers were largely oblivious.

- *Probability and How to Use It.* With the ascension of quantum mechanics, understanding probability became essential to understanding physics – although it is fair to say that probability should have been essential since the discovery of least squares by Legendre. Popper, who did not count himself a logical empiricist, proposed treating probability as a fundamental physical feature and proposed reformulations so that measure zero events could be conditioned on. Reichenbach's doctoral thesis was on the foundations of probability, exploiting ideas due to Poincaré, but made no technical advance. His later work, mixing attempts at a frequency theory with "probability logic", was a mess, and the statistical parts of his *Theory of Probability* were likely written, or at least sketched out, by Valentine Bargmann, Einstein's assistant at the Institute for Advanced Study and an old friend of Reichenbach's.[4] Carnap's logical theory of probability was utterly inapplicable, and his attempt to characterize entropy was so antipathetic to physics that von Neumann urged him not to publish it (Köhler 2001). Frank Ramsey's brief behaviorist account of subjective probability had a large influence on statistics but only because of development of the idea by de Finetti.

- *Quantum Logic.* Reichenbach's attempt at a logic for quantum phenomena was inept, unoriginal, and eclipsed by Birkhoff and von Neuman's earlier work.
- *The Relation between Probability and Causality.* Specifying this relation was the most original part of Reichenbach's *The Direction of Time*, and Reichenbach got it badly wrong, likely because he could not actually compute joint probabilities of two variables conditional on their common effect.
- *The Logic of Discovery.* Reichenbach (1938, 114–129) posed the problem of how "unobserved things" can be discovered. He gave a simple, vivid example, the "Cube World", in which inhabitants are to figure out the causes of images on the walls. His solution was a wave of the hand at common causes.
- *The Generation of Novel Concepts.* In their neo-Kantian youth, Russell and Carnap sketched logical constructions of concepts from data. The constructions were not carried out. Carnap characterized the question of whether such logical constructions correspond to anything real as a "pseudo-question". The related question of how useful scientific concepts can be generated from data was never broached. (This was the critique against the *Aufbau* of Eino Kaila and Reichenbach as well.)

Each of these examples was a foundational issue of physics or an instance of one, and some of them are examples of generic problems in philosophy of science. Others, mathematicians, logicians, and physicists, contributed to our understanding of these problems, but the logical empiricists did not. When they tried, as did Reichenbach, their work was inferior to contemporary efforts by others on the same topics or, as with Carnap's theory of probability, scientifically useless.

The logical empiricists had minor scientific influences outside of physics. Hempel, who had taken a doctorate with Reichenbach but attached himself to Carnap's views and projects, collaborated with Paul Oppenheim, who made original contributions to linguistics (as did Reichenbach). One of Carnap's graduate students, Herbert Simon, fathered artificial intelligence, and Bruce Buchanan, Hempel's graduate student's graduate student, embedded Nicod's and Hempel's accounts of confirmation in the first practical machine learning program. Carnap's efforts at a semantic characterization of modals contributed to the development of modern modal logic.

12.4. Why?

A more thorough search might find crumbs in the logical empiricists' bag of influences on physics, but it is pretty much empty. So the question is: why? I suggest four causes:

1. The Kantian conception of the role of philosophy;
2. The influence of Einstein;
3. The model of Frege's and Hilbert's theories of proof;
4. Limited mathematical competence.

Quine (1951) famously construed the logical empiricist program as two dogmas, a distinction between analytic and synthetic propositions and the "reducibility" of scientific concepts to those about the observable. Earlier, Russell (1950) gave an even more trenchant summary and criticism, which seems to have gone unnoticed. These summaries were only nutshells. A fuller, more accurate account of what was going on with logical empiricism seems first to have been given by Bella K. Milmed (1961).[5] But she, along with Russell and Quine, left out half the story.

There are two interwoven threads in the modern philosophical tradition, often sported by the same people. On the one side, from Descartes, Leibniz, Locke, Berkeley, and Hume, the concern was with a speculative – Hume called it "experimental" – cognitive psychology and with how one could know the workings of everyday thought and experience, some of which involved what nowadays we would perhaps call the analysis of concepts. Hume's account of the idea of causation is the most famous example. On the other side, there was methodology, general principles, sometimes speculative, sometimes quasi-algorithmic, for fathoming nature and engagement to improve the sciences with them. Descartes offered deduction from the "natural light of reason" and claimed to put it to work in his geometry, physics, and elsewhere. Leibniz proposed a universal calculus for scientific discovery, never realized. Bacon expounded a method that presaged modern causal analysis and used it to provide a correct theory of heat. Whewell championed "consilience", and while he discovered nothing I know of, he at least named some scientific kinds, anion and cation. Mill provided a more or less systematic handbook of methodology, central parts of it borrowed from Bacon. Boole developed an algebra of logic intended as ancillary to the discovery of causes. Richard Price offered Bayesian reasoning as an antidote to Hume's skepticism and as a method for the sciences. Helmholtz mixed philosophy and mathematics in a critique of Kantian ideas about space. Peirce introduced randomization in experimental design.

In the "protocol sentence" debate in the Vienna Circle, Neurath and Carnap rejected the very idea of a relation between the language of science and an independent external world in virtue of which what is said is true or false. Thus Carnap supplemented his "Logical Construction of The World" with "Pseudo-Problems in Philosophy". Whether a construction of a concept denotes something in the world is a pseudo-problem. The result of these discussions for Hempel was that the notion of truth has no place in science. More generally, the result was antipathy to questions of the reliability – the truth guidingness – of methods – and a replacement of the search for truth by surrogates: explanation and confirmation. Reichenbach, in Berlin, somehow remained largely free of this attitude. By the thirties, his epistemology began with our knowledge of everyday things, and the business of science was to discover general truths about them and about equally real unobserved things.

Logical empiricism thus inherited a divided conception of the relation of philosophy and the sciences. On the first side, represented most vividly by Carnap, philosophy is a meta-logical enterprise remote from contributions to any science or to methodology. On the second side, represented by Reichenbach, ideas about scientific language were combined with articulating methodology and put to work in more concrete scientific efforts in physics, statistics, and elsewhere, generally without success. Nominal allies, Carnap and Reichenbach represented two radically different versions of the job of philosophers, one as passive observers of the life of science, the other as active contributors. Carnap's disposition, if not his syntactic anthem, carried the day.

12.4.1. Kant

Kant was the most proximate and most important philosophical influence on the logical empiricists. Kant's first *Critique* announced its problem to be how synthetic a priori knowledge is possible. That was a teaser. The work can be more broadly viewed as addressing a problem of inference in the face of apparent radical underdetermination: how are we, from the buzzing and booming of sensation, the "matter of experience", to know an external world of things and processes and relationships ordered in time, located in places and occurring with causal regularity? On a psychological reading of Kant, which I endorse (since I don't really understand the alternatives), the pocket answer is that one's mind creates the "cognizable" "manifold of experience" – the things and processes and relationships ordered in time and located in places and related by causal connections. That construction embodies general constraints that cannot be contradicted by experience – synthetic a priori truths that include Euclidean geometry, the elements of Newtonian physics, the universality of causal connections, and, in one edition of the *Critique of Pure Reason*, the framework of phlogiston theory. Kant's view was that scientific concepts and claims must be cashed out exclusively through the meshing of a priori functions with sensations.

Kant's engagement with Newtonian physics was real and serious but not technical or mathematical and, except for rather opaque references to the argument of Book III of Newton's *Principia*, uninterested in how physical theories can be tested. Indeed, in Kant's view, the central assumptions of Newtonian mechanics and celestial mechanics could not be tested, because necessarily no conceivable experience could contradict them. Kant, who made minor but genuine contributions to science, had no evident interest in the details of scientific method; his engagement with Newtonian dynamics focused on proposing a priori foundations and arguing that those foundations are the cynosure of the relations of mind and experience. Kant pointed philosophy of science away from creating new paradigms to stir the sciences and towards "reconstruction" and "justification".

The 19th and early 20th century wreaked havoc with Kant's framework. Not long after Kant had declared the a priori certainty of physical Euclidean geometry, Gauß attempted to measure the deviation of the internal angles of triangles from 180 degrees. After non-Euclidean geometries came proofs that hyperbolic geometry is consistent if Euclidean geometry is. Well before the theories of relativity, there was work to do saving Kant and work to do reconsidering which, if any, of Kant's arguments were sound.[6]

Reichenbach (1915/2008), Carnap (1921/2019), and Russell (1897) all wrote doctoral theses devoted to revising, extending, or saving Kantian epistemology from mathematical and scientific developments in the late 19th century. Russell and Carnap focused on the geometry of physical space. Russell's thesis offered an a priori demonstration of the "axiom of free mobility" – what we call constant curvature; Carnap argued that the topology of space is known a priori. Reichenbach's aim was to make the theory of probability synthetic a priori or at least a necessary condition for empirical knowledge. Russell's and Reichenbach's proposals were not particularly original. Helmholtz had given arguments for constant curvature, and Reichenbach's thesis was straight out of Poincaré but with a Kantian twist. Judging from Kant's remarks on "monadology" in the *Metaphysical Foundations of Natural Science*, the nascent logical empiricists would have had more to do saving Kantianism if Kant had lived long enough to learn of Dalton's atomic theory.

Kant's most important influence was not in prompting this kind of backing and filling. It was in framing the very enterprise of philosophy of science. Among the logical empiricists, "reconstruction" replaced the innovation Friedman later envisioned and that philosophy had traditionally tried to provide. Kantian quasi-psychology was replaced by an account of how conventions of language did the same job for science. The a priori became the analytic by convention. The logical empiricists transformed Kant's fixed a priori into a more flexible linguistic mode, with "coordinating definitions", and "correspondence rules" somehow connecting words and things but beyond empirical testing. They were a priori because the language of scientific practice made them so. They could be abandoned but not refuted. They were the "pragmatic" a priori. Having thus identified the conventional or stipulative or analytic, the remainder of science was revealed by Nature, empirically. A principal philosophical task became doing this sorting for one or another piece of science, separating the "analytic" from the "synthetic". Some of Reichen-bach's writing about relativity is of this kind, and Carnap's alleged mas-terpiece, *The Logical Syntax of Language*, articulated the program in a straightjacket: philosophy is properly about logical syntax.

12.4.2. Einstein

Whatever bit of methodological thread Kant left to the logical empiri-cists was cut shorter by Einstein. By the later part of the 19th century, a tenable opinion was that physics was basically done, settled. John Tyn-dall, the eminent physicist and popularist, lectured that physics was fully formed; all that remained was the discovery and formulation of fur-ther forces beyond gravity and electromagnetism fitting into the New-ton scheme, if any exist. In that setting, the Theory of Relativity was an epistemological shock. Neo-Kantians of various kinds, Cassirer and Schlick, were all over it. Schlick leaned on it for his epistemology and wrote a "philosophical" exposition, which Einstein praised. But most importantly, what it meant for the logical empiricists and their allies was that no matter how much evidence might accrue for a theory, the theory could still be false. The collapse of Newtonian theory brought back the skeptical lesson Hume took from *The Meno*. The moral most of the logical empiricists drew was that science could not be an enterprise whose business is to discover general truths about the world, because that is impossible. Instead, Popper claimed science is an enterprise for tormentors of Sisyphus – ever offering new rocks of theory to carry up in search of verification and making sure that each rock fell, smashed to pebbles. Hempel, who claimed the notion of truth is useless to science, offered ersatz goals: explanation and confirmation. The unknowability of the *ding an sich* was promoted to the unknowability of generaliza-tions about the empirical world. Which is why, to the surprise of many who misunderstood what they were about, Carnap and Hempel found a kindred spirit in Thomas Kuhn.

Einstein was almost an oracle for the logical empiricists. Einstein, the determinist, held that science is a "free creation". Science could only spring from human genius. There could be no recipe, no algorithm, for scientific discovery.[7] The logical empiricists signed on, mostly by ignoring the very possibility, even as regression and factor analysis were advanc-ing as algorithmic methods in psychology and elsewhere. With a fine instance of an enduring philosophical argument – *I cannot imagine how to do X, therefore doing X is impossible* – Hempel (1988/2000) explicitly denounced the possibility of algorithmic contributions to scientific dis-covery. Science, he argued, requires the introduction of new theoretical quantities that are not explicitly definable from observed quantities, and no algorithm could ever do that. Only Reichenbach, with his Principle of the Common Cause, half-demurred.

12.4.3. Frege and Hilbert

Frege provided a model of logical achievement. He provided an explicit mathematical theory intended to capture the very idea of a demonstrative proof, a theory that seemed adequate for understanding much of mathematical practice. Carnap studied with him, and Russell famously corresponded. His efforts were remedied and improved by Hilbert, with whom Hempel studied. Gödel attended Vienna Circle meetings. Having dismissed the very idea of science as a search for special kinds of truths, the logical empiricists attempted to do for their surrogates – explanation and confirmation – what Frege and Hilbert had done for proof. Formal theories of explanation and confirmation were cast as logical relations.

Neither Frege nor Hilbert had attempted to give mathematical accounts of notions such as "mathematically interesting" or "mathematical explanation". They were wise enough to know that some notions are too diffuse and subjective for useful logical formalization. The logical empiricists were not.

Logic was a clumsy tool for real physics, which required more specialized mathematics. Methodological problems that were pertinent to science fit within formal logic only with a complexity that made them intractable. Hilbert, a great mathematical physicist, gave an axiomatic theory of geometry that could be formalized in second-order logic, but he did not do his physics in any order logic. The logical empiricists missed the message. So, for example, the logical empiricists proposed accounts of "empirical meaning" intended to separate empirically meaningful predicates and sentences from everything else but said nothing about the nearly isomorphic problems in physics that came later to be known as "gauge symmetries", already present in the scalar potential of Maxwell's electrodynamics or the similar problems for linear models that came in economics to be known as "identification problems".

12.4.4. Competence

Few human beings have the mathematical capacities of von Neumann or Gödel or Hilbert. But less gifted minds, well-trained and serious and imaginative, can and do make important contributions. Of the logical empiricists, after the physicists and mathematicians, Carnap was perhaps the most mathematically able, but his efforts were entirely – albeit often ingeniously, as with his theories of probability and modality – confined to developing reconstructions within a logical formalism. Except for Philip Frank, whose philosophical contributions were chiefly reportorial, Reichenbach was the logical empiricist who most engaged physics and who best attempted to contribute to its conceptual problems. In his early years, he was sufficiently competent with field theories to produce a technical argument that they could all be "geometrized", an argument with which Einstein seems to have agreed.[8] But outside of that setting, he lacked the mathematical knowledge and skills to carry out some of his bold and imaginative ideas. Although he wrote extensively about probability, I very much doubt he could have defined or calculated a correlation coefficient. He did not understand the relevant properties of independence and conditional independence. In consequence, his attempt at a macroscopic causal/statistical characterization of the direction of time was a nest of confusions. Away from relativity, Reichenbach was a mathematical wood-pusher.

12.5. The Legacy

Philosophy's most important legacy from the logical empiricists is the rejection of Friedman's manifesto: furthering the search for truth; providing new methods, new insights into the structure of going theories; jolting the sciences with a new track is none of the business of philosophy of science. Philosophy of science lost to psychology any claim to fathom the mind and abandoned methodology to statistics. The result is that, with occasional excellent counter-examples, philosophy of science has become bizarre journalism, or fan fiction, or exposition of esoterica (e.g. string theory) with "philosophical" commentary that could as well have been obtained from high school chemistry or just silliness. So, in recent philosophical books, we read that scientists cannot discover theories they have not thought of (Stanford 2006), and we read that Darwin's reasoning to the theory of evolution was produced by changes in neural connections in Darwin's brain (Churchland 2012), and we read that Newton's argument for universal gravitation was really all about estimating the gravitational constant in that law (Harper 2011).

Notwithstanding, many of the problems of physics and other sciences that the logical empiricists passed by or dealt with poorly were taken up in the late 20th and 21st centuries by their heirs, wittingly or not. And there are novel frameworks the logical empiricists did not imagine. I will give just a few examples.

1. *Confirmation and learning.* Confirmation theory became a matter of assigning numbers to sentences and claiming their relations captured patterns of "support" or confirmation. In contrast, a body of mathematical work by philosophers, inspired by Hilary Putnam's (1965) work in the early 1960s, has generated a theory of learnability from data that parallels recursion theory. It applies pretty directly to aspects of physics, for example, to inferences based on the post-Newtonian representation of metrical theories of gravitation. Most remarkably, it has been used to show that the practice of postulating novel fundamental particles, their quantum numbers, and conservation laws is exactly in accord with the norms of formal learning theory (see Kelly 1996; Schulte 2000).

2. *Identifiability in relativity.* Einstein observed that the choice of surfaces of simultaneity with respect to an unaccelerated observer is "conventional". Reichenbach, and then Adolf Grünbaum, made a huge deal of that. Philosophers have since shown that the simultaneity surfaces are uniquely determined by the geometric structure of Minkowski space-time (see Malament 1977; Spirtes 1982). But philosophers also demonstrated another indeterminacy, the global topology of space, and hence of space-time in the most likely physical models of the universe (see Glymour 1972, 1977; van Fraassen 1970).

3. *Meaning of quantum theory.* In works too numerous to cite, philosophers of science have pursued issues of measurement, non-locality, existence of joint probability distributions, and consistency with relativity.

4. *Theoretical equivalence.* A "purely syntactic" criterion for equivalence of theories in different languages was produced, giving the required results that Euclidean geometry and real closed field theory are equivalent and Euclidean geometry and elliptic geometry are not (see Glymour 1970; Barrett and Halvorson 2016). Thanks to Beth's theorem, there is a corresponding model theoretic equivalence for first-order languages. The criterion was informally applied to various gravitational theories (Spirtes and Glymour 1982).

5. *Computability and knowability.* Philosophers (and others) adapted the theory of learning to computationally bounded learners, with results that contradict several hoary methodological dogmas (see Osherson, Stob and Weinstein 1986).

6. *The relation between probability and causality.* Statisticians and computer scientists first articulated the "Markov Condition" for graphical models and its mathematical properties. Philosophers gave the representation a causal interpretation and showed that it corresponds to the relations between causality and probability in experimental interventions (Spirtes, Glymour and Scheines 1993).

7. *Quantum logics.* Abstracting from von Neumann's Hilbert Space representation, philosophers and physicists provided "logics" of quantum mechanics based on the operator algebra (Engesser, Gabbay and Lehmann 2011).

8. *Probability and how to use it.* Beyond the debates over the "interpretation of probability", philosophers have contributed to novel foundations and results (Kadane, Schervish and Seidenfeld 1999) and detailed analyses of statistical debates (Mayo 2018).

9. *The discovery of "unobserved things".* Using the graphical causal model framework, philosophers produced a procedure, the FCI algorithm, that distinguishes between variables that are not confounded by unrecorded common causes and those that are (Spirtes, Glymour and Scheines 1993). Other algorithms correctly clustered measured variables with a common cause. Reichenbach's "Cube World" problem is

easily solved with these algorithms, and they allow a "macroscopic statistical" characterization of the direction of time (Field 2003).

10. *The generation of novel concepts.* The discovery of an unobserved common cause can be the generation of a novel concept. More directly, philosophers and collaborators have produced algorithms for generating concepts directly from time series of "sense data" – time series of pixel firings, for example. The methods can be used for real physical data (Chalupka, Perona and Eberhardt 2015; Chalupka, Bischoff, Perona and Eberhardt 2016).

There are novel scientific frameworks. Extensions of David Lewis' ideas have led to a theory of language creation through simple learning mechanisms applied to signals (Skyrms 2010). Homotopy type theory, a new branch of mathematics, was created in part by philosophers (Awodey and Warren 2009). The project of algorithms for discovering causal relations from non-experimental data was a philosophical creation, now with contributions from hundreds of statisticians and computer scientists. And this is only a brief survey. For each reference, there are a hundred others equally relevant, which I do not have the space or energy to give.

There are ironies. Early in their careers, Russell and Carnap sketched proposals for constructing from primitive data what Kant might have called the manifold of experience. Both Russell and Carnap proposed applying higher-order logic to representations of the data of experience – sense data in Russell's case, similarities of elementary experiences in Carnap's – to construct applicable concepts of ordinary properties, things, events, and processes. Only Carnap attempted actual constructions, and he did not get very far, nor did he do it very well. Recently, Frederick Eberhardt, a philosopher at the California Institute of Technology, and his collaborators developed an algorithm for generating causally relevant, novel, higher-order variables from time series of data, such as pixels on a screen or firings of an array of retinal nerves. They applied their algorithm to data on sea surface temperatures and pressures, and out came the ENSO phenomenon. Eberhardt presented these results at a recent meeting to celebrate and discuss Reichenbach's work (*All Things Reichenbach*, LMU-Munich, 22–24 July 2019). Unsatisfied with the example, a philosopher demanded to know whether the variables the algorithm would generate are the real variables in the world. I directed him to Carnap, *Pseudo Problems in Philosophy*.

Notes

1. It should be noted that Friedman gives no 20th-century example at all.
2. Philipp Frank is discussed, mainly from a philosophical and sociological point of view, by Don Howard and Adam Tamas Tuboly in this volume. On von Mises, see Maria Carla Galavotti's chapter.

3. A somewhat refined picture is about Carnap's ideas on categoricity is provided by Awodey and Carus (2001).
4. Reichenbach gratefully acknowledged Bargmann's help both in *Theory of Probability*'s original German edition (Reichenbach 1935, 423, n2 and 438) and in his *Philosophic Foundations of Quantum Mechanics* (1944, vii), where he thanks Bargmann for "advice in mathematical and physical questions".
5. Milmed's insightful book was damned with faint praise by Charles Parsons. Her account is similar to the more detailed view later offered by Friedman.
6. I have never found a concerted analysis by any of the logical empiricists of the defects in any of Kant's *arguments,* for example, the transcendental deduction of the categories of the understanding or the antinomies of reason. Perhaps I should read more.
7. The relation of Einstein to logical empiricism is taken up by Thomas Ryckman and Don Howard in their chapters.
8. On Reichenbach's ideas about geometry and relativity, see Marco Giovanelli's chapter in this volume.

References

Awodey, S. and Carus, A. W. (2001), 'Carnap, Completeness, and Categoricity: The Gabelbarkeitssatz of 1928', *Erkenntnis,* 54 (2): 145–172.

Awodey, S. and Warren, M. A. (2009), 'Homotopy Theoretic Models of Identity Types', *Mathematical Proceedings of the Cambridge Philosophical Society,* 146 (1): 45–55.

Barrett, Th. W. and Halvorson, H. (2016), 'Glymour and Quine on Theoretical Equivalence', *Journal of Philosophical Logic,* 45 (5): 467–483.

Carnap, R. (1921/2019), 'Der Raum/Space', in A. W. Carus et al. (eds.), *The Collected Works of Rudolf Carnap, Volume 1: Early Writings.* Oxford: Oxford University Press, pp. 21–208.

——— (1934/1937), *Logical Syntax of Language.* London: Kegan Paul, Trench, Trubner & Co. Ltd.

Chalupka, K., Bischoff, T., Perona, P., and Eberhardt, F. (2016), 'Unsupervised Discovery of El Nino Using Causal Feature Learning on Microlevel Climate Data', in *Proceedings of the Thirty-Second Conference on Uncertainty in Artificial Intelligence,* Corvallis: AUAI Press, pp. 72–81.

Chalupka, K., Perona, P., and Eberhardt, F. (2015), 'Visual Causal Feature Learning', in *Proceedings of the Thirty-First Conference on Uncertainty in Artificial Intelligence,* Corvallis: AUAI Press, pp. 181–190.

Churchland, P. M. (2012), *Plato's Camera: How the Physical Brain Captures a Landscape of Abstract Universals.* Cambridge, MA: MIT Press.

Engesser, K., Gabbay, D. M., and Lehmann, D. (eds.) (2011), *Handbook of Quantum Logic and Quantum Structures: Quantum Structures.* Amsterdam: Elsevier.

Field, H. (2003), 'Causation in a Physical World', in D. Zimmermann (ed.), *Oxford Handbook of Metaphysics.* Oxford: Oxford University Press, pp. 435–460.

Friedman, M. (2001), *Dynamics of Reason.* Stanford, CA: CSLI Publications.

Glymour, C. (1970), 'Theoretical Realism and Theoretical Equivalence', in R. C. Buck and R. S. Cohen (eds.), *PSA 1970: Proceedings of the Biennial Meeting of the Philosophy of Science Association.* Dordrecht: D. Reidel, pp. 275–288.

——— (1972), 'Topology, Cosmology and Convention', *Synthese,* 24 (1): 195–218.

——— (1977), 'Indistinguishable Space-Times and the Fundamental Group', in J. Earman, C. Glymour, and J. Stachel (eds.), *Foundations of Space-Time Theories: Minnesota Studies in the Philosophy of Science, Volume 8*. Minneapolis, MN: University of Minnesota Press, pp. 50–60.

Harper, W. L. (2011), *Isaac Newton's Scientific Method: Turning Data into Evidence about Gravity and Cosmology*. Oxford: Oxford University Press.

Hempel, C. G. (1988/2000), 'On the Cognitive Status and the Rationale of Scientific Methodology', in R. Jeffrey (ed.), *Carl G. Hempel: Selected Philosophical Essays*. Cambridge: Cambridge University Press, pp. 199–228.

Kadane, J., Schervish, M., and Seidenfeld, T. (1999), *Rethinking the Foundations of Statistics*. Cambridge: Cambridge University Press.

Kelly, K. (1996), *The Logic of Reliable Inquiry*. Oxford: Oxford University Press.

Köhler, E. (2001), 'Why von Neumann Rejected Carnap's Dualism of Information Concepts', in M. Rédei and M. Stöltzner (eds.), *John von Neumann and the Foundations of Quantum Physics*. Dordrecht: Kluwer, pp. 97–134.

Malament, D. (1977), 'Causal Theories of Time and the Conventionality of Simultaneity', *Noûs*, 11 (3): 293–300.

Mayo, D. G. (2018), *Statistical Inference as Severe Testing*. Cambridge: Cambridge University Press.

Milmed, B. K. (1961), *Kant and Current Philosophical Issues*. New York: New York University Press.

Neurath, O. (1930/1983), 'Ways of the Scientific World Conception', in M. Neurath and R. S. Cohen (eds.), *Otto Neurath: Philosophical Papers 1913–1946*. Dordrecht: D. Reidel, pp. 32–47.

Osherson, D. N., Stob, M., and Weinstein, S. (1986), *Systems That Learn: An Introduction to Learning Theory for Cognitive and Computer Scientists*. Cambridge, MA: MIT Press.

Putnam, H. (1965), 'Trial and Error Predicates and the Solution to a Problem of Mostowski', *The Journal of Symbolic Logic*, 30 (1): 49–57.

Quine, W. V. O. (1951), 'Two Dogmas of Empiricism', *Philosophical Review*, 60: 20–43.

Reichenbach, H. (1915/2008), *The Concept of Probability in the Mathematical Representation of Reality*. Chicago and La Salle, IL: Open Court.

——— (1930), 'Zur Einführung', *Erkenntnis*, 1: 1–3.

——— (1935), *Wahrscheinlichkeitslehre. Eine Untersuchung über die Logischen und Mathematischen Grundlagen der Wahrscheinlichkeitsrechnung*. Leiden: Sijthoff.

——— (1938), *Experience and Prediction*. Chicago: University of Chicago Press.

——— (1944), *Philosophic Foundations of Quantum Mechanics*. Berkeley and Los Angeles: University of California Press.

Russell, B. (1897), *An Essay on the Foundations of Geometry*. Cambridge: Cambridge University Press.

——— (1950), 'Logical Positivism', *Revue Internationale de philosophie*, 4 (11): 3–19.

Schulte, O. (2000), 'Inferring Conservation Laws in Particle Physics: A Case Study in the Problem of Induction', *The British Journal for the Philosophy of Science*, 51 (4): 771–806.

Skyrms, B. (2010), *Signals: Evolution, Learning, and Information*. Oxford: Oxford University Press.

Spirtes, P. L. (1982), *Conventionalism and the Philosophy of Henri Poincaré*. Ph.D. Thesis. University of Pittsburgh.

Spirtes, P. L. and Glymour, C. (1982), 'Space-Time and Synonymy', *Philosophy of Science*, 49 (3): 463–477.

Spirtes, P. L., Glymour, C., and Scheines, R. (1993), *Causation, Prediction and Search*. Dordrecht: Springer.

Stadler, F. (2011), 'The Road to Experience and Prediction from within: Hans Reichenbach's Scientific Correspondence from Berlin to Istanbul', *Synthese*, 181: 137–155.

Stanford, P. K. (2006), *Exceeding Our Grasp: Science, History, and the Problem of Unconceived Alternatives*. Oxford: Oxford University Press.

van Fraassen, B. C. (1970), *An Introduction to the Philosophy of Time and Space*. New York: Random House.

Weyl, H. (1922), *Space-Time-Matter*. Translated by Henry L. Brose. London: Methuan & Co. Ltd.

Part 3
General Philosophy of Physics

13 Probability Theory as a Natural Science
Richard von Mises' Frequentism

Maria Carla Galavotti

13.1. About von Mises

In his opening lecture delivered at the "Colloque consacré à la théorie des probabilités" held in 1937 in Geneva, Maurice Fréchet (1938, 19) praised von Mises for "having awakened the interest to questions previously addressed in a fragmentary way". This is evidence of the influence exerted by von Mises' views on probability, which by that time had been entrusted to a number of articles published in prestigious journals like the *Mathematische Zeitschrift* (see, among others, von Mises [1919a, 1919b]), *Zeitschrift für Angewandte Mathematik und Mechanik* and *Statistik und theoretische Physik* and the books *Wahrscheinlichkeit, Statistik und Wahrheit* (1928) and *Wahrscheinlichkeitsrechnung und ihre Anwendung in der Statistik und theoretischen Physik* (1931). As suggested by the last title (in English: *Probability Theory and Its Application to Statistics and Theoretical Physics*), von Mises' interest in probability focuses on its applications to statistics and physics.

Born in Lvov in 1883, after studying mathematics, physics and engineering at the Technische Hochschule in Vienna and obtaining a Ph.D. in 1907 from the local technical university, von Mises was awarded his *habilitation* from Brno, qualifying as a lecturer in engineering and machine construction. He then served as lecturer and professor at the universities of Strasbourg, Frankfurt and Dresden and in 1920 was appointed professor of applied mathematics and director of the newly founded Institute for Applied Mathematics in Berlin. After Hitler came to power in 1933, von Mises, being of Jewish descent, moved to Istanbul, where he became the chair of pure and applied mechanics and the director of the mathematics institute. Finally, in 1939, he moved to Harvard University to become professor of aerodynamics and applied mathematics and died in Boston in 1953.

Memoirs by a number of authors, including Philipp Frank and Harald Cramér, portray von Mises as an extremely energetic man with a very broad spectrum of interests ranging from mathematics, physics, mechanics and engineering to probability, philosophy and poetry – among other things, he was a fervent admirer of the poet and novelist Rainer Maria Rilke and a "recognized authority" (Goldstein 1963, xiii) on his work. His collection of Rilke's writings is reputed to be one of the best. Despite such a wide range of interests, von Mises' scientific production is characterized by great internal coherence. However, as observed by Philipp Frank, one of his best friends and co-author of the Frank-Mises handbook of mathematical physics:[1]

> it would be a misinterpretation of his work if it were considered as the output of a versatile man who split his interests because he was casually attracted by many topics. Actually, von Mises chose the topics according to a definite viewpoint, determined by his ideas about the essence and method of every thoroughly scientific research.
>
> (1954, 823)

Von Mises left outstanding contributions in many fields, including fluid dynamics, the theory of plasticity and geometry. His work on probability and statistics is no less important and covers technical results as well as original views on the foundations of probability, which are the object of this chapter.

13.2. von Mises' Program

von Mises' concern revolves around scientific knowledge, whose most sophisticated representation he found in physics. His attention is not so much focused on theories but rather on "the relationship between the direct sense of observation of the experimental physicist and the conceptual system of science" (Frank 1954, 823). In other words, the problem at the core of von Mises' work is the connection between the data coming from the observations and experiments made by scientists and the abstract concepts belonging to scientific laws. He identified the proper tool for analyzing such a connection as applied mathematics, and more precisely statistics, which provides the means for connecting large sets of experimental data to the abstract entities belonging to theories, such as length (in classical mechanics) or electromagnetic field (in Maxwell's theory of electromagnetism).

As Frank (1954, 823) remarks, the "*problem of connection* between sense observations and abstract principles has always been the critical point in the philosophy of science" (original emphasis). As reported by Friedrich Stadler (2001, 51), von Mises was close to the group of intellectuals, which included, among others, Philipp Frank, Otto Neurath, and Hans Hahn, often associated with the "First Vienna Circle",[2] and later took part in the meetings and discussions of the Berlin group of scientist-philosophers who belonged to the *Gesellschaft für empirische Philosophie*, the association founded in 1928, the same year when in

Vienna the *Verein Ernst Mach* – related to the Vienna Circle – was officially launched.

Like its Viennese counterpart, the Berlin group patronized the attitude towards philosophy, knowledge, and human activities in general summarized by the expression "scientific conception of the world" and which appeared in the title of the "manifesto"[3] of logical empiricism – or logical positivism, as von Mises used to call it. Of note is the title "Scientific Conception of the World: On a New Textbook of Positivism" of a paper by von Mises published in *Erkenntnis* in 1940 (reprinted in von Mises [1964a, 524–529]), in which he summarizes the main theses contained in the book *Kleines Lehrbuch der Positivismus* (1939), later published in English under the title *Positivism*. After he moved to the United States, von Mises was part of the group of scientists and philosophers who used to meet at Harvard, there reviving the logical positivist movement. He also took part in the activities of the Institute for the Unity of Science, founded by Frank in 1947, "somewhat along the lines of the Ernst-Mach-Verein" (Holton 1993, 62).

It has been pointed out that "the Berliner's plan was to explore philosophical problems with scientists and mathematicians in their specific disciplines", while "the objective of the Vienna Circle . . . was to advance specific theories" (Milkov 2013, 5).[4] von Mises' work is fully in tune with such a description. A scientist by training, von Mises considered philosophy not only an important part of his research work but an essential component of his engagement as an intellectual and man of his time. In the preface of *Positivism*, the book in which his philosophical standpoint is spelled out at length, von Mises traces back his own version of positivism to Comte, Poincaré, and especially Mach, whose work "has been continued and in essential points amplified by the so-called Vienna Circle, Carnap, Frank, Hahn, Neurath, Schlick. It is *this* development which provides the basis for all comments on the various problems presented in the following text" (von Mises 1951, v–vi, original emphasis).[5]

A major element of novelty attributed to the Vienna Circle movement is the focus on language and more in particular the stress on the need "to face the language that is used with unceasing logical criticism" (von Mises 1964a, 524). Ernst Mach is the author who exercised the strongest influence on von Mises, who described himself "a devoted disciple of Mach" (von Mises 1964a, 524) and credited Mach with having singled out the gist of the "positivistic" attitude, considered as requiring "a continual readiness to give up a judgment once made or to change it if new experiences require" (von Mises 1951, 1). An essential component of such an attitude is a radical version of empiricism according to which "the majority of all statements which play any role in practical life and in science originate in observations and are continually tested by experience" (von Mises 1951, 6), together with the view that scientific theories are descriptions of connections among phenomena subject to continuous revision in the light of new data. More generally: "science consists of a continually progressing *adaptation of ideas to facts*" (1951, 7, original emphasis). Also part of the same attitude is the tenet that scientific statements are predictions about future experiences and that the criterion for acceptance of scientific theories and laws is given by their utility.

Inspired by these philosophical convictions, von Mises engaged in the program of endowing probability theory with a solid foundation "starting from the thesis that *probability theory should be regarded as the mathematical theory of a group of observed phenomena, in the same way as, for example, geometry and theoretical mechanics*" (Cramér 1953, 658, original emphasis). Observable phenomena form the object of probability, whose task is "to provide a mathematical model of the statistical regularities observed in cases where a given experiment or observation may be repeated a large number of times under similar conditions" (Cramér 1953, 658). By taking such a viewpoint, von Mises is led on the one hand to place physics center stage and on the other to maintain that "probability theory can never become a part of the mathematical theory of sets" (von Mises 1957, 100). Probability, continues the previously quoted passage, "remains a natural science, a theory of certain observable phenomena, which we have idealized in the concept of a collective". The notion of *collective* provides the building block of von Mises' theory of probability. It represents a most lucid and uncompromising version of frequentism, which exerted a great influence on generations of scientists.

13.3. The Nature of Probability

Von Mises is not interested in the intuitive notion of probability or in its everyday use but rather in developing a *rational* notion liable to be defined unambiguously.[6] His first step in this direction amounts to restricting the domain of probability "to problems in which either the same event repeats itself again and again, or a great number of uniform elements are involved at the same time" (1957, 11). To assume that the object of probability theory is mass phenomena, or indefinitely repeatable events, implies that "in order to apply the theory of probability we must have a practically unlimited sequence of uniform observations" (1957, 11). Typical fields of application of probability are games of chance; insurance matters such as determining the premium to be paid for a travel fortuity or accidental death; and physical phenomena of the kind described by the theory of gases, Brownian motion, and radioactivity. Such mass phenomena are called *collectives* by von Mises. Besides representing the object of the theory of probability, collectives also delimit its scope of application. As emphasized by the author:

> The principle which underlies the whole of our treatment of the probability problem is that a collective must exist before we begin

to speak of probability. The definition of probability which we shall give is only concerned with "the probability of encountering a certain attribute in a given collective".

(1957, 12)

As an immediate consequence of this standpoint, probability cannot refer to single events, such as a specific coin throw or the behavior of one gas molecule. von Mises puts it very clearly: "the phrase 'probability of death', when it refers to a single person, has no meaning at all for us. This is one of the most important consequences of our definition of probability" (1957, 11). To say that the probability of single events has "no meaning" is in tune with the terminology adopted by logical positivists, who labeled "meaningless" the concepts belonging to metaphysics.

As stated, collectives are mass phenomena, or repeatable events, which allow for several series of observations that can in principle be prolonged indefinitely. Repeated observations of a given attribute exhibit frequency values that must tend to a limit if these series can qualify as collectives. In addition to being a requirement that must be fulfilled by collectives, the existence of limiting frequencies is the essence of von Mises' frequentism, according to which probability is the limit of the relative frequency of a given attribute observed in the initial part of an indefinitely long sequence of repeatable events. In other words: a collective is an infinite sequence of elements that exhibit a certain attribute whose relative frequency, determined on the basis of the observation of a sample of n elements, is m/n. The probability of such an attribute in the sequence is put as equal to the limit to which the relative frequency m/n tends when $n \rightarrow \infty$. Aware that the limit assumption is a strong one, von Mises requires it all the same, holding that it can be justified in a great number of cases belonging both to science and everyday life. By way of justification, von Mises appeals to Poisson, who "recognized that this assumption is an appropriate picture of reality, at least in very many applications" (von Mises 1951, 168). In addition, he argues that the limit assumption can be retained in the three paradigmatic cases we mentioned earlier, namely games of chance, insurance problems, and the theory of gases.

The limit assumption is necessary but not sufficient to define a collective; *randomness* is also required. von Mises invites us to consider the case of a

road along which milestones are placed, large ones for whole miles and smaller ones for tenths of a mile. If we walk long enough along this road, calculating the relative frequencies of large stones, the value found in this way will lie around 1/10. The value will be exactly 0.1 whenever in each mile we are in that interval between two small milestones which corresponds to the one in which we started. The deviations from the value 0.1 will become smaller and smaller as the number of stones passed increases; that is, the relative frequency tends towards the limiting value 0.1. This result may induce us to speak of a certain "probability of encountering a large stone".

(von Mises 1957, 23)

However,

the sequence of observations of large or small stones differs essentially from the sequence of observations, for instance, of the results of a game of chance, in that the first sequence obeys an easily recognizable law. Exactly every tenth observation leads to the attribute "large", all others to the attribute "small". After having just passed a large stone, we are in no doubt about the size of the next one; there is no chance of its being large.

(1957, 23)

As shown by the example, sequences that qualify as collectives must be lawless or unpredictable; that is, they obey the requirement of randomness. Part of von Mises' program is to provide a definition of the notion of randomness that restates in rigorous terms the intuitive idea that an event is random when it is unpredictable and cannot be accounted for in causal terms.

13.4. Randomness

Von Mises defines randomness in an operative fashion based on the notion of *place selection*. This amounts to selecting members from a sequence to form a sub-sequence according to a rule that, for every element of the sequence, states unambiguously whether it ought to be included in the sub-sequence. The decision to include in the sub-sequence the pth term of the original sequence may depend on the number p and on the attributes of the $(p - 1)$ preceding elements but not on the attribute of the pth term itself or any other attribute characterizing the following elements of the sequence. Examples of sub-sequences obtained by place selection mentioned by von Mises (1957, 25) are those "formed by all odd numbers of the original sequence, or by all members for which the place number in the sequence is the square of an integer, or a prime number, or a number selected according to some other rule, whatever it may be".

Having defined the notion of place selection, von Mises goes on to stipulate randomness as *insensitivity to place selection* to the effect that the limiting values of the relative frequencies in a given sequence should not be affected by any of all possible selections that can be performed on it. Moreover, the limiting values of the relative frequencies in all sub-sequences obtained by place selection must be equal to those of the original sequence. In other words, "the limiting values of the relative frequencies in a collective must be independent of all possible place selections" (1957, 25). von Mises calls this condition the *Principle of randomness*, or the *Principle of the impossibility of a gambling system*, because it entails the impossibility of developing a system allowing for a sure gain (or loss) in whatever game of chance. It is precisely on the failure of all the attempts made in order to devise gambling systems that von Mises grounds the empirical foundation of the notion of a random collective:

> Everybody who has been to Monte Carlo, or who has read descriptions of a gambling bank, knows how many 'absolutely safe' gambling systems, sometimes of an enormously complicated character, have been invented and tried out by gamblers; and new systems are still being suggested every day. The authors of such systems have all, sooner or later, had the sad experience of finding out that no system is able to improve their chances of winning in the long run, i.e., to affect the relative frequencies with which different colors or numbers appear in a sequence selected from the total sequence of the game. This experience forms the experimental basis of our definition of probability.
>
> (1957, 25)

Von Mises draws an analogy between the impossibility of finding a gambling system and the impossibility of constructing a perpetual-motion machine, calling attention to the fact that both are empirically based. The impossibility of perpetual motion is based on the law of the conservation of energy, which

> is nothing but a broad generalization – however firmly rooted in various branches of physics – of fundamental empirical results. The failure of all the innumerable attempts to build such a machine plays a decisive role among these. . . . There is no question of proving the law of conservation of energy – if we mean by "proof" something more than the simple fact of an agreement between a principle and all the experimental results so far obtained.
>
> (1957, 26)

This analogy serves the purpose of showing that probability theory can be built in a scientific fashion, in all respects like the way in which physics is developed through a process of generalizing from observations. As von Mises (1957, 26) puts it:

> by generalizing from the experience of gambling banks, deducing from it the Principle of the Impossibility of a Gambling System, and

including this principle in the foundation of the theory of probability, we proceed in the same way as did the physicists in the case of the energy principle.

Notably, von Mises pushes the analogy even further by equating the principle of randomness with the laws of nature, to be regarded simply as expressing restrictions on our expectations based on past experience, a view whose paternity he ascribes to Ernst Mach.

13.5. Probability Is a Theory of Collectives

Having spelled out the notion of collective, von Mises faces the task of building probability as a theory of collectives. In the first place, he reaffirms that once taken in accordance with the frequency approach, the theory of probability is in all respects similar to all other sciences: it "starts from observations, orders them, classifies them, derives from them certain basic concepts and laws and . . . draws conclusions which can be tested by comparison with experimental results" (1957, 31). In the theory of probability, as well as in science at large, one moves from given probabilities to probabilities obtained from the former ones. In the author's collective-based perspective, the purpose of the theory of probability is identified with the derivation of probabilities by shifting from one or more collectives to a new collective derived from the initial ones.

Before clarifying how such a derivation can be made, von Mises calls attention to the fact that once probability is taken as the limiting value of the relative frequency observed in a collective, the values 1 and 0 do not correspond respectively to certainty and absolute impossibility but rather to the "practical certainty" that a given attribute will or will not be observed again. It can happen that an attribute is so rare that its frequency converges to 0 when the sequence to which it belongs is prolonged enough or, by contrast, that some attribute is nearly always present. In such cases, limiting frequencies would correspond to extremely rare or almost certain events, but that would not mean that an attribute could not recur at all in a given sequence or, conversely, that it necessarily ought to be found in every possible observation. In von Mises' (1957, 34) words: "the indeterminate character of all statements of the probability theory is maintained in the limiting cases as well".

The next step towards building the theory of probability in terms of collectives amounts to the definition of the notion of *distribution*. A distribution denotes

> the whole of the probabilities attached to the different attributes in a collective. . . . If, for instance, the numbers 1/5, 3/5, and 1/5 represent the distribution in a collective with three attributes A, B, and C, the probabilities of A and C being 1/5 each, and that of B being 3/5,

then in a sufficiently long sequence of observations we shall find the attributes A, B, and C "distributed" in such a way that the first and third of them occur in 1/5 of all observed cases and the second in the remaining 3/5.

(von Mises 1957, 35)

The notion of continuous distribution is then introduced in order to take care of the cases involving a continuum of attributes. As an example, von Mises mentions the distribution of a certain mass along a rod of non-uniform thickness, describable by indicating the mass density at each point of the rod. To handle such cases, he introduces the notion of "probability density" and with reference to the previous example, clarifies that "the distribution is described by a function representing the probability density per unit of length over the range of the continuous variable" (1957, 37).

The notion of distribution plays a crucial role within von Mises' theory of probability, whose purpose "is to calculate the distribution in the new collective from the known distribution (or distributions) in the initial ones" (1957, 37). He goes on to operationally define the basic properties of probability in terms of four operations called *selection*, *mixing*, *partition*, and *combination*, which make it possible to derive one collective from another, and shows that by means of such operations, the whole theory of probability can be re-stated in terms of collectives. The first operation of *selecting* consists in obtaining new collectives from a given one by the method of place selection. The fact that, as we saw, place selection must obey the condition of randomness ensures that the new collectives obtained by selecting have the same probability distribution as the original one.

The second operation of *mixing* corresponds to additivity, the third operation of *partition* corresponds to conditional probability, and the fourth operation of *combination* corresponds to the multiplication rule. Together with the two requirements imposed on collectives, namely (1) the existence of the limit of relative frequencies and (2) the invariance under place selection (randomness), the four operations provide the building blocks of von Mises' theory of probability, which he saw as standing on a solid empirical basis, like any other science. This machinery provides a way of measuring probabilities on the basis of frequencies. In this way, probability is operationally reduced to a measurable quantity: the genuinely operational character of von Mises' approach should not pass unnoticed.

It is worth mentioning that in von Mises' writings, the *basic assumptions* (1) and (2) are often called *axioms*, or *postulates* – for instance, they are introduced as axioms in the article "Grundlagen der Wahrscheinlichkeitsrechnung" (see von Mises 1919b), where his frequency theory of probability is spelt out for the first time, and in various passages of his later production he refers to them sometimes as axioms and sometimes as postulates. von Mises is aware of the fact that his treatment is not in line with the "orthodox" approach to mathematical axiomatics because it combines mathematical elements, such as the notion of limit, with empirical observations. In line with his conviction that – as already observed – probability theory must be built inductively from reality, that is, from empirical data, he maintains that "the 'mixture of empirical and theoretical elements' is . . . unavoidable in mathematical science" (von Mises 1964b, 45). In that connection as well, von Mises grounds his claim on a comparison with mechanics:

> When in the theory of elasticity we introduce the concepts of strain and stress, we cannot content ourselves by stating that these are symmetric tensors of second order. We have to bring in the basic assumptions of continuum mechanics, Hooke's law, etc., each of them a mixture of empirical and theoretical elements.
>
> (1964b, 45)

In Chapter 9 of *Positivism*, entitled "The Exact Theories", von Mises maintains that mathematics "like any other science, has a tautological and an empirical side" which has been often neglected to concentrate only on its tautological part, but wrongly so because *the relation between tautological systems and (extramathematical) experiences . . . is its very purpose, i.e., to make this relation a part of the mathematical system itself*" (1951, 135, original emphasis). In other words, logic and axiomatic systems are important for the sake of bringing clarity into the matters under discussion, provided that their connection with experience is not left behind in favor of their purely symbolic aspect: "the axiomatization is always a secondary activity which follows the actual discovery of the pertinent relations and puts them in a precise form" (1951, 113). One should keep in mind that von Mises was not interested in pure, but in *applied*, mathematics, and the primary task he pursued was to address the "fundamental philosophical and practical problems of the application of mathematics to reality" (Siegmund-Schultze 2004, 346). In pursuing that task, von Mises was led to work out his theory of probability by the awareness of the ubiquitous presence of stochastic notions in science, from measurement to abstract theories. As reported by Siegmund-Schultze (2004, 358), von Mises attributed a "unifying function" to stochastic notions and meant such a function to be realized primarily in physics, albeit in no way limited to this realm.[7]

13.6. Criticism of Other Theories of Probability

Von Mises reckoned his own theory a decisive improvement over the classical definition of probability, to which he objects that: "equally possible cases do not always exist, e.g., they are not present in the game with a biased die, or in life insurances" (von Mises 1957, 80). When one deals with the probability of the outcome of the throws of a biased die, or the probability of death, one is compelled to resort to frequencies. At that point, however – so goes von Mises' discussion of the matter – most of those who tacitly assume the classical approach apply the laws of probability defined in terms of equally possible cases to probability assignments representing frequencies, thereby "passing, as if it were a matter of no importance, from the consideration of a priori probabilities to the discussion of cases where the probability is not known a priori" (1957, 70). With "extraordinary intrepidity", the same authors justify such a move by an appeal to the Law of Large Numbers, "which is supposed to form a bridge between the concept of a priori probabilities and the determination of probabilities from observations" (1957, 70). In von Mises' eyes, such a way of proceeding is ill founded and should be abandoned in favor of explicit acceptance of the frequency approach. In his words:

> [U]p to the present time, no one has succeeded in developing a complete theory of probability without, sooner or later, introducing probability by means of the relative frequencies in long sequences. There is, then, little reason to adhere to a definition which is too narrow for the inclusion of a number of important applications and which must be given a forced interpretation in order to be capable of dealing with many questions of which the theory of probability has to take cognizance.
>
> (von Mises 1957, 70)

The relationship between the laws of large numbers and frequentism is discussed at length in Chapter 4 of *Probability, Statistics and Truth*. Without going deeper into von Mises' very detailed account of several versions of Bernoulli's, Poisson's, and Bayes' results, it is worth mentioning his conclusion, which amounts to the claim that the frequency theory is neither implied by nor contradicts them, as is sometimes held. Between those results and his version of frequentism, there is a fundamental difference, which amounts to the randomness assumption. That said, von Mises is deeply convinced that if combined with the notion of probability in terms of collectives, all the laws of large numbers acquire a "clear and unambiguous meaning free from contradictions" (1954, 134).

Von Mises is also critical of John Maynard Keynes, Harold Jeffreys, and Rudolf Carnap, all of whom he calls "subjectivists".[8] By claiming that probability should be based on logical grounds, they were unable, according to von Mises, to establish a solid link between the propositions expressing probability values and the observational data supporting them. The general objection moved against the "subjectivist" attitude is worth mentioning, because it highlights the analogy between probability and physics that is the leitmotiv of von Mises' approach. It goes thus:

> It would not be impossible to carry out a detailed psychological investigation into the foundations of our subjective probability estimations, but its relation to probability calculus is similar to that of the subjective feeling of temperature to scientific thermodynamics. Thermodynamics had its starting point in the subjective impression of hot and cold. Its development begins, however, when an objective method of comparing temperatures by means of a column of mercury is substituted for the subjective estimate of the degree of warmth. Everyone knows that objective temperature measurements do not always confirm our subjective feelings. . . . These discrepancies certainly do not impair the usefulness of physical thermodynamics, and nobody thinks of altering thermodynamics in order to make it agree with subjective impressions of hot and cold. . . . [R]epeated observations and frequency determinations are the thermometers of probability theory.
>
> (von Mises 1957, 76)

The comparison with science, and more particularly with physics, is at the core of von Mises' criticism of George Polya's theory of plausibility statements[9] and the viewpoint of the so-called "nihilists", who deny the need to give a precise definition of probability and maintain that the task of the theory of probability is simply to determine the exact values of probability in connection with specific problems. According to von Mises (1957, 97), the upholders of a similar view

> completely misunderstand the meaning of exact science. . . . [E]verywhere, from the most abstract parts of mathematics to the experimental physical sciences, in so far as they are treated theoretically, the exact definition of concepts is a necessary step which precedes the statement of propositions or goes parallel to it.

In the same spirit, von Mises (1957, 100) rejects the claim that probability is part of the theory of sets, because – as mentioned in Section 13.1 – probability theory is "a natural science, a theory of certain observable phenomena".

13.7. Methodological Issues

Von Mises' program to build the theory of probability on purely empirical grounds raises some objections of which the author is aware and which he rebuts by appealing to the analogy with physics. A first questionable point is the definition of probability as the limit of the relative frequency in an infinite sequence. To the objection that infinite sequences are not to be found in nature, von Mises answers that the use of such idealized notions is not different from what happens in science. In fact, "such questions have been solved completely for physics since the days of Leibniz and Newton" (von Mises 1951, 168), who made use of limiting assumptions in order to define notions like velocity and density. von Mises then argues that his own definition of probability "follows methodically the same lines as geometry or mechanics or similar branches of science" (1951, 169).

That said, there remains an issue of applicability of the limiting frequency definition to real phenomena. Von Mises proceeds as follows: first, a repeatable event, or mass phenomenon, is assumed to be random and to display stability of the relative frequency; deductive derivations are then made in the mathematical theory to reach conclusions which must be tested against experience through retranslation from the mathematical theory. According to von Mises, there are no fixed rules of retranslation from theory to experience. Instead, the criteria of application are contained in the inductive process leading from experience to an exact theory. In the case of probability, the inductive process consists of the recognition of the stability of relative frequency and insensitivity to selection of sub-sequences. Once an inductive process had led to an exact theory, it is treated in the hypothetico-deductive fashion. Again, von Mises resorts to the analogy with physics to support his methodological approach:

> The original conjecture is first of all tested with the phenomena out of which it evolved. If it not only passes these tests but also holds good in other cases, of which the discoverer did not think originally, faith in it grows. At a certain stage of the process of testing, the guess becomes a physical theorem, a law. By collecting a suitable group of such laws . . . an axiomatic system is created. This entire procedure . . . is known as the method of *induction*. . . . Once a system of axioms . . . is established, deductive treatment . . . may begin. In mechanics, and most other parts of physics one has to use essentially mathematical analysis; differential equations are integrated, their possible solutions are investigated, and so on.
>
> (1951, 141–142)

von Mises emphasizes that the process of induction "has two sides: it is to a certain extent determined by observations, but it also leaves a certain measure of freedom of decision to the scientist" (1951, 142). While most of those who addressed the issue focused on the first of these aspects, von Mises credits Poincaré for having called attention to the second, emphasizing the conventional character of the laws of physics. In von Mises' words: "One can see that conventionalism comprehends correctly one aspect of the process of induction, an aspect that is significant also from our point of view" (1951, 142). In other words, the relationship between an abstract theory and observable phenomena cannot be captured by a logical relation. Insofar as probability is concerned, "the notion of the infinite collective can be applied to finite sequences of observations in a way which is not logically definable, but is nevertheless sufficiently exact in practice" (von Mises 1957, 85).

Equally problematic is the notion of randomness in terms of insensitivity to all possible place selections put forward by von Mises. As we saw, his choice of embracing an absolute, unrestricted notion of randomness was philosophically motivated by the urge to secure an objective foundation to probability. However, soon after it was proposed, von Mises' definition raised serious objections because it turned out to be unsatisfiable, except for the trivial case of sequences whose attributes have probability 0 or 1. Attempts to improve on von Mises' view were made by the statistician Abraham Wald (1937) and the logician Alonzo Church (1940). Church formulated the requirement that the functions fixing selection rules be effectively calculable, in the sense of being recursively enumerable. The condition that selection rules should be expressed by enumerable functions had already been proposed by Wald, who had been able to prove that if an enumerable set of place selection is taken, the set of sequences insensible to place selections associated with a probability value other than 0 or 1 is a continuum. von Mises' work inspired also the French mathematician Jean Ville (1936, 1939), who developed a theory of randomness based on the notion of martingale, and Per Martin-Löf (1966), Claus Peter Schnorr (1977), and Andrej Kolmogorov, who proposed a notion of randomness in terms of complexity.[10]

Hans Reichenbach, the other major representative of the frequency theory of probability within the logical empiricist movement, regarded von Mises' unrestricted notion of randomness as far removed from scientific practice and proposed a weaker notion, relative to a restricted domain of place selections. Unlike von Mises, Reichenbach holds that

> the significance of the problem of the definition of random sequences should not be overestimated. . . . All types of probability sequences are found in nature. A mathematical theory of probability should not be restricted to the study of one specific type of sequence but should include suitable definitions of various types, chosen from the standpoint of practical use.
>
> (1971, 151)

Reichenbach's concern for all sorts of practical applications prompted the introduction of the notion of "practical limit" for sequences that "in dimensions accessible to human observation, converge sufficiently and remain within the interval of convergence" (1971, 347). For similar reasons, Reichenbach made an attempt to extend the frequency theory to single-case probabilities.[11] For his part, von Mises did not accept a similar restricted notion of randomness, retained the infinite character of collectives, and stood on the view that probability is a theory of repetitive events.

It should not pass unnoticed that the philosophical motivation underpinning von Mises' view of randomness lies with his indeterminism. In fact, von Mises pushes his positivistic view of probability to its extreme consequences by embracing indeterminism and rejecting the principle of causality. Regarding causality, he claims that:

> It now appears inevitable that we must abandon another cherished notion that has its origin in everyday life and pre-scientific thought and has been elevated to the rank of an eternal category of thought by overly zealous philosophers: the naive concept of causality.
>
> (1957, 210)

In his eyes, philosophers have given so many and so vague formulations of the principle of causality that "it is not at all easy, perhaps hardly possible, to contradict this 'law'" (1957, 210). This makes the principle itself useless. The developments occurring within physics, where recourse to statistical methods has gradually superseded causal talk, point in the same direction:

> [W]hen physics, and more generally natural science based on observations, shall have completely assimilated the methods and arguments of statistical theory and shall have recognized them as essential tools, the feeling will disappear that these methods and theories contradict any logical need, any "necessity of thought", or that they leave some philosophical requirement unfulfilled. In other words, the principle of causality is subject to change and it will adjust itself to the requirements of physics.
>
> (von Mises 1957, 211)

von Mises vigorously rejects the idea that statistical theories will eventually be superseded by deterministic ones: "The point of view that *statistical theories* are merely preparatory explanations in contrast to the final deterministic ones is a prejudice, which is bound to disappear with increasing understanding" (von Mises 1964a, 527, original emphasis).

13.8. Applications to Physics

For von Mises, physics is not only the paradigm of good science but represents the privileged field of application of the frequency theory. In the last chapter of *Probability, Statistics and Truth*, titled "Statistical Problems in Physics", von Mises discusses the application of frequentism to physics and maintains that probability can be applied to phenomena belonging to physics insofar as they can be reduced to "chance mechanisms" having the characteristics of collectives. Among the areas of physics where, according to the author, the frequency theory can be applied in a straightforward manner, he mentions the kinetic theory of gases, Brownian motion, radioactivity, and Planck's theory of black-body radiation. von Mises discusses the matter in great detail to show that the phenomena that are the object of such theories can be reduced to collectives and treated probabilistically in the frequency fashion.

For instance, in connection with the kinetic theory of gases, he maintains that: "It appears to be appropriate, i.e., in accordance with experimental findings, to consider the molecules as elements of a collective, and to apply to this collective the rules of probability calculus" (1957, 181). In similar fashion, von Mises argues that the frequency notion of probability applies to the phenomena belonging to the previously mentioned theories, where what the theory predicts is "not the exact result of a single sequence of observations but the outcome of the great majority of identical experiments (each experiment consisting of a large sequence of observations), repeated a very large number of times" (1957, 199).

Finally, von Mises (1957, 211) discusses "the new quantum statistics created by de Broglie, Schrödinger, Heisenberg and Born" and maintains that the frequency theory can be extended to this field as well. von Mises looks with special favor at Heisenberg's principle of uncertainty, which he welcomes as urging the withdrawal of both causality and the deterministic worldview traditionally associated with it. Moreover, von Mises welcomes Heisenberg's principle as a possible basis for unifying the old and the new physics on probabilistic grounds. After the advent of Boltzmann's statistical mechanics in the second half of the nineteenth century, it was conceded that "the predictions of classical physics are to be understood in the sense of probability statements of the type of the Laws of Large Numbers" (1957, 217–218), although at that stage "the usual assumption was that the atomic processes themselves, namely the motions of single molecules, are governed by the exact laws of deterministic mechanics" (1957, 218). Quantum mechanics put an end to that kind of dualism: "the rise of quantum mechanics has freed us from this dualism which prevented a logically satisfactory formulation of the fundamentals of physics" (1957, 218). These developments made it possible for the whole edifice of science to rest on a statistical conception of nature, thereby granting indeterminism the same credibility traditionally attached to determinism. This is a conviction that von Mises shared with some of the fathers of quantum mechanics, including, besides Heisenberg, Max Born, who in "Statistical Interpretation of Quantum Mechanics"[12] deems determinism "an article of faith" (Born 1967, 97) created by the success of Newtonian mechanics and its developments by Laplace.

Another point of agreement between von Mises and the physicists of the Copenhagen interpretation of quantum mechanics lies with a radical form of empiricism according to which physics can only speak of what is observable, and "the term 'happens' is restricted to observation" (Heisenberg 1990, 40). For those physicists, like for von Mises, such a principle immediately leads to measurement, and this in turn leads to probability; as put by Heisenberg (1990, 40): "what one deduces from an observation is a probability function". The conclusion reached by von Mises is that

> the result of *all* measurements are collectives. In the realm of macrophysics the objects of measurement are themselves statistical conglomerates. . . . In microphysics, where we are concerned with measurements on a single elementary particle, the inexactness is introduced by the statistical character of the light quanta striking the particle during and through the very act of measuring. In both cases we are faced with the indeterministic nature of the problem as soon as we inquire more closely into the concrete conditions of the act of measuring.
>
> (1957, 215, original emphasis)

Notwithstanding von Mises' conviction, the application of the frequency theory to quantum mechanics is not free from difficulties, mostly caused by the single case problem. Indeed, within quantum mechanics, it is common to talk about single case probabilities – for instance, the probability that a single atom is in a certain state – but as we have seen, such attributions are not admitted by von Mises. This difficulty is widely acknowledged in the literature, starting with Karl Popper, who in the 1950s proposed a new interpretation of probability, namely the propensity theory, devised for application within quantum mechanics.[13] The topic nurtures an ongoing debate in which the upholders of different approaches confront each other. To put it briefly, the controversy on the interpretation of probabilities occurring within quantum mechanics is all but settled. Besides this problem, after von Mises, the frequency theory of probability became quite popular among physicists and is still the object of broad consensus.

13.9. Final Remarks

Probability, Statistics and Truth ends with the following passage, meant by von Mises to illustrate the main thesis of the book, as reflected by the three words appearing in the title: "Starting from a logically clear concept of *probability*, based on experience, using arguments which are usually called *statistical*, we can discover *truth* in wide domains of human interest" (1957, 220, original emphases). The idea is that statistical methods making use of the frequency notion of probability provide a powerful heuristic tool for the construction of knowledge not only in science but also in all fields of human activity. In other words, in von Mises' hands, the frequency view becomes the key to the investigation of all aspects of reality, to be carried out in accordance with the attitude he calls "positivist". It is not out of place to recall what being a positivist meant for the author. In a paper entitled "The Role of Positivism in the XX Century", containing the text of an address given in 1953 at a meeting of the Society for the Unity of Science (published posthumously in von Mises (1964a)), von Mises maintains that "*He is a positivist who, when confronted by any problem in life, reacts in the manner in which a typical contemporary scientist deals with his problems of research*" (von Mises 1964a, 539, original emphasis). This involves in the first place judging "*on the ground of experience*" (1964a, 539, original emphasis).

In the same paper, von Mises makes it clear that the version of positivism he favors matches the philosophy heralded by the Vienna and Berlin Circles: it is "logical positivism, as it is often called to-day" (1964a, 541). As already observed, von Mises identifies with the critical attitude towards language the most important innovation brought about by logical positivism. This involves the tenet that "*[a] statement is meaningful if and only if it ultimately leads to sentences that relate only to observational terms*" (1964a, 543, original emphasis) and the consequent rejection of metaphysics. The latter is condemned by von Mises as a powerful tool for deception: "Any type of metaphysics, whether with or without religious shading, leads by its very nature to intolerance and injustice, makes a peaceful life and the pursuit of happiness for the whole of mankind impossible" (1964a, 547). By contrast, the positivist invites us to behave like the scientist, who "does not believe that any scientific theory is definitive or unchangeable" (1964a, 545). This is part of the lesson taught by Mach, together with the conception of scientific laws as connections among phenomena "set up in a constructive manner as conjectures . . . subject to continual testing by new observations" (von Mises 1951, 369).

Physics is the major source of inspiration of von Mises' positivism, as well as the privileged field of application of his frequentism, namely the pillar on which his positivism is made to rest. Moreover, he regards physics as the paradigm of a fruitful way of addressing problems not only in science but also in every field, including history, sociology, and the so-called "humanistic sciences", which, according to von Mises, should not be seen as opposed to the natural sciences. No dividing line should be traced between human and natural sciences, because: "In every case the scientific procedure is essentially the same; we start with observations, try preliminary generalizations, attempt a theoretical statement and eventually build up a more or less axiomatic system" (von Mises 1964a, 527). Following this path will bring great benefits to mankind: "What mankind needs is: Less loose talk and more criticism of language, less emotional acting and more scientifically disciplined thinking, less metaphysics and more positivism" (von Mises 1964a, 547).

Notes

1. The genesis and content of this work is illustrated in Siegmund-Schultze (2007).
2. On the birth and development of logical positivism, from the First Vienna Circle onwards, including the influence exerted by the movement on the philosophical milieu, see Haller (1993) and Stadler (2001/2015), where Stadler deems von Mises one of "those proponents of the empiricist conception of science whose work has been undervalued" (p. xxiv). On the First Vienna Circle, in addition to the previously mentioned works by Haller and Stadler, see Uebel (2015, 2017). See also Limbeck-Lilienau (2018), where the importance and the very existence of the First Vienna Circle are questioned.
3. The manifesto, published originally in 1929, was called "The Scientific Conception of the World: The Vienna Circle"; see Stadler and Uebel (2012) containing the reprint of the original text followed by translations in English, French, Spanish, and Italian, together with two studies of its origin and reception by Henk Mulder and Thomas Uebel.
4. The quotation is from a chapter appearing in the volume *The Berlin Group and the Philosophy of Logical Empiricism* (Milkov and Peckhaus 2013), which strangely enough does not include a chapter on von Mises, who is hardly mentioned in the references included in some chapters.
5. Siegmund-Schultze (2004, 348) conjectures that von Mises was not very active in the Berlin circle, and "deliberately chose a certain distance" with the Vienna Circle. To the present writer, such a conjecture looks at odds with von Mises' repeated pronouncements in favor of logical positivism to be found in his writings – see Section 13.9, and the previously quoted passage.
6. The following sections borrow some passages from Galavotti (1995/1996, 2005).
7. For an excellent discussion of von Mises' theory of probability, see Gillies (2000).
8. As is well known, these authors are the major representatives of the logical interpretation of probability. See Galavotti (2005, Ch. 6) for more on their conception of probability.
9. In the years 1919–20, von Mises and Polya entertained a correspondence. See Siegmund-Schultze (2006).
10. For a survey of the problems raised by von Mises' notion of randomness, see Martin-Löf (1969). The theories advanced by Ville, Martin-Löf, Schnorr, Kolmogorov, and others are examined in detail in Bienvenu, Shafer and Shen (2009). Kolmogorov's approach, together with the reception of von Mises' work, is discussed in von Plato (1994).
11. For more on Reichenbach's version of frequentism, see Galavotti (2005, 2011).
12. The paper appeared in *Science* in 1955 and is reprinted with other writings of the author in the volume *Physics in my Generation* (Born 1967).
13. See Galavotti (2005, Ch. 5) and Gillies (2000) for more on the propensity theory. See also Galavotti (1995/1996, 2001) for more on probability in physics.

References

Bienvenu, L., Shafer, G., and Shen, A. (2009), 'On the History of Martingales in the Study of Randomness', *Journal Electronique d'Histoire des Probabilités et de la Statistique*, 5 (1), www.jehps.net/juin2009/BienvenuShaferShen.

Born, M. (1967), *Physics in My Generation*. New York: Springer.

Church, A. (1940), 'The Concept of a Random Sequence', *Bulletin of the American Mathematical Society*, 46: 130–135.

Cramér, H. (1953), 'Richard von Mises' Work in Probability and Statistics', *The Annals of Mathematical Statistics*, 24: 657–662.

Frank, Ph. (1954), 'The Work of Richard von Mises: 1883–1953', *Science*, 119: 823–824.

Fréchet, M. (1938), 'Les principaux courants dans l'évolution récente des recherches sur le calcul des probabilités', *Colloque consacré a la théorie des probabilités, Première partie. Actualités scientifiques et industrielles*, 734: 19–23.

Galavotti, M. C. (1995/1996), 'Operationism, Probability and Quantum Mechanics', *Foundations of Science*, 1: 99–118.

—— (2001), 'What Interpretation for Probability in Physics?', in J. D. Bricmont et al. (eds.), *Chance in Physics: Foundations and Perspectives*. Berlin, Heidelberg and New York: Springer, pp. 265–269.

—— (2005), *Philosophical Introduction to Probability*. Stanford, CA: CSLI Publications.

—— (2011), 'On Hans Reichenbach's Inductivism', *Synthese*, 181 (1): 95–111.

Gillies, D. (2000), *Philosophical Theories of Probability*. London and New York: Routledge.

Goldstein, S. (1963), 'Richard von Mises. 1883–1953', in Ph. Frank et al. (eds.), *Selected Papers of Richard von Mises: Volume I: Geometry, Mechanics, Analysis*. Providence: American Mathematical Society, pp. ix–xiv.

Haller, R. (1993), *Neopositivismus. Eine historische Einführung in die Philosophie des Wiener Kreises*. Darmstadt: Wissenschaftliche Buchgesellschaft.

Heisenberg, W. (1990), *Physics and Philosophy*. London: Penguin Books.

Holton, G. (1993), 'From the Vienna Circle to Harvard Square: The Americanization of a European World Conception', in F. Stadler (ed.), *Scientific Philosophy: Origins and Developments*. Dordrecht: Kluwer, pp. 47–72.

Limbeck-Lilienau, Ch. (2018), 'The First Vienna Circle: Myth or Reality?', *Hungarian Philosophical Review*, 62 (4): 50–65.

Martin-Löf, P. (1966), 'The Definition of Random Sequences', *Information and Control*, 9 (6): 602–619.

—— (1969), 'The Literature on von Mises' Kollektivs Revisited', *Theoria*, 35: 12–37.

Milkov, N. (2013), 'The Berlin Group and the Vienna Circle: Affinities and Divergencies', in Milkov and Peckhaus (2013), pp. 3–32.

Milkov, N. and Peckhaus, V. (eds.) (2013), *The Berlin Group and the Philosophy of Logical Empiricism*. Dordrecht: Springer.

Reichenbach, H. (1971), *The Theory of Probability*. Berkeley and Los Angeles: University of California Press (Original German edition 1935; first English edition 1949).

Schnorr, C. P. (1977), 'A Survey of the Theory of Random Sequences', in R. E. Butts and J. Hintikka (eds.), *Basic Problems in Methodology and Linguistics*. Dordrecht: Reidel, pp. 193–211.

Siegmund-Schultze, R. (2004), 'Non-Conformist Longing for Unity in the Fractures of Modernity: Towards a Scientific Biography of Richard von Mises (1883–1953)', *Science in Context*, 17 (3): 333–370.

——— (2006), 'Probability in 1919/20: The von Mises-Polya Controversy', *Archive for History of Exact Sciences*, 60: 431–515.

——— (2007), 'Philipp Frank, Richard von Mises, and the Frank-Mises', *Physics in Perspective*, 9: 26–57.

Stadler, F. (2001/2015), *The Vienna Circle: Studies in the Origins, Development, and Influence of Logical Empiricism.* 2nd ed. Dordrecht: Springer.

Stadler, F. and Uebel, Th. (eds.) (2012), *Wissenschaftliche Weltauffassung. Der Wiener Kreis.* Wien: Springer.

Uebel, Th. (2015), 'American Pragmatism and the Vienna Circle: The Early Years', *Journal for the History of Analytical Philosophy*, 3 (3): 1–35.

——— (2017), 'American Pragmatism, Central-European Pragmatism and the First Vienna Circle', in S. Pihlström, F. Stadler, and N. Wiedtmann (eds.), *Logical Empiricism and Pragmatism.* Cham: Springer, pp. 83–102.

Ville, J. (1936), 'Sur la notion de collectif', *Comptes rendus*, 203: 26–27.

——— (1939), *Étude critique de la notion de collectif.* Paris: Gauthier-Villars.

von Mises, R. (1919a), 'Fundamentalsätze der Wahrscheinlichkeitsrechnung', *Mathematische Zeitschrift*, 4: 1–97. Reprinted in von Mises (1964a), pp. 35–56.

——— (1919b), 'Grundlagen der Wahrscheinlichkeitsrechnung', *Mathematische Zeitschrift*, 5: 52–99. Reprinted in von Mises (1964a), pp. 57–105.

——— (1951), *Positivism: A Study in Human Understanding.* Harvard: Harvard University Press (Original German edition 1939).

——— (1957), *Probability, Statistics and Truth.* New York: Dover (Original German edition 1928; first English edition 1939).

——— (1964a), *Selected Papers of Richard von Mises: Volume II: Probability and Statistics, General.* Edited by P. Frank et al. Providence: American Mathematical Society.

——— (1964b), *Mathematical Theory of Probability and Statistics.* New York: Academic Press.

von Plato, J. (1994), *Creating Modern Probability.* Cambridge: Cambridge University Press.

Wald, A. (1937), 'Die Widerspruchsfreitheit des Kollectivsbegriffes der Wahrscheinlichkeitsrechnung', *Ergebnisse eines mathematischen Kolloquiums*, 8: 38–72.

14 From Physical Possibility to Probability and Back

Reichenbach's Account of Coordination

Flavia Padovani

14.1. Introduction: Reichenbach's Axiomatic Approach

Reichenbach's two major areas of contribution are space-time and probability theories, for both of which he attempted to provide an axiomatization (in 1924/1969, as well as in 1932 and 1932–33/1949, respectively), which did not meet fully enthusiastic reactions, to put it mildly.[1] Reichenbach's fascination with an axiomatic method is not surprising, given his early academic interests and the list of outstanding figures whose lectures he could benefit from during his university years, including Hilbert. Overall, a leitmotif of Reichenbach's early philosophical reflections is the role played by mathematics with respect to our representation of reality. He is not interested in discussing the foundations of mathematics but rather in how we can use it to describe the world, notably in physics.[2] In other words, Reichenbach's concern is mainly the problem of application, that is, how formal systems consisting of non-interpreted symbols and formation rules can be used to tell us something about reality. Physical and, in general, any scientific laws express relations about quantities in the world that are formulated in mathematical terms. The problem is whether these relations actually capture the relations between objects/features of the world and the mathematical symbols that would stand for them in the law. In his first writing on the topic (1916), this issue is embedded in a broader discussion on the difference between mathematical judgments and physical judgments, and it already includes the idea of coordination, namely whether the mapping between the set of mathematical symbols and the set of objects/features of the world is well grounded.

As is well known, Reichenbach's initial epistemological project is guided by a Kantian approach, which becomes particularly interesting in 1920, when he starts looking more closely at the foundations of different theories and at the way these become incompatible with subsequent theories considered as a whole. To be able to describe reality, a formal system must be grounded in a variety of theory-specific principles that ultimately enable the application of that system to reality. In most cases, these principles are foundational relative to a theory (i.e., they are constitutive of meaning), and they can be revised in light of future evidence. This typically occurs if the coordination no longer allows for a consistent interpretation and some discrepancy appears in our description of reality. Coordinating principles also provide certain "rules of application" at the level of our basic access to phenomena, particularly in relation to the quantities that appear in physical equations. As we will see in Section 14.2, Reichenbach's discussion of coordination constantly oscillates between these two distinct levels when addressing the constitutive role of principles (and later definitions) within a scientific construction. The fact that these principles are revisable makes them appear to be quite akin to conventions. And, in fact, after his first two volumes and a famous exchange with Moritz Schlick, Reichenbach mostly drops the language of "constitution" and "principles" and embraces a terminology (and an approach) aligned with that of conventionalism (Section 14.3), with a shift that my appear to be not necessarily unproblematic given the original background of his reflections. It is especially Reichenbach's reliance on the indispensable role of probability that challenges this shift, and it is also probability that explains how coordination can de facto be achieved in his work, as we will conclude in Section 14.4.

14.2. Cognition as Coordination in Reichenbach's Early Writings

14.2.1. *Physical Judgments and Approximation*

Reichenbach starts dealing with the problem of coordination[3] in his first scientific publication, his 1915 doctoral thesis *The Concept of Probability in the Mathematical Representation of Reality*, published as (1916/2008). This thesis revolves around the nature of probability statements and the justification of the necessity of their use in physics, as well as, more generally, in any empirical science. The motivation behind this work is to develop an objective interpretation of probability equally covering statistics, physics, and, broadly, any important question dealing with uncertainty in the sciences and our lives. One of the main targets of this thesis is the account promoted by the physiologist Johannes von Kries in his *Principien der Wahrscheinlichkeitsrechnung* (1886), which for Reichenbach makes too many concessions to epistemic elements such as those implied by the principle of indifference.[4] For Reichenbach, an overarching account of probability should express not so much how the occurrence of an event relates to our subjective knowledge (or lack thereof) but rather how this occurrence is connected with an objective (i.e., true) state of affairs in the actual world. In Kries's logico-objective account of probability, objective, physical possibilities appear to be the ones that are allowed (or at best not ruled out) by laws of nature.[5] On the other hand, for Kries, these possibilities may not necessarily appear as instantiated, and this leads to a confusion, in Reichenbach's view, about the concept of possibility, namely that "the universe is the one real world among many possible worlds" (1916/2008, 121). To be sure, for Reichenbach this is non-sensical and one should explain why only one world happens to be instantiated. In fact, for him, there exists only one possible physical world – the one in which we are in – and the way we understand it.

Even setting the general many-worlds issue aside, Reichenbach argues, what Kries's account implies is that a process that is merely defined by laws of nature, that is, nomologically, may still be considered indeterminate if its realization does not take place. For Reichenbach, however, this would entail that it should not even be called "a process" in the first place. As he puts it:

> Such a description of a process by purely nomological features is always a generalized summary of a class. This is the actual meaning of the distinction. The class is given by the form of the differential equations and the constants that occur in them that already must have been numerically determined; the integration constants are left unspecified and their particular values correspond to one instance of the class. But an instance is not real because the constants have assumed particular values; and it is not possible just because the values have been left undetermined. Lack of determination is not the criterion for possibility, and determination is not the definition of reality.
>
> (Reichenbach 1916/2008, 121)

So how are we supposed to operate within this tension? To start with, for Reichenbach, physics relies on the methodological assumption (which, he emphasizes, does *not* correspond to an empirical claim) that mathematics is applicable to real objects. An empirical claim, instead, "*asserts the validity of a particular mathematical structure of reality*" (emphasis in the original). Furthermore, since no process in nature is closed, and "real events are always governed by an infinity of equations", a mathematical structure will only provide an approximate representation of reality. The application of mathematical concepts to reality consequently hinges on an understanding of how to manage the unavoidable component of approximation involved in this process (1916/2008, 109–111).

In this connection, Reichenbach illustrates two directions in which empirical research takes place. The first direction goes downwards, that is, from the law to the individual case. When we look at particular equations that are valid for reality, what we notice is that they will only truly determine the quantities that are numerically measurable of the real objects they represent "down to a small error". Other aspects or influences pertaining to those events might be neglected in those equations, which means that those equations alone do not capture reality in full but only limited to a class of objects that are deemed appropriate or basically selected through experience. As an example of this approach, Reichenbach first discusses Maxwell's theory of electricity, whose four basic equations and derivative laws should capture all phenomena described by the theory. However, he points out, the system of equations of this theory only provides a mathematical framework to understand real events but that obviously cannot

> fully account for reality. The theory becomes natural law when its claim to *validity in reality* can be asserted, i.e., that it represents the real processes to numerical approximation. . . . The possibility of physical knowledge has thereby been traced back to the assertability of numerical approximations.
>
> (Reichenbach 1916/2008, 111–113)

Reichenbach then goes on to discuss the relationship between general law and individual case using the example of the application of Boyle's law. In this case, there is an element of idealization, in that this law only applies to a subclass of gases for which R is taken to be constant. This is obviously not the case for non-ideal gases, for which R would become a "complicated function that itself would have to be determined anew". This is an indication, according to Reichenbach, that there are "numerical determinations that do not apply to all of nature". So, the task of physics in this case would be to improve our characterization of objects of empirical intuition by "increasing numbers of unmeasured variables appearing as letters in formulas, whose values must be determined numerically". In principle, this would eventually allow us to fully capture the individual case – if it weren't that a complete determination can never be reached anyway. It would be an infinite task.

A second direction of research follows a different path, namely the formulation of more general laws from particular equations and laws. The tendency in this case is to "resolve constants into functions, to find more and more general laws that contain the previous law as a special case". The idea here is to consider how theories have emerged in a subsequent scientific stage. For instance, a constant given in specific laws is "brought into connection with completely different quantities, so that it appears a function whose specific value in the previous laws is only attained under special circumstances". As an example, Reichenbach mentions the expression of Galileo's constant in the law of falling bodies which, in Newton's theory, becomes "a function of the distance to the centre of the earth" (1916/2008, 115). As it is impossible to fully grasp the individual case, it is equally "impossible to subsume the whole universe under one physical equation", so even this research direction cannot be exhausted. Although the approach towards more general laws is different, the task of physics remains the same in both cases, enabling "*the coordination of a class, which is initially given as a system of mathematical theorems, with objects of empirical intuition: this includes the numerical determination of constants*" (1916/2008, 117–119; emphasis in the original). The implication of this notion of the two directions of research is not entirely fleshed out in this context, with respect to coordination, as it will be in the 1920s, as we will soon see.

For Reichenbach, any mathematical representation of reality ultimately implies the use of approximation, which is the distinctive feature of physical judgments as opposed to mathematical ones. Without coming to grips with approximation, we would not be able to apply (i.e., to coordinate) any judgments expressed in mathematical terms to reality, that is, we would not be able to articulate any physical judgments, and thus we would be incapable of formulating any physical laws. As he writes:

> Physical knowledge consists in the coordination of equations, and consequently of numbers, to classes of objects of empirical intuition. The equality of these numbers for a number of actual objects of the class cannot be asserted, but only the approximate equality. The reason for this approximation is the existence of a law for the distribution of values. While mathematical judgments determine variables in such a way that they are the same for all their individual objects in all places at all times, the variables in a physical judgment are not equal for all individual objects in their class, but rather subject to a law of distribution in space and time. Instead of the general validity of mathematical claims, we have in the case of physical judgments the subsumption under the law of distribution.
>
> (Reichenbach 1916/2008, 127)

Given this state of affairs, Reichenbach introduces a new principle to complement the traditional Kantian principle of causality, a principle that he indeed calls "the principle of lawful distribution of values".[6] The essential function of this principle is to guarantee that the frequency distribution of those values converges to a limit. In Reichenbach's view, this principle is tantamount to a synthetic a priori principle in that it embodies a condition for the very possibility to express – and even conceive of – physical laws.

14.2.2. *The Kantian (Dis)connection*

As we have seen, Reichenbach's doctoral thesis is an attempt to integrate probability into a Kantian framework without changing Kant's philosophical system at its roots. However, Einstein's lectures, which Reichenbach attended in Berlin in 1917–1918, had a significant impact on his views and prompted him to amend his own previous approach to the principles of knowledge. For Reichenbach, the theory of relativity showed that in the evolution of a theory, some principles that appear to be essential in a previous stage can be contradicted by the fundamental assumptions on which a subsequent theory is based and so those principles can eventually be dismissed. In his reflections about the importance of some of those principles, Reichenbach reaches the conclusion that it is no longer possible to retain the whole set of principles characterizing Kant's philosophy, which was designed in light of a previous (now overcome) physical theory, and that the two systems of principles considered as a whole are incompatible.[7] Reichenbach's *Habilitation* thesis, *The Theory of Relativity and A Priori Knowledge* (1920), is devoted to discussing this incompatibility and its consequences in an effort to salvage at least what Reichenbach regards as the core message of Kant's philosophy, the idea that some principles do play a foundational role in determining our knowledge of the world, at least relative to a theory.

Reichenbach's solution is to separate two distinct meanings within the notion of "a priori" – namely "necessarily true", in absolute terms, and "constituting the concept of the object", which are conflated in Kant's work – and retain only the second one. Along with what Cassirer had already pointed out in *Substance and Function* (1910), Reichenbach emphasizes that a priori means "*before* knowledge", but not "for all time" and not "independent of experience" (1920/1965, 48).[8] This distinction opens the way to the consideration that all principles of knowledge, previously deemed a priori according to both meanings of the term, can now be simply viewed as a priori only qua preconditions of knowledge. In this sense, they can still be held to be "constitutive" principles, albeit temporarily, as they can well be rejected or their role revised in light of new evidence.

To better articulate this idea and how it works when it comes to theory change, Reichenbach introduces another distinction, one that involves the idea of coordination. As we saw, in his doctoral thesis, Reichenbach delineates two different directions of research, one aimed at capturing the individual case and one oriented towards formulating progressively more and more general laws. In line with this second direction, in 1920, he draws a distinction between what he calls "axioms of coordination" and "axioms of connection". Axioms of coordination are essential because they enable an empirical interpretation of physical laws (the axioms of connection), which would otherwise simply be pieces of mathematics. They are a priori principles of coordination, for they function as preconditions of knowledge that are specific to a theory; hence, they are not immutable.[9] In the shift from one stage of physics to the next, some coordinating principles may change status and become axioms of connection, that is, simple laws of physics. For Reichenbach, the metric provides an emblematic example of this shift in the conceptual role of axioms. As he explains, within "traditional physics", the Euclidean metric merely indicated the "relation to which space points combine to form extended structures independently of their physical quality", whereas in Einsteinian physics, the metric has a different function, "namely the characterization of a *physical state*". In other words, "*[t]he metric is no longer an axiom of coordination but has become an axiom of connection*" (1920/1965, 53, 98–100, original emphasis).

With respect to what we saw in Reichenbach's doctoral thesis, the notion of coordination is now more articulated but still introduced through the distinction between mathematical and physical judgments. Now, coordination is more explicitly presented as a form of mapping between two sets. As opposed to the objects belonging to the world of mathematics, physical objects cannot be determined by axioms and definitions, as they are "things" of the real world. The distinctiveness of physical coordination is that the formal or conceptual side is fully determined, while the side that provides the "contribution" of reality is yet to be determined, being dependent on perception. However, Reichenbach explains – and it is worth citing the whole passage:

> perceptions do not furnish *definitions* of what is real. . . . [S]ince perceptions do not define the elements of the universal set, one side of the cognitive process contains an undefined class. Thus, it happens that individual things and their order will be defined by physical laws. The coordination itself creates one of the sequences of elements to be coordinated. One might be inclined to dismiss this difficulty simply by declaring that only the ordered set is real, while the undefined one is fictitious, a hypostatized thing-in-itself. . . . But such a view is certainly false. There remains the peculiarity that the defined side does not carry its justification within itself; its structure is determined from outside. Although there is a coordination to undefined elements, it is restricted, not arbitrary. This restriction is called "the determination of knowledge by experience". We notice the strange fact that it is the defined side that determines the individual things of the undefined side, and that, vice versa, it is the undefined side that prescribes the order of the defined side. *The existence of reality is expressed in this mutuality of coordination.* . . . This mutuality attests to what is real.
>
> (1920/1965, 42, original emphases)

Following Kant, Reichenbach emphasizes that perception only provides the material of which the object is made through "an act of judgement", which ultimately is "a subordination into a determinate schema", so that, "depending on the choice of the scheme, either an object or a certain type of relation will result" (1920/1965, 48–49). While this could look like a top-down model, it is not, and in trying to articulate his view of how physical coordination is supposed to work, Reichenbach maintains that the set that is not determined is actually the one that "prescribes the order" to the other set and that genuine knowledge is only possible thanks to this "mutuality of coordination". What this means is not fully explained in this context, but it is clear that Reichenbach envisages some sort of regulative-limiting role to be performed on the side of perception, one that will induce some constraints on the relationship between the two sides of the coordination and that, through an iterative procedure of some kind, will eventually lead to a more refined representation.[10] More importantly, the limits to the arbitrariness of coordination that a top-down approach would seem to imply are explicitly highlighted by Reichenbach, which is the reason he initially rejects conventionalism. For him, in fact – at least in this framework – physics has clearly determined that "the theorems of Euclidean geometry do not apply to our physical space" (1920/1965, 4).

The idea of coordination as mapping operates on two different levels, calling to mind of the motives that prompted Reichenbach to identify two main directions of research in 1915. From a wider perspective, there is coordination between scientific theories as a whole (each with its own particular most general law) and experience; from a narrower point of view, there is coordination between restricted parts of the theory and individual objects/features of the world, that is, how "physical things are coordinated to equations. Not only," as he puts it, "is the totality of real things coordinated to the total system of equations, but *individual* things are coordinated to *individual* equations" (1920/1965, 37, original emphases).[11] The twofold function of coordinating, constitutive principles is clearly intertwined, in the sense that the entire system (roughly, a "paradigm", as we would rather call it today) functions like a whole, where all the different parts of the theory are interrelated. The fact that some of these constitutive principles are strictly peculiar to some theory and might well be dismissed in the case of theory change does not imply that all of the coordinating principles would be dismissed en bloc should that change occur. To be sure, the principles that enable our access to phenomena are the ones that seem to be less easy to discharge. One of these is the one Reichenbach refers to as the "principle of probability", which is a refined version of the "principle of lawful distribution of values" that we have seen previously. The principle of probability is in charge of ensuring that the quantities that appear in physical equations are more and more accurate. We can represent magnitudes by the same numerical value obtained out of different empirical data by approximation because we can rely on this principle.

As we saw, the idea that probability is so crucial at the basic level of our scientific constructions is already at the core of Reichenbach's doctoral thesis, where it is the unifying feature of the two approaches to scientific inquiry. However, in 1920 probability enjoys an even clearer central role. One of the distinctive traits of this newly elaborated version of coordination is the integration of elements derived from Schlick's

General Theory of Knowledge (1918), in particular, the idea of univocal (*eindeutig*) coordination as indicating true (physical) knowledge.[12] In their meaning-constituting function, axioms of coordination are supposed to lead to a univocal coordination, which is obtained when "all chains of reasoning lead to the same number for the same phenomenon" (1920/1965, 43). The criterion of truth in terms of univocal coordination appears to primarily concern the determination of individual quantities, but it is of course also relevant when a whole system is applied to reality. Thus, it is the ultimate function of the coordinating principle of probability to enable the mapping between our formal representations and the corresponding interpretation. Having rejected the first meaning of Kant's concept of "a priori", now Reichenbach maintains that every assumption – even the ones that we deem fundamental – could be discharged one day. Thus, like any other assumption, the feature of *Eindeutigkeit* is not immune to revision and could in principle be dismissed in a future state of knowledge.[13]

One could raise an objection to this very idea here and suggest that the very tool that we should use to detect whether inconsistencies arise (univocal coordination) is the one that we would also be using to dismiss it (if necessary). So, one could be tempted to say that if experience results from constitutive principles, and we formulate our laws on the basis of empirical data, then any interpretation in disagreement with those data should be rejected. Yet, Reichenbach argues, it is "*admissible*" and not "logically *inconsistent*" to find such contradictions.[14] One of the main shortcomings of Kant's approach to the principles of knowledge is indeed that it surreptitiously presupposes the validity of the "hypothesis of the arbitrariness of coordination", as Reichenbach calls it, that is, the idea that a system of coordination is characterized by "self-evidence" and that therefore contradictions between constitutive principles and experiences cannot arise. Instead, for Reichenbach, the theory of relativity has clearly shown that "*there exist systems of coordinating principles which make uniqueness of the coordination impossible; that is, there exist implicitly inconsistent systems*" (1920/1956, 67, original emphasis). If this were not the case, we would not have scientific progress.

14.3. From Coordinating Principles to Coordinative Definitions

As is well known, Reichenbach's overall attitude towards the status of coordinating principles underwent an important shift towards conventionalism[15] after a famous and often-cited correspondence with Schlick.[16] The most evident consequence of this exchange is in the novel approach displayed by Reichenbach soon after 1920, an approach visible both in his "Report" (1921) and more extensively in *The Axiomatization of the Theory of Relativity* (1924/1969). In this monograph, Reichenbach makes a clear-cut separation between conventional and factual components of the physical theory, namely between what he now calls "coordinative definitions" (which are indeed conventional) and "empirical axioms", those that can (at least ideally) make direct contact with experimentally testable facts. Despite being "empirical", axioms cannot be easily confirmed. To be sure, as he underlines, an axiom can only be confirmed indirectly, and "depending on the confirmation of its consequences, it may be called true or false with a certain degree of probability" (1924/1969, 4). The purpose of this "constructive axiomatization" is to show how "abstract conceptualization" is actually derivable from the "observable facts", which are chosen in a way such that they can be "grasped in accordance with pre-relativistic physics" (1924/1969, 5–7). What is new is their different combination within the conceptual system of the new theory, which is also a way to show how progress can be obtained also building on concepts belonging to a previous scientific stage.

As in the previous two volumes, Reichenbach appeals to the difference between mathematics and physics to articulate how definitions differ in both. Differently than in previous work, however, the *Axiomatization* deliberately avoids using terminology with a potential Kantian connotation and indeed introduces a definition that derives from Hilbert's axiomatization of geometry. Mathematical definitions are now defined as "conceptual definitions", that is, implicit definitions, in the sense that they elucidate the meaning of a concept through other concepts. Physical definitions, instead, take "the meaning of a concept for granted" and then apply it "to a physical thing". What this means is that physical definitions *are* coordinative definitions, because they "consist in the coordination of a mathematical definition to a "piece of reality"; one might call them *real definitions*" (1924/1969, 8). However, "the physical thing that is coordinated is not an immediate perceptual experience but must be constructed from such an experience by means of an interpretation", and this is a construction that goes "beyond perception". Coordinative definitions are thus necessary to assign a meaning to different mathematical concepts, while conceptual/mathematical definitions are presupposed in order to determine a physical concept.

As an example, Reichenbach briefly discusses the mathematical definition of unit of length as opposed to the physical, coordinative definition which "designates the Paris standard meter as the unit of length. In this physical definition, the mathematical definition of the concept is presupposed" (1924/1969, 8). This is not tantamount to saying that the meaning of those concepts is completely determined by those definitions. A preliminary coordination provides a condition for the possibility of those concepts to be meaningful (and eventually be used in our description of the world), but it is merely a necessary condition, and of course we have a certain freedom in relation to the choice of coordinative definitions.[17] Still, there is a restriction in the arbitrariness of this choice in that coordinative definitions need to lead to univocal coordination in conformity with the "facts" laid down by the axioms.[18]

A substantial part of Reichenbach's discussions around conventionalism relate to geometry and its use in our physical theories. In one of his most popular books, *The Philosophy of Space and Time* (1928), Reichenbach devotes an entire chapter to articulating the nature of geometry as a theory of relation and to better explain what is it that we coordinate to the physical sphere (1928/1958, §14). Purely abstract and basic concepts such as "'element', 'relation', 'univocal coordination',[19] 'implication', etc." are not definable, while all the other concepts, which geometry as a relational structure is composed of, are represented "as functions of these basic concepts". As Shapiro pointed out, what is coordinated to elements of the "physical system" are indeed these structure-constituting concepts and not the basic, abstract ones.[20]

An additional feature of this broader view of coordination takes shape in connection with a reflection on the importance of visualization (1928, §15). Mathematical concepts, which are defined by implicit definitions, enable our "control (*Beherrschung*) of natural phenomena", but they are not tied to a unique form of visualization. Any piece of knowledge having a logical structure that could suitably be "coordinated to *mathematical* geometry" could very well "also be coordinated to *physical* geometry and represented by means of diagrams". Coordinative definitions could therefore be established for, say, "direct currents, or increases in the tension of a stretched rod", not only for rigid measuring rods and the sort. At the same time, geometrical axioms could be employed to visualize other objects, such as "compressed gases, electrical phenomena or mechanical forces". As Reichenbach elucidates:

> We are so accustomed to the coordination of rigid bodies to mathematical geometry as a theory of relations that we no longer notice that there exists a duality. . . . On the one hand we have the mathematical system *A* of relations and on the other hand the physical system *a* of rigid bodies. Every assertion about *A* can be translated into an assertion about *a* and it is customary to use assertions about *a* alone which are symbolic of assertions about *A*. This is called *visual* geometry. The system *a* is the *visual* space of *A*. In contrast, the content of *A* cannot be visualized. . . . This consideration also clarifies the term *pure visualization*. We do not think of the system *a* as a system of natural objects, but objects exemplifying the relations of Euclidean geometry: then the system *a* of things is a *space of pure visualization*.
>
> (1928/1958, 103, original emphases)

It is only for pragmatic reasons that rigid measuring rods have become our preferred tools of measurement and that in our graphical representations we make, accordingly, extensive use of geometrical relations. Graphical representations, however, are merely expressing a coordination of an abstract system to a physical system, if this can realize the conceptual system. The use of a symbolic language merely responds to our practical needs and to the limits of our thinking capacity, but the choice of a system will not determine the content expressed by that system. As he puts it:

> Thinking completely without symbols seems to be impossible. However, this fact should not lead to the mistaken impression that the chosen symbol is essential for the content of the thought. It is as irrelevant as is the color of the beads of an abacus for the arithmetical operations they represent. By content in the logical sense is meant only the system of relations common to a given set of symbolic systems. The fact that we can think of a system of relations only in terms of concrete objects does not change its independent and purely logical significance.
>
> (Reichenbach 1928/1958, 107)

Now, as much as this discussion appears to be linear and unproblematic, it seems to be mainly applicable to systems whose interpretations have already been sufficiently developed. However, when it comes to our description of the real world, we primarily need to be able to establish whether our interpretations are actually adequate descriptions of reality. Even more, considering scientific progress and how theories constantly expand, a pressing issue would be to know what it is that allows us to achieve a satisfactory mapping between formal representations and their interpretations. This is where the centrality of probability lies.

14.4. Coordinative Definitions, Probability, and Induction

In broad terms, the same approach Reichenbach uses in discussing geometry as a theory of relation applies to probability, and it is featured in 1932, in the first paper in which Reichenbach elaborates on the foundations of probability in connection to inductive logic. The calculus of probability is presented as a collection of formulas which can be regarded as "implicit definitions of the concept of probability". The same general considerations that apply to geometry also apply to probability as a system. From a formal point of view, within the axioms system, operations can be carried out without establishing an interpretation of the sign "$P( )$". On the other hand, an interpretation can be given to this sign compatible with the properties expressed within the axiomatic system. In this case, when we assign an interpretation to that sign, we give a *coordinative definition*. According to Reichenbach, this "double manipulation of the axiomatic system" turns out to be extremely important for the calculus of probability, as it enables us to construe probability as a frequency and base the entire axiomatic system on the frequency interpretation of probability (1932/33/1949, 307–308).[21]

As we have seen, at the beginning of his career, Reichenbach strongly emphasizes the essential role played by the principle of probability in granting the validity of mathematical structures for the description of reality. The problem at the core of the application of mathematics, which is intrinsic to coordination, is also at the heart of any scientific theorizing. In order to make predictions, for instance, science consistently selects the simplest curve resulting from measurements and assumes that this curve will approximate the outcomes of future measurements. This selection can be made only if we have a valid procedure in place that justifies this way of making extrapolations. At the same time, we also need to differentiate between well-grounded theories and arbitrary interpretations that do not properly describe the physical world. In both cases, which correspond to the twofold role of coordination (and to the two directions of research that Reichenbach envisaged in his doctoral thesis), we make use of an inductive inference.[22] In Reichenbach's frequency interpretation, an inductive inference asserts that the values that we observe in a finite sequence represent, "with certain limit of exactness, the value of the limit for the whole sequence" (1949, 351). Thus, every empirical statement, even every empirical law, will eventually rely on inductive inferences and so on induction (and ultimately on probability). Unsurprisingly, Reichenbach will try to solve the problem of induction throughout his entire career, specifically while attempting to provide a solid foundation to his probability theory.[23]

In the early 1930s, in the wake of the discussions around the status of core principles, Reichenbach addresses the issue of the nature and the justification of the principle of induction, assessing whether it could be interpreted as a convention or alternatively as an extra-scientific rule. He dismisses the second possibility, claiming that induction is the only tool we can use to ascertain empirically adequate theories as opposed to improper ones, which amounts to determining what the degree of probability of scientific theories is after all.[24] He rejects the first option even more decisively by arguing that if the principle of induction were a mere convention, we would not be able to justify it (and so the whole scientific construction would appear to be built on a quite unstable footing). As is well known, Reichenbach's justification of induction is based on the assumption that if a limit of the observed frequency of an event does exist, this limit will be detected by applying what is now known as "the straight rule". The idea of this rule is that the inductive method is convergent, that is, that the reiterated use of induction will eventually find – to a sufficient degree of approximation – convergence if a limit exists at all (1949, 474).

Providing a logical foundation for this type of inductive reasoning is obviously crucial in Reichenbach's work, especially considering that without proper foundation, it would not be clear how his axiomatic constructions could ever support a specific choice of coordination (or a set of coordinative definitions) with respect to a "piece of reality". In fact, as we have seen, the "physical thing" that is coordinated is already the result of a construction, which necessarily employs that very type of inductive reasoning which Reichenbach's entire philosophical approach to scientific reasoning revolves around. Reichenbach famously argued that the inductive procedure is a *"self-corrective method"* that will "automatically lead to success in a finite number of steps" if the straight rule is consistently implemented (1949, 446). Inductive inferences will eventually be assessed through and within other inferences, being constantly adjusted or revised if needed. This practice culminates in a system of higher-level inductions that Reichenbach calls "cross induction", which leads to an advanced state of knowledge, where inferences are combined with several other inferences in a network. This "method of *concatenation*" among inferences is what grants the success of induction (1949, 430–31) and, for Reichenbach, ultimately enables scientific reasoning.

The focal point of Reichenbach's early philosophical project was an analysis of how we can use mathematics to successfully describe the world (and why we need probability to do so). The role of probability, initially viewed as complementing the Kantian principle of causality, was featured as part of the problem of coordination and later examined in the framework of an original interpretation of constitutive principles. Reichenbach's shift towards conventionalism resulted in his attempt to axiomatize the theory of relativity, and later the theory of probability. As we saw, the prominent role of probability lies in its indispensability in all empirical reasoning. It is probability that explains how coordination can effectively be attained and that is why probability cannot be interpreted in a conventionalist fashion. Addressing the complexities of Reichenbach's approach to probability and induction would be beyond the scope of this chapter, but it is clear that probabilistic considerations cannot be dispensed with,[25] neither as far as basic scientific operations and activities are concerned nor in relation to the assessment of whole theories. Reichenbach's account of coordination, far from becoming an expression of his conventionalist turn, rather exemplifies one more limitation on reading him as a conventionalist.

Notes

1. Weyl's dismissive reaction to Reichenbach (1924) is fully documented, among others. See, in particular, Ryckman (2005) and literature therein. Reichenbach's overall approach to probability was also met with quite a lot of criticism. Cf. Eberhardt and Glymour (2011).
2. In connection with this, as we shall see subsequently, Reichenbach carefully distinguishes the formal system from the relational structure it defines, which is what matters to him.

3. As far as the notion of "coordination" is concerned, in this volume, Reichenbach's main philosophical inspiration is Cassirer's *Substance and Function* (1910/1923). See Padovani (2011), as well as Ryckman (1991) and Friedman (2007).
4. On this issue, see Eberhardt and Glymour's Introduction to Reichenbach (1916/2008) as well as Eberhardt (2011).
5. In this respect, Reichenbach argues, they are on a par with Kant's notion of "metaphysical possibilities" (Reichenbach 1916/2008, 119). Cf. also Padovani (2010).
6. Reichenbach justifies this principle by what he calls "a transcendental deduction". Cf. Eberhardt (2011) as well as Eberhardt and Glymour (2011).
7. Cf. in particular Chapters 2 and 3 of Reichenbach (1920/1965).
8. As Cassirer (1910/1923, 269) writes, "[a] cognition is called a priori not in any sense as if it were prior to experience, but because and in so far as it is contained as a necessary premise in every valid judgment concerning facts". See also Ryckman (1991, 85).
9. Literature on constitutive principles is now abundant and often referred to by using the expression coined by Friedman (2001), "relativized a priori". See also Friedman (2012) and, more recently, Stump (2015).
10. In (1923/1932), Reichenbach presents a possible elaboration of this idea and how it should work from a more practical point of view. See Padovani (2015b).
11. This narrower sense of coordination also appears to ultimately cover the relation between quantity terms and measurement procedures, in line with what Reichenbach presented in his doctoral thesis. See Padovani (2015a, 2017).
12. Cf. Ryckman (2005, Ch. 2). In this little volume, Reichenbach heavily criticizes Schlick for considering the uniqueness (*Eindeutigkeit*) of coordination a demand instead of a criterion that could, like any other assumption, be dispensed with in the future. Cf. (1920/1965, footnote 27 on p. 85).
13. To be sure, Schlick considers the criterion of *Eindeutigkeit* an intrinsic and necessary feature of any act of designation (if it aspires to truth). For Reichenbach, instead, *Eindeutigkeit* is rather an assumption from below, some sort of pragmatic condition of possibility for any empirical statement, if that statement has to provide an accurate description of reality. In (1920), Reichenbach associates this criterion with "truth", but in his later writing, it will be rather associated with an idea of empirical adequacy. Cf. Section 14.4.
14. Despite the insistence on contradictions, and an image of scientific theories that is in many respects reminiscent of Kuhn's paradigms, Reichenbach envisages the succession of scientific theories as a form of "expansion" (*Erweiterung*). Contradictions do emerge when a system has expanded sufficiently, but when the new system kicks in, the old theory is eventually included into the new one by means of a generalization of those principles that have become inconsistent with the old "paradigm". As he writes: "The contradiction that arises if experiences are made with the old coordinating principle by means of which a new coordinating principle is to be proved disappears on one condition: if the old principle can be regarded as an approximation [*Näherung*] for certain simple cases. Since all experiences are merely approximate laws [*Näherungsgesetze*], they may be established by means of the old principles; this method does not exclude the possibility that the totality of experiences inductively confirms a more general principle. It is logically admissible and technically possible to discover inductively new coordinating principles that represent a successive approximation of the principles used until now. We

can call such a generalization 'successive' (*stetig*) because for certain approximately realized cases the new principle is to converge toward the old principle with an exactness corresponding to the approximation of these cases. We shall call this inductive procedure the method of successive approximations [*Verfahren der stetigen Erweiterung*]" (1920/1965, 68–69).

15. A more articulated reflection on conventionalism plays an important role both in the way Reichenbach uses it to lay out the foundations of a theory and in the way axiomatic systems are applied to reality. On all these issues, see Ben-Menahem (2006).

16. This exchange is available at http://echo.mpiwg-berlin.mpg.de/content/modernphysics/reichenbach1920-22. See also Ryckman (2005, 55ff).

17. In this chapter, I will not address the implications of a choice of coordinative definitions. In particular, the idea that we can choose among different sets lends itself to an important discussion involving equivalence, namely whether the different descriptions of the world that would derive from a different set of coordinative definitions could be deemed equivalent. Reichenbach argues that different descriptions of the same matter of fact must be equivalent because no matter what "forms of expression" we pick, they must eventually "have the same content" (cf. especially the whole discussion in *Experience and Prediction* [1938]). However, to establish exactly what "empirical equivalence" means and whether it can actually be achieved through different sets of coordinative definitions is not an easy business. To be sure, as has long been pointed out, Reichenbach's notion of equivalence through coordinative definitions is quite defective, and his account of equivalent descriptions does not hold. See, for instance, Mühlhölzer (1991).

18. Consider, in fact, what Reichenbach writes in *Experience and Prediction* about geometrical conventionalism: "if in different worlds the definitions of coordination are settled in the same way, the resulting geometry may be different. Geometrical conventionalism is accordingly a misleading idea" (1938, 371).

19. In the English edition, "*eindeutige Zuordnung*" is translated as "one-to-one correspondence", which does not seem to capture the extent of Reichenbach's approach to coordination (1928/1958, 92–93).

20. As he puts it, these concepts are "best taken (albeit anachronistically) to represent the types of objects specified in the definition of a class of *mathematical structures* whose instantiations in each structure satisfy the postulates specified in this definition" (1994, 291).

21. See also Reichenbach (1933/34). In a footnote to the English translation, Reichenbach points out that in (1932/1933/1949), he did not mention the fact that in order to "make probability an analogue of truth a certain change in the interpretation of probability is necessary: probability then is not regarded as a property of events, or things, but as a property of linguistic expressions. In other words, the calculus of probability is then incorporated in the metalanguage. This transition is easily achieved for the frequency interpretation, since the numerical value of a probability will be the same whether we count events or the corresponding sentences about the occurrence of the events. Although the paper correctly employs the second interpretation, it would have been advisable to mention the duality of interpretation" (1932/1933/1949, 308).

22. Whether this approach is successful is obviously another story. See, again, Eberhardt and Glymour (2011).

23. As he puts it in *Experience and Prediction* (1938, §38), "the theory of probability involves the problem of induction, and a solution to the problem of probability cannot be given without an answer to the question of induction".

24. As we know, this was one of the points at issue in the discussion between Reichenbach and Popper. Reichenbach devotes an entire chapter of his *Theory of Probability* to further discuss the possibility to assess the probability of hypotheses (1949, §85). See also Eberhardt and Glymour (2011).
25. This idea appears to be in contrast to the reference to "truth" (relative to a given stage of knowledge) that seems to be entailed by Reichenbach's initial approach to univocal coordination. In fact, especially after 1930, Reichenbach claims that the dichotomous true-false of classical logic must abandoned in favor of a continuous scale of probability.

References

Ben-Menahem, Y. (2006), *Conventionalism: From Poincaré to Quine*. Cambridge: Cambridge University Press.

Cassirer, E. (1910/1923), *Substance and Function*. Chicago: Open Court.

Eberhardt, F. (2011), 'Reliability via Synthetic A Priori: Reichenbach's Doctoral Thesis on Probability', *Synthese*, 181 (1): 125–136.

Eberhardt, F. and Glymour, C. (2011), 'Hans Reichenbach's Probability Logic', in D. Gabbay, S. Hartmann, and J. Woods (eds.), *Handbook of the History of Logic X: Inductive Logic*. Amsterdam: Elsevier, pp. 357–389.

Friedman, M. (2001), *Dynamics of Reason*. Stanford, CA: CSLI Publications.

⸻ (2007), 'Coordination, Constitution, and Convention: The Evolution of the A Priori in Logical Empiricism', in A. Richardson and Th. Uebel (eds.), *The Cambridge Companion to Logical Empiricism*. Cambridge: Cambridge University Press, pp. 91–116.

⸻ (2012), 'Reconsidering the Dynamics of Reason', *Studies in History and Philosophy of Science*, 43: 47–53.

Mühlhölzer, F. (1991), 'Equivalent Descriptions', *Erkenntnis*, 35: 77–97.

Padovani, F. (2010), 'Statistical or Dynamical Lawfulness? Reichenbach and Schlick on the Laws of Nature', in F. O. Engler and M. Iven (eds.), *Moritz Schlick: Ursprünge und Entwicklungen seines Denkens*. Berlin: Parerga Verlag, pp. 225–255.

⸻ (2011), 'Relativizing the Relativized A Priori: Reichenbach's Axioms of Coordination Divided', *Synthese*, 181 (1): 41–62.

⸻ (2015a), 'Measurement, Coordination, and the Relativized A Priori', *Studies in History and Philosophy of Modern Physics*, 52: 123–128.

⸻ (2015b), 'Reichenbach on Causality in 1923: Scientific Inference, Coordination, and Confirmation', *Studies in History and Philosophy of Science*, 53: 3–11.

⸻ (2017), 'Coordination and Measurement: What We Get Wrong about What Reichenbach Got Right', *European Studies in Philosophy of Science*, 5: 49–60.

Reichenbach, H. (1916/2008), *The Concept of Probability in the Mathematical Representation of Reality*. Chicago: Open Court.

⸻ (1920/1965), *The Theory of Relativity and A Priori Knowledge*. Berkeley and Los Angeles: University of California Press.

⸻ (1921), 'Bericht über eine Axiomatik der Einsteinschen Raum-Zeit-Lehre', *Physikalische Zeitschrift*, 22: 683–686.

⸻ (1923/1932), 'Die Kausalbehauptung und die Möglichkeit ihrer empirischen Nachprüfung', *Erkenntnis*, 3 (1): 32–64.

—— (1924/1969), *The Axiomatization of the Theory of Relativity*. Berkeley and Los Angeles: University of California Press.

—— (1928), *The Philosophy of Space and Time*. New York: Dover.

—— (1930), 'Kausalität und Wahrscheinlichkeit', *Erkenntnis*, 1 (2–4): 158–188.

—— (1932), 'Axiomatik der Wahrscheinlichkeitsrechnung', *Mathematische Zeitschrift*, 34 (4): 568–619.

—— (1932/1933/1949), 'The Logical Foundations of the Concept of Probability', in H. Feigl and W. Sellars (eds.), *Readings in Philosophical Analysis*. New York: Appleton-Century-Crofts, pp. 305–323.

—— (1938), *Experience and Prediction*. Chicago: University of Chicago Press.

—— (1949), *The Theory of Probability*. Berkeley: University of California Press.

Ryckman, T. A. (1991), 'Conditio Sine Qua Non? Zuordnung in the Early Epistemologies of Cassirer and Schlick', *Synthese*, 88 (1): 57–95.

—— (2005), *The Reign of Relativity: Philosophy in Physics 1915–1925*. New York: Oxford University Press.

Schlick, M. (1918), *Allgemeine Erkenntnislehre*. Berlin: Springer.

Shapiro, L. S. (1994), '"Coordinative Definition" and Reichenbach's Semantic Framework: A Reassessment', *Erkenntnis*, 41: 287–323.

Stump, D. (2015), *Conceptual Change and the Philosophy of Science*. New York: Routledge.

von Kries, J. (1886), *Die Principien der Wahrscheinlichkeitsrechnung. Eine logische Untersuchung*. Freiburg i. B.: Mohr.

15 Two Constants in Carnap's View on Scientific Theories

Sebastian Lutz

15.1. Introduction

The received view on the history of Rudolf Carnap's logical empiricist view on the structure of scientific theories is an almost pitiful one. Initially, he assumed that theories can be formalized in predicate logic, with all theoretical terms defined in observational terms. The latter assumption Carnap himself had to weaken almost immediately and repeatedly until he had completely abandoned the demand that all concepts be introduced successively, starting from observational terms. What was left was the hope that observational terms could be defined in theoretical terms. Carnap's assumption that theories could be formalized in predicate logic was abandoned by almost everyone else, because it would render any actual formalization of theories inconvenient and far removed from actual science. Almost nothing of Carnap's original approach remains, as it has now been effectively replaced by the semantic view of, for instance, Bas van Fraassen. This story is very briefly recounted in Section 15.2.

Against this received view, I argue that the history of Carnap's view on the structure of scientific theories is not as tragic as it appears. First, from his earliest writings on the topic (which, I argue in Section 15.3, should be considered contributions to the logical empiricist view on theories), Carnap advocated for a description of physical theories as restrictions on mathematical spaces (Section 15.4.1). This is essentially the state-space approach advocated by van Fraassen (Section 15.4.2). Second, Carnap's conception of correspondence rules was not weakened in repeated concessions. Rather, his final conception, the possibility of defining maybe all observational terms in theoretical terms and the impossibility of the converse, was already present in his earliest writings on the general structure of scientific theories, albeit at times in contradiction to his professed position (Section 15.5.1). The same conception is also assumed by van Fraassen in his state-space approach (Section 15.5.2).

Carnap's view on the structure of scientific theories thus emerges as surprisingly resilient, surviving first as an undercurrent beneath his overly positivistic tendencies during the time of the Vienna Circle and later as a core idea of one of its alleged replacements.

15.2. The Decline and Fall of Carnap's Early View on the Structure of Scientific Theories

With respect to the interpretation of scientific terms, Carnap (1963, §9) himself describes in his "Intellectual Autobiography" the development of his view on scientific theories as a gradual liberalization. Initially, in *Der logische Aufbau der Welt* (Carnap 1928/1967, §35, *Aufbau* from now on), every meaningful sentence is supposed to be translatable into a sentence about experiences, and this means that "the concepts of science are explicitly definable on the basis of observation concepts" (Carnap 1963, 59).

In "Testability and Meaning", Carnap (1936, 1937) relaxes this claim, because he has come to the conclusion that it is impossible to define disposition terms. Instead, he suggests that a new term should be introduced by a pair of reduction sentences, that is, one necessary and one sufficient condition for the new term.

In the *Foundations of Logic and Mathematics*, Carnap (1939) describes two ways of relating theoretical terms (he calls them "abstract") and observational ("elementary") terms. There is, first, the construction of theoretical terms from observational terms by reduction sentences and, second, the construction of observational terms from theoretical terms. In both methods, only the observable terms are directly interpreted. Carnap (1939, 206) conjectures that in the second method, "so it seems at present, explicit definitions will do", while the first method does not allow for such a strict relation.

In two later works, Carnap does not even mention the first method of interpreting theoretical terms. In "The Methodological Character of Theoretical Concepts", he states that the correspondence rules (*C*-rules)

> specify the relation *R* which . . . relates to an observable space-time region *u*, e.g., an observable event or thing, a class *u'* of coordinate quadruples which may be specified by intervals around the coordinate values *x*, *y*, *z*, *t*.
>
> On the basis of these *C* rules for space-time designations, other *C*-rules are given for [theoretical terms]. These rules . . . hold for any space-time location. They will usually connect only very special kinds of value distributions of the theoretical magnitudes in question with an observable event. For example, a rule might refer to two material bodies . . . observable at locations *u* and *v*. . . . Another rule may connect the theoretical term "temperature" with the observable predicate "warmer than" in this way: "If *u* is warmer than *v*, then the temperature of *u'* is higher than that of *v'*.
>
> (Carnap 1956, 47)

Thus observable space-time regions can be assigned to specific space-time coordinates, and observable terms are assigned to specific values of theoretical magnitudes. Carnap also emphasizes the asymmetry between observable terms and theoretical magnitudes when noting that some, but not all, distributions of theoretical magnitudes are assigned an observable event.[1] For instance, Carnap (1950/1962, 12–13) notes that it is not required that "if x is not warmer than y (in the prescientific sense), then the temperature of x is not higher than that of y", because "[w]hen the difference between the temperatures of x and y is small, then, as a rule, we notice no difference in our heat sensations".

In "Beobachtungssprache und theoretische Sprache", Carnap repeats the construction from "The Methodological Character of Theoretical Concepts" and notes that there is an asymmetry not only between theoretical magnitudes and observable predicates but also between space-time coordinates and space-time regions:

> We can . . . introduce a space-time coordinate system in which each small body at any time point can be assigned an ordered quadruple of real numbers. Then by generalization, every such quadruple can be regarded as representative of a space-time point (therefore as an unobservable theoretical object.) Then physical magnitudes, such as mass-density, may be introduced, which have a value for every space-time point e.g., a real number. A function of this sort is construed in our system by a function F of quadruples of real numbers.
>
> (1958/1975, 81)

This passage is very much in keeping with the previous one (Carnap 1956, 47). The space-time point u (rather than the region) is given by a small body, and the theoretical magnitudes are the physical magnitudes. In this passage, however, Carnap is even clearer about the impossibility of translating every theoretical statement into an observation statement.

Carnap's *Philosophical Foundations of Physics: An Introduction to the Philosophy of Science* (1966), based on a seminar taught in 1958 but updated to include later results, serves as a stopping point for this brief history of the decline of Carnap's view on scientific theories. In this introductory textbook, he repeats the claim that, while it may be possible to define all observational terms in theoretical terms, the reverse is impossible (1966, 234). In connection with measurement and with the correspondence rules between theoretical and observational terms, Carnap (1966, 57, 72, 266) also references a monograph on concept formation by Hempel (1952, 684, §7), who repeatedly stresses the importance of

> *theoretical constructs*, i.e., the often highly abstract terms used in the advanced stages of scientific theory formation, such as "mass", "mass point" . . ., "volume", "Carnot process" . . ., "proton", "ψ-function", etc. . . . Terms of this kind are not introduced by definitions or reduction chains based on observables; in fact, they are not introduced by any piecemeal process of assigning meaning to them individually. Rather, the constructs used in a theory are introduced jointly, as it

were, by setting up a theoretical system formulated in terms of them and by giving this system an experiential interpretation.

One way of giving an experiential interpretation to the theoretical system consists of defining further concepts with the help of the theoretical constructs and interpreting those further concepts directly (1952, 686–687, §7). This is Carnap's second method of giving empirical meaning to theoretical terms (Carnap 1939); Hempel's treatment of "mass" as such a theoretical term (Hempel 1952, §12) was even suggested by Carnap himself (1952, 738, n. 72). Hempel (1952, §§6–7) considers the use of theoretical constructs another step in the liberalization of empiricism: the claim that all terms are explicitly definable in observational terms is the narrower thesis of empiricism (1952, 676); the claim that all terms are reducible to observational terms (Carnap 1936) is the liberalized thesis of empiricism (Hempel 1952, 683). The need for reduction sentences shows that the narrower thesis is false, and the need for theoretical constructs shows that the liberalized thesis is false.

While Carnap saw value in the concept of correspondence rules until the end, Hempel (1970, §6) extended his critique of explicit definitions and reduction sentences to theoretical constructs and dismissed the whole concept of correspondence rules as misguided. Further criticisms of correspondence rules (Suppe 1972, §II) and the use of axiomatizations in general (Hempel 1970, §3, Suppe 1974, §IV) contributed to the downfall of Carnap's view on the structure of scientific theories. Hempel (1969, 1970), for example, abandoned it completely. In its stead, the semantic view on scientific theories became an important framework for the reconstruction of theories, possibly the dominant one (Suppe 1989, 3). The semantic view assumes set theoretical descriptions of scientific theories (Suppes 1967) or descriptions in phase space (van Fraassen 1970), both of which are considered closer to actual science than Carnap's view.

So much for the received view on Carnap's view. To arrive at a more positive assessment, it suffices to look at Carnap's earliest works on the structure of scientific theories.

15.3. Three Early Works of Carnap

Both Suppe (1974, 12) and Feigl (1970, 3) cite "Über die Aufgabe der Physik und die Anwendung des Grundsatzes der Einfachstheit" ("On the Task of Physics and the Application of the Principle of Maximal Simplicity"; Carnap 1923/2019, "Aufgabe" from now on) as Carnap's earliest exposition of the logical empiricist's view on scientific theories (Feigl also mentions Campbell 1920). Mormann criticizes this classification (in a footnote, without elaborating further):

> Feigl once went so far to trace back the essentials of the Logical Empiricist account of empirical theories to an early (pre-Vienna)

paper of Carnap that may well be classified as belonging to his neo-Kantian period. . . . This stance betrays, to put it mildly, that Feigl did not pay too much attention to the amendments that had taken place since then.

(2007, 159, n. 13)

However, even without consulting the "Aufgabe" itself, there are some reasons to consider it a contribution to logical empiricism's view of theories. For one, Mormann's claim of a discontinuity rests on the text's belonging to Carnap's neo-Kantian period, but even though Carnap wrote the article while holding neo-Kantian views (Carus 2019a, note b), this does not mean that they are manifest in the article. Carnap himself, for example, did not seem to think so. When discussing the influence of Kant's views on his own work in his autobiography, Carnap (1963, 12) mentions his doctoral dissertation (Carnap 1922), specifically the chapter on intuitive space (Carnap 1963, 4), but not the "Aufgabe". As influences for this article, Carnap (1963, §13, 77–78) rather lists Poincaré and Hugo Dingler and at another point Hilbert, Poincaré, and Duhem.

Furthermore, Carnap (1963, 15, §2) classifies the "Aufgabe" as written in the same "period" as the monograph *Physikalische Begriffsbildung* (*Physical Concept Formation*; Carnap 1926/2019), which was written in Vienna after the main work on the *Aufbau* had concluded (Carus and Friedman 2019, lxi) and whose main points were much later taken up by both Hempel (1952) and Carnap (1966).

Perhaps most importantly, Carnap begins §13 of his autobiography, entitled "The Theoretical Language", with the "Aufgabe" (followed by his monograph on concept formation) and ends it with his article on the methodological character of theoretical terms (Carnap 1956). This is relevant because §13 occurs in Part II of Carnap's autobiography, entitled "Philosophical Problems", where "[i]n each section, a certain problem or complex of problems [is being] dealt with". So Carnap himself considers the "Aufgabe" a starting point of the development that led to one of his core articles on scientific theories.

It is revealing that Carnap ends the overview that begins with the "Aufgabe" with his article on theoretical terms, which is a discussion of the formal structures of correspondence rules and the formal requirements they must fulfill so that theoretical terms and sentences count as empirically significant. Such formal aspects can survive radical changes in meta-theoretical perspective. For instance, Carnap's discussion of empirical significance showed no significant break when he included explicitly semantic considerations in his analysis (cf. Lutz 2017) and neither did his reliance on higher-order logic (Lutz 2012, §2). Thus, even if Carnap wrote the "Aufgabe" from a Kantian perspective, all other aspects, and especially the technical ones on the structure of scientific theories, can already anticipate the logical empiricist view.

In his autobiography, Carnap summarizes the "Aufgabe" already in his later terminology:

> I imagined the ideal system of physics as consisting of three volumes: The first was to contain the basic physical laws, represented as a formal axiom system; the second to contain the phenomenal-physical dictionary, that is to say, the rules of correspondence between observable qualities and physical magnitudes; the third to contain descriptions of the physical state of the universe for two arbitrary time points. From these descriptions, together with the laws contained in the first volume, the state of the world for any other time-point would be deducible (Laplace's form of determinism), and from this result, with the help of the rules of correspondence, the qualities could be derived which are observable at any position in space and time. The distinction between the laws represented as formal axioms and the correlations to observables was resumed and further developed many years later in connection with the theoretical language.
>
> (1963, 15)

Here, then, Carnap also points out the continuity between the "Aufgabe" and later discussions of theoretical terms.

In the "Aufgabe" itself, Carnap introduces his main point with a reference to Poincaré and Dingler, but not Kant:

> It is the main thesis of the conventionalism expounded by Poincaré and further developed by Dingler that in the construction of physics we have to make stipulations that are subject to our free choice. . . . But the choice among these stipulations ought not therefore to be made arbitrarily, rather it should follow certain methodological principles – and in the end the principle of maximal simplicity has to decide.
>
> (1923/2019, 90/211, emphases removed)

The laws of physics in the first volume can be chosen according to the principle of maximal simplicity, Carnap (1923/2019, 90/211) states, because they are not determined by experience. Whether, for example, the world has a Euclidean or non-Euclidean geometry depends on the objects that are chosen to be rigid bodies (an example that Carnap will repeat in his introductory textbook [1966] when discussing the conventional elements in the concept of length). Therefore, Carnap concludes, the laws of physics are

> *synthetic a priori propositions*, although not exactly in the Kantian transcendental-critical sense. For that would mean that they expressed necessary conditions of the objects of experience, themselves

conditioned by the forms of intuition and thought. . . . Actually, though, [this volume's] construction is in many ways left to our choice. . . . As an identifying description of the first volume, the concept of a "hypothetico-deductive system" . . . is therefore to be preferred to the Kantian concept of the a priori.

(Carnap 1923/2019, 97/233)

This is the only passage in the "Aufgabe" that contains an explicit reference to Kant, and it is far from an endorsement of the Kantian doctrine of the synthetic a priori but rather, just as Carnap (1963) states in his autobiography, an endorsement of Poincaré's conventionalism. One could, of course, argue that Poincaré's conventionalism is neo-Kantian, but then Carnap never gave up neo-Kantianism in this sense, since he endorsed the conventionality of theoretical constructions to the end.

The connections between the "Aufgabe" and Carnap's later view on scientific theories are obvious. He makes a strict distinction between, first, the domain of perception and, second, the domain of physical theories, a distinction that "cannot be emphasized sharply enough". The contents of perception do not occur at all in theoretical physics, which is not obvious only because terms like "pressure" and "heat" are used in both domains (Carnap 1923/2019, 99/227). The connection between the two domains is given, "in a way, through a kind of dictionary that indicates which objects (elements, complexes, processes) in the second domain correspond to the particular ones of the first". This dictionary is just a metaphor for correspondence rules. Carnap gives the following examples for dictionary entries:

> "Such and such a shade of blue (designated, for example, according to Ostwald's color system) corresponds to a certain periodical movement of electrons (denoted by the frequency of oscillation)" . . . "Such and such a pungent smell (the smell of chlorine; [smells] lack a classificatory system) corresponds to a certain mixture of peculiarly structured electron complexes (C1-atoms)"; "A certain temperature sensation ([these also] lack a classificatory system) corresponds to a certain average kinetic energy of a number of electron complexes (atoms or molecules)".
>
> (1923/2019, 99–100/227–229)

The two examples for correspondence rules that Carnap would later give in his introductory textbook read almost like translations of these dictionary entries:

> An example for such a rule is: "If there is an electromagnetic oscillation of a specified frequency, then there is a visible greenish-blue color of a certain hue". Here something observable is connected with a nonobservable microprocess.

Another example is: "The temperature (measured by a thermometer, and therefore, an observable in the wider sense explained earlier) of a gas is proportional to the mean kinetic energy of its molecules". This rule connects a nonobservable in molecular theory, the kinetic energy of molecules, with an observable, the temperature of a gas.

(1966, 233)

Here, the Ostwald color-system is substituted by the hue of the colors, and the missing system of designation for heat experiences is circumvented by using an observational term in the wider sense, the temperature according to a thermometer. The connection to heat experiences has to be given through the further correspondence rules between temperature and the observational relation "warmer than" already noted previously (Carnap 1956, 12–13).

In an article published shortly after the "Aufgabe", entitled "Dreidimensionalität des Raumes und Kausalität. Eine Untersuchung über den logischen Zusammenhang zweier Fiktionen" ("Three-Dimensionality of Space and Causality: An Investigation of the Logical Connection Between Two Fictions"), Carnap (1924/2019, 107–108/241) relies, as in the "Aufgabe", on the distinction between the domain of perception and the domain of physical theories. He speaks of experiences of the first level, which are the phenomena, and experiences of the second level, which can be those of physics or, since this level is subject to conventional choices, also the ordinary experiences involving everyday concepts. The contents of the experiences on the first level are called the "primary", those of the second level the "secondary world", which can again be the "world of physics" or the "ordinary world" (1924/2019, 108–109/253–255). The connection between the secondary and the primary world is, with a reference to the "Aufgabe", given through a "relation of coordination" (Carnap 1924/2019, 108/253).

In his monograph on concept formation, Carnap (1926/2019, 60/417) also gives another example of correspondence rules: specific electron configurations are assigned specific atoms or crystals, say, chloride and sodium or sodium chloride. These configurations are then assigned the qualities "white" and "salty". These correspondence rules are very close to the ones in his introductory textbook,[2] but they are even closer to an example that Carnap gives somewhat earlier:

[L]et us imagine a calculus of physics is constructed, according to the second method, on the basis of primitive specific signs like "electromagnetic field", "gravitational field", "electron", "proton", etc. The system of definitions will then lead to elementary terms, e.g. to "Fe", defined as a class of regions in which the configuration of particles fulfils certain conditions, an "Na-yellow" as a class of space-time regions in which the temporal distribution of the electromagnetic field fulfils certain conditions. The semantic rules are laid down stating that "Fe" designates iron and "Na-yellow" designates a specific yellow color. (If "iron" is not accepted as sufficiently elementary, the rules can be stated for more elementary terms.)

(1939, 207)

When it comes to the detailed description of the relation of the observational and theoretical terms, the "Aufgabe", the monograph on concept formation, this much later text, and the introductory textbook show barely a change at all.

Finally, the *Aufbau* puts the discussions of the three early works in a wider perspective. Here, Carnap aims at developing a constitution system in which all statements are translatable into statements that contain only one primitive relation (although more are possible) (Carnap 1928/1967, §§1–2, §35, §156). The lower levels of the system constitute the autopsychological objects from the primitive relation, the middle level the physical objects from the autopsychological objects, and the upper levels the heteropsychological and cultural objects from the physical objects. The correspondence rules of the "Aufgabe" and the monograph on concept formation are found in the middle level.

The autopsychological objects are first used to constitute "the entire space-time world, with the assignment of sense qualities to the individual world points, [which] we call the perceptual world" (Carnap 1928/1967, §133). The perceptual world then allows the constitution of the "physical world" (1928/1967, §136).[3] When constructing the physical world, Carnap refers to his three earlier works for further elaboration.[4]

In the construction of the world as described in the *Aufbau*, then, the method of interpreting theoretical terms that Carnap (1923/2019) describes in the "Aufgabe" is not involved in the assignment of sensory qualities to space-time points but covers the whole of the second step, the correlation of numbers as values of state magnitudes to these sensory qualities. The Kantian notions of Carnap's doctoral dissertation would occur somewhere in the construction leading to the assignment of sensory qualities to space-time points. So whatever Kantian notions about synthetic a priori statements Carnap had at the time of writing, they would not have been relevant for the construction developed in the "Aufgabe". This is the final reason Mormann's claim that the "Aufgabe" cannot be considered a contribution to the logical empiricist's theory of science is incorrect.

In fact, a bigger break in Carnap's reasoning about the structure of scientific theories seems to have come after he had written his monograph on concept formation: the finitism of Wittgenstein's *Tractatus* seemed to make a theoretical language impossible (Carus and Friedman 2019, xli). Accordingly, there is a long gap of Carnap's writings on the philosophy of science until after the *Logische Syntax der Sprache* (Carnap 1934/1967), which deals with this problem (and others).

15.4. The Formalization of Scientific Theories

In one of his criticism of the logical empiricists, van Fraassen writes:

> The scholastic logistical distinctions that the logical positivist tra-
> dition produced – observational and theoretical vocabulary, Craig
> reductions, Ramsey sentences, first-order axiomatizable theories, and
> also projectible predicates, reduction sentences, disposition terms,
> and all the unholy rest of it – had moved us mille milles de toute
> habitation scientifique, isolated in our own abstract dreams.
>
> (1989, 225)

I will discuss in the next section how much worse Carnap's logical
empiricist view of the relation between theory and observation is than
van Fraassen's semantic view. In this section, I will focus on how much
worse Carnap's view on the formalization of theories is. Before I do so,
it bears repeating that the alleged fixation of logical empiricists on first-
order logic was invented from whole cloth by critics of logical empiricism
(Lutz 2012, §2) and that Carnap saw higher-order logic as including all
of mathematics (Carnap 1939, §14), which is a quite plausible position
(Andrews 2002, vi–xiv). The question is, then, how logic and mathematics were supposed to be used for describing theories according to Carnap.

15.4.1. *Carnap's Purely Numerical Structures*

In his article on three-dimensionality and causality, Carnap repeats the
point of the "Aufgabe" that the physical world is free from perceptual
qualities and contains

> only spatial and temporal magnitudes, together with certain nonsen-
> sible state-magnitudes. In its purest form, moreover, these three types
> of magnitudes have no character comparable with spatiality, tempo-
> rality, or sensible qualities, but are rather mere numerical determina-
> tions, i.e., relational terms.
>
> (1924/2019, 108/253)

The laws in this pure physical world are also free from the concept of causation as it occurs in the ordinary world (Carnap 1924/2019, 108/253,
emphasis in the original): "The processes in the physical world *do not act*
on one another; rather they are governed by a dependency that is to be
conceived of as a purely mathematical, functional relation". The use of
"mere numerical determinations" with a dependency that is considered
"a purely mathematical, functional relation" to describe scientific theories is the first major constant in Carnap's view on scientific theories.

Carnap (1924/2019, 120/271) notes that instead of describing the
value of each magnitude at each point in space for two different times,

one can equivalently describe the value and the derivative of each magnitude at each point in space for just one time. In the "Aufgabe", Carnap (1923/2019, 101/229) notes that, for logical reasons, using two different times is more correct (Carnap refers to Russell and Hausdorff (Carus 2019a, note f) for the proof that derivatives cannot be considered instantaneous magnitudes); the equivalence, however, enables the use of the more expedient description at one time:

> If it is possible to calculate, from the coordinate values of the n elements at time t_0 and the $3n$ components of their velocities, their $3n$ coordinates at time t_1, then we only need to think of the $3n$ equations specifying these coordinates as solved for the $3n$ velocity components in order to see that these components are then also determined by their coordinates at times t_0 and t_1.

While Carnap in general in the "Aufgabe" speaks of the laws of scientific theories without further specification as a set of formal axioms,[5] this more concrete elaboration is a description of phase space that is only missing the explicit classification as a geometrical space.[6] (This classification, however, brings with it substantial formal gains; Scheck 2010, §1.18.)

The explicit classification as a geometrical space is first mentioned in the article on three-dimensionality and causality, where Carnap (1924/2019, II.c) claims that the physical world is four-dimensional with the qualification that this "concerns only spatial and temporal dimensions, not qualitative dimensions" (1924/2019, 113/261). Here Carnap is conceptualizing physical values as fixing one dimension of an abstract space.[7] Laws of theories are described as restrictions on abstract numerical functions:

> If any element of a class depends on other elements in such a way that it is uniquely determined as soon as a certain subclass of the remainder is fixed, then we call the relation of dependency a "*determining law*" and the class "*determined*". . . .
> We call laws of dependency that do not result in unique determinacy for any element, even if all the rest are determined, but still limit the possibilities for this element, "*constraining laws*".
> (Carnap 1924/2019, 118/269)

This abstract view on scientific theories is elaborated in Chapter III of Carnap's monograph on concept formation, entitled "Abstract Stage: the Four-Dimensional Universe". Here, Carnap first suggests identifying space-time points by four-tuples so that a physical description of the world just consists of assigning the values of the basic physical magnitudes to each point in four-dimensional space-time. But, Carnap notes, it is also possible to go beyond this geometrical description and let descriptions of the world consist of sets of tuples, with the first four values being the space-time points, the rest being the values of the physical magnitudes. This, then, is the purely numerical description of his article on three-dimensionality and causality. As in that article, laws of nature are in both cases restrictions on the possible descriptions: restrictions on the possible assignments of values to space-time points in the geometrical case and restrictions on the possible sets of tuples in the numerical case (Carnap 1926/2019, 58/415). In §136 of the *Aufbau*, finally, the "physical world" is accordingly considered a "(purely numerical) structure".

In the monograph on concept formation, Carnap (1926/2019, 55–56/411-413) is explicit that he is thinking of field theories,[8] noting that it is at this point not yet clear whether all of physics can be so described. A classical field theory is the infinite dimensional analogue to the finite dimensional phase space (Scheck 2010, §§7.1, 7.4, 7.6).

Carnap himself considered this aspect of his monograph closely connected to his later work:

> I described the world of physics as an abstract system of ordered quadruples of real numbers to which values of certain functions are co-ordinated; the quadruples represent space-time points, and the functions represent the state-magnitudes of physics. This abstract conception of the system of physics was later elaborated in my work on the theoretical language.
>
> (1963, 15–16)

Of course, these points of the monograph essentially recapitulate the same points in the article on three-dimensionality and causality. The formalism suggested in these two works would later be used again (Carnap 1956, 1958/1975), with the same stress on the purely mathematical character of the space-time tuples and the assignment of other numbers to those tuples as physical magnitudes at those space-time points.

15.4.2. Van Fraassen's State-Space

In the semantic view, theories are not formalized as sets of postulates in predicate logic but rather through the structures that are models of such sets (Suppe 1989, 4) or through restrictions on an abstract state-space (van Fraassen 1980, 44). When they rely on structures, the formalizations of the semantic view can be expressed in higher-order logic (Halvorson 2013; Lutz 2014b; Da Costa and Chuaqui 1988; Hudetz 2017).[9] This leaves van Fraassen's state-space approach as a possible competitor to Carnap's view.

In an outline of his semantic view that is endorsed, for example, by Suppe (1974, 221–222), van Fraassen focuses on

> the formal structure of *nonrelativistic* theories in physics. . . . A physical system is conceived of as capable of a certain set of *states*, and these states are represented by elements of a certain mathematical space, the *state-space*. . . . To give the simplest example, a classical particle['s] . . . state-space can be taken to be Euclidean 6-space, whose points are the 6-tuples of real numbers $(q_x, q_y, q_z, p_x, p_y, p_z)$.
> (1970, 328–329)

Van Fraassen (1970, 330) then distinguishes between laws of coexistence, laws of succession, and laws of interaction. In the non-statistical case (1970, §5.1), laws of coexistence select the physically possible subset of the state-space (1970, 330); laws of succession select, in the instantaneous state picture, the physically possible trajectories in the state-space (1970, 331); and laws of interaction, at least in principle, reduce to the previous (1970, 332).

van Fraassen's phase space example recalls Carnap's description of an *n*-particle system in the "Aufgabe". And van Fraassen's schema for the formalization of scientific theories is essentially that described in Carnap's article on three-dimensionality and causality. Both assume a purely mathematical description of the possible states of a system and consider scientific theories as restrictions on this space. The main difference is that Carnap focuses on systems that contain continuous fields and thus have infinitely many degrees of freedom, while van Fraassen focuses on systems that contain a finite number of particles with accordingly finitely many degrees of freedom.[10]

Formally, the main difference between Carnap and van Fraassen is that the former assumes that the description of the theory is given in higher-order logic, and the latter assumes that it is given in the semi-formal meta-language of an elementary language to be discussed subsequently. But since both Carnap and van Fraassen take their respective languages to include all of mathematics and allow as much or as little explicit formalization as expedient (Lutz 2012, §§2–3), this difference is merely verbal (Lutz 2014b, 1489).

15.5. The Relation Between Observational and Theoretical Terms

The formalizations assumed in Carnap's view on theories did not move us "mille milles de toute habitation scientifique", at least not more so than van Fraassen's view. But how about the "observational and theoretical vocabulary . . . and all the unholy rest of it"? Specifically, is Carnap's view, and especially his early view, on the rules connecting theory and observations indeed too far removed from science?

15.5.1. Carnap's Correspondence Rules

In his introductory textbook, Carnap writes:

> Different writers have different names for these rules. I call them "correspondence rules". P.W. Bridgman calls them operational rules. Norman R. Campbell speaks of them as the "Dictionary". Since the rules connect a term in one terminology with a term in another terminology, the use of the rules is analogous to the use of a . . . dictionary. . . .
>
> There is a temptation at times to think that the set of [correspondence] rules provides a means for defining theoretical terms, whereas just the opposite is really true. A theoretical term can never be explicitly defined on the basis of observable terms, although sometimes an observable can be defined in theoretical terms.
>
> (1966, 233–234, footnote removed)

This is Carnap's final view on correspondence rules: Observational terms can sometimes be explicitly defined in theoretical terms but not vice versa. As I will argue, it is the second constant in Carnap's view on scientific theories, identifiable in all of his major works on the structure of scientific theories, starting from the "Aufgabe". There, Carnap states that any statement in observational terms can be described in theoretical terms but not vice versa:

> The dictionary can be used in both directions: it serves to translate a phenomenal state of affairs to the corresponding physical one and vice versa. However, we have to note that *the correlation is unique only in the second case*; a particular sensory content corresponds not to just a single particular physical state of affairs but to an infinite number of them.
>
> (1923/2019, 100/229, original emphasis)

Carnap's first reason for the lack of (unique) translatability into the phenomenal language are the multiple microscopic realizations of physical macrostates in, for example, thermodynamics. His second reason is the "psycho-physical fact of the threshold of sensitivity" (1923/2019, 100/229). As recapitulated previously, Carnap (1950/1962, 13) will later argue that because of this threshold of sensitivity, it cannot be required that there is a difference in temperatures only if there is a difference in heat sensations. Carnap (1924/2019, 126/281) also adduces the perception threshold in his article on three-dimensionality and causality and adds two more reasons: First, a sensation does not uniquely identify the location of its physical source. Second, a sensation, say, of a color, corresponds to a multitude of physical states, say, a multitude of frequency distributions of electromagnetic waves. Thus, phenomenal states are multiply realized by physical states, which, if they are physical macrostates (as in thermodynamics), are themselves multiply realized by physical microstates.

In his monograph on concept formation, Carnap repeats the claim that observational states are uniquely determined by physical states:

> The retranslation of the pure number statements of abstract physics into qualitative statements is possible because a particular distribution of values of particular state-magnitudes is uniquely co-ordinated with certain physical qualities, ultimately sensory qualities. A certain character of a set of 14-tuples of numbers is e.g. to be interpreted as a certain motion of electrons within a certain spatial configuration. And this in turn is to be interpreted e.g. as a chlorine atom or a sodium atom or a sodium chloride crystal, i.e. table salt. A set of 14-tuples so constituted is then co-ordinated with the qualities white and salty. Another such set is to be interpreted as, say, another motion of a set of electrons, and this in turn as a particular periodic distribution of air molecules, hence a sound wave of a particular frequency; a set of 14-tuples of numbers so constituted is then associated with a tone of particular pitch, timbre and volume, or a precisely characterized sound.
>
> (1926/2019, 60/417)

Another example would be the translation of the quadruples of space-time numbers with their assigned momenta into average kinetic energies, from there into temperature, and finally into heat experience (Carnap 1958/1975, 1966, 1950/1962).

Note that "uniquely co-ordinated" here means that each state-magnitude is assigned exactly one sensory quality, so that "retranslation" means that each description of state-magnitudes is assigned exactly one description of physical qualities but not necessarily vice versa; this is the converse of the modern use of "translation" in logic and definition theory (cf. Creath and Richardson 2019, note dd)

In the monograph, Carnap does not explicitly state that physical qualities cannot be uniquely co-ordinated with state-magnitudes but repeats his claim from the "Aufgabe" and his article on three-dimensionality and causality that there is a perception threshold:

> As a first attempt, we could . . . assign a higher temperature to one of the bodies if it evokes a stronger sensation of warmth. . . . This

method of assignment, however, would turn out not to work, owing to the fact of "heat transfer". For if two bodies are brought into contact with each other [and] only one has a perceptible change, then we ascribe to the other a complementary change of imperceptible magnitude (for the sake of later laws of nature in connection with the concept of specific heat).

(Carnap 1926/2019, 17/364-365)[11]

This illustration of the perception threshold led Carnap in the earlier two articles to conclude that the theoretical states are not uniquely determined by the observational states, and Carnap does nothing to counter this conclusion here. However, he does not give a perfectly clear endorsement of the thesis; this clear endorsement is given in the *Aufbau*, of all places.

In §136 of the *Aufbau*, the "physicoqualitative correlation" consists, first, of a one-to-one correspondence between the world points of physics (the space-time points from the monograph on concept formation) and the world points of the perceptual world and, second, a many-one relation between the physical magnitudes and the perceptual qualities, as in the "Aufgabe". While each physical state can be assigned a perceptual state,

> in the opposite direction, the correlation is not unique; the assignment of a quality to a world point in the perceptual world does not determine which structure of state magnitudes is to be assigned to the neighborhood of the corresponding physical world point of the world of physics; the assignment merely determines a class to which this structure must belong.

Implicitly, this position entails that observational terms can be determined in theoretical terms but not vice versa, because, as in the "Aufgabe" (Carnap 1923/2019, 99/227), Carnap assumes a strict distinction between the concepts applying to the physical states (which are named by theoretical terms) and the concepts applying to the perceptual states (which are named by observational terms).

In modern terminology, the relation between phenomenal and physical states can be phrased syntactically or semantically. (Carnap could not easily have made that distinction, since it did not become clear to him until 1930; see Reck 2007, 189–191.) Syntactically, Carnap claims that every sentence in observational terms can be translated into a sentence in theoretical terms with the help of the correspondence rules but not vice versa. From the theory of definition, it is known that then the correspondence rules entail an explicit definition for each observational term in theoretical terms, while the converse does not hold. Semantically, Carnap claims that a structure for the theoretical vocabulary can be expanded at most in one way to a model of the correspondence rules, while at least some structures for the observational vocabulary can be expanded to more than one model of the correspondence rules. This means that the correspondence rules do not entail an explicit definition for each theoretical term (according to Padoa's theorem) and that the correspondence rules do entail an explicit definition for each observational term (according to Beth's theorem) for correspondence rules in first-order logic and (according to theorem 3 by Tarski 1935) for a finite number of correspondence rules in finite type theory (cf. Leivant 1994, §5.1).

In other words, already in the "Aufgabe", his very first publication on the structure of theories, as well as in his article on three-dimensionality and causality, his monograph on concept formation, and in the *Aufbau*, Carnap indirectly claims that theoretical terms are interpreted through observational terms according to the second method described in the *Foundations of Logic and Mathematics* (Carnap 1939), although without the later qualification that explicit definability *seems* possible given the contemporary state of science. Of course, up to the 1920s, definition theory and formal semantics were not yet developed enough to phrase this consequence so clearly. Nonetheless, this is what Carnap's position entails.[12]

In his autobiography, Carnap notes how the part of the *Aufbau* in which he claims the indefinability of theoretical terms relates to his later work:

> For the construction of the world of physics on the basis of the temporal sequence of sensory qualities, I used the following method. A system of ordered quadruples of real numbers serves as the system of co-ordinates of space-time points. To these quadruples, sensory qualities, e.g., colors, are assigned first, and then numbers as values of physical state magnitudes. . . . In general, I introduced concepts by explicit definitions, but here the physical concepts were introduced instead on the basis of general principles of correspondence, simplicity, and analogy. It seems to me that the procedure which is used in the construction of the physical world, anticipates the method which I recognized explicitly much later, namely the method of introducing theoretical terms by postulates and rules of correspondence.
>
> (1963, 19)

Carnap's discussion of his attempt to define all theoretical terms in observational terms here follows Quine (1969, 76–77) and focuses on the first step at which the attempt fails: the construction of the phenomenal world through simplicity and analogy (cf. Friedman 1992, n. 9; Friedman 1999, 160–162). But Carnap discusses the step to the phenomenal world together with the step to the physical world and thereby brushes over the point I have made here: even if Carnap had succeeded in explicitly defining the concepts of the perceptual world, the physical magnitudes and anything that relies on physical magnitudes for its constitution would not be explicitly definable in perceptual terms, according to Carnap's own position in his three earlier works and the *Aufbau* itself.

Again it seems that the technical details of Carnap's view on the structure of scientific theories survived changes in his meta-theoretic views: Carnap (1928/1967) clearly thinks of his technical analysis in his three early works as correct, since he refers to them for the details of his construction and explicitly restates their central point about correspondence rules. It is just that he also has meta-theoretic views (here: overly positivistic ones) that in this case conflict with his technical analysis.

Contrary to this result, Friedman (1992), in his discussion of §136 of the *Aufbau*, claims that Carnap does give a method for arriving at explicit definitions in spite of the one-to-many relation between phenomenal and physical facts. Specifically, Friedman (1992, 21–22) states:

> Although Carnap [claims] that the coordination between "phenomenal facts" and corresponding state-magnitudes is only unique . . . in the direction from the latter to the former, he . . . outlines a procedure for nonetheless approximating to a unique assignment of physical state-magnitudes by focusing on a small neighborhood of a given phenomenally characterized space-time point and working back and forth using the laws of physics (1923, pp. 102–03). The crucial point is that the laws of physics, together with an unambiguous determination of phenomenal qualities from physical state-magnitudes, provide a methodological procedure for narrowing down the ambiguity in the assignment of physical state-magnitudes: in principle, a unique assignment is thereby constructed after all.

This is not Carnap's claim. The method to which Friedman refers is described in a passage in the "Aufgabe" in which Carnap relaxes the idealizing assumptions about the third volume of an ideal physics, the complete knowledge of the state of the world. "Then the task is to calculate from the observed state of a restricted region, namely our own spatio-temporal neighborhood, the state of another space–time region" (Carnap 1923/2019, 102/231). As a technical problem of this task, Carnap notes that to calculate even an arbitrarily small area just for one second would demand knowledge of the state of the world in a 300,000-km radius. The bigger problem is that in principle, the physical state of an area cannot be uniquely determined from observations, because the dictionary contains only one-many relations.

The method to come to predictions is given in the following passage:

> The reason why a physics that is as yet far from even this more modest fiction is nevertheless able to make predictive calculations on the

basis of observations is the following. To be sure, an infinite set of physical states of the region corresponds to a given observational result, and thus also an equipollent set of such states for the future moment to be calculated. . . . But with the inverse translation of this infinite set of physical states back into sensory contents, there often turns out to be a relatively small set of sensory contents that in favorable cases forms a continuous domain of qualities (e.g., a domain of similar shades of color). What we try to do is, first of all, to carry out the observations in such a way that several unconnected qualitative domains do not result for the future point in time and, second, to narrow the boundaries of the one qualitative domain. The two defects of prediction, ambiguity and imprecision, can thus be reduced more and more as science progresses. In special cases, for time intervals that are not too long, they can be entirely eliminated, i.e., unambiguous prediction can be achieved. . . . In contradistinction, science always remains infinitely far removed from the unambiguous prediction of physical states, even for arbitrarily small time intervals.[13]

(Carnap 1923/2019, 102–103/233)

So, contrary to Friedman's claim, Carnap does not give a method for explicitly defining the terms of physics (and there is no working back and forth). Rather, he points out that for some cases and small regions of space-time, exact prediction of a future observation, but not of a future physical state, is possible. His argument against the prediction of physical states rests on the one-many relation between observations and physical states, the assumptions that physical states can be determined only through observations (this is implicit), and that for each current physical state, there is exactly one physical state in the future (a set of physical states evolves over time into a set of physical states of the same cardinality). Since at any point in time, one can only determine an infinite set of physical states, any prediction can therefore also only determine an infinite set of physical states.[14] (Carnap does not consider the possibility of using observations at more than one point in time to narrow down the set of physical states.)

   Friedman was probably led astray because he assumed that Carnap's exposition in §136 is compatible with the central claims of the *Aufbau*. Clearly, Friedman is correct about Carnap's intentions regarding the definability of theoretical terms, but he is mistaken about the implications of Carnap's three early works and their recapitulation in §136, just as Carnap was mistaken when he referenced and recapitulated them.

15.5.2. Van Fraassen's Satisfaction Function

In Carnap's view, then, theories are restrictions on mathematical spaces, and the correspondence rules "uniquely co-ordinate . . . with certain physical qualities, ultimately sensory qualities" (Carnap 1926/2019, 60/417). Thus, according to Carnap, volumes of the mathematical spaces are mapped to physical qualities, which in turn are mapped to phenomenal qualities. Basically this account of theories was suggested about 45 years later by van Fraassen.

In his current account, van Fraassen (1980, 64) describes the connection through model theory, specifically through embeddings. This account can be expressed directly in Carnap's view on scientific theories (Lutz 2014a). In van Fraassen's earlier account (cf. van Fraassen 1989, 365, n. 34), the relation between physical theories and observations relies on "satisfaction functions":

> Besides the state-space, the theory uses a certain set of *measurable physical magnitudes* to characterize the physical system. This yields the set of *elementary statements* about the system (of the theory): each elementary statement U formulates a proposition to the effect that a certain such physical magnitude m has a certain value r at a certain time t. (Thus we write $U = U(m, r, t) \ldots$.)
>
> . . . For each elementary statement U there is a region $h(U)$ of the state-space H such that U is true if and only if the system's actual state is represented by an element of $h(U)$. . . . The mapping h (the satisfaction function) is the third characteristic feature of the theory. . . . The exact relation between $U(m, r, t)$ and the outcome of an actual experiment is the subject of an auxiliary theory of measurement, of which the notion of "correspondence rule" gives only the shallowest characterization.
>
> (1970, 328–329)

The disparaging remark about measurement and correspondence rules references an article by Suppes (1967), which is odd considering that both Hempel (1952, §§11–14) and Carnap (1966, §§6–10) extensively discuss the representational theory of measurement championed by Suppes.

Be this as it may, van Fraassen's account thus assumes a purely mathematical description of the physical system whose states are mapped to values of physical quantities, which are in turn related to measurement results and finally observations. That is exactly what happens in Carnap's account.

15.6. Conclusion

The structure of correspondence rules as it was presented in Carnap's three early works and in §136 of the *Aufbau* did not change much: Specifically, the one-many relation between observational states and theoretical states that Carnap first described in the 'Aufgabe" was still present in his last works on the structure of scientific theories. In his later works, however, Carnap (e.g., 1939) phrased it in terms of the explicit definability of observational terms in theoretical terms. The description of the world as a mathematical space and the conceptualization of scientific theories as restrictions on this mathematical space as conceived in Carnap's article on three-dimensionality and causality, his monograph on concept formation, and in §136 of the *Aufbau* also remained in Carnap's view on scientific theories to the end.

That Carnap uses the results of his three earlier works in the *Aufbau* and rediscovers them in his later works (Carnap 1956, 1958/1975) shows that these works belong to Carnap's logical empiricism. That the results survived Carnap's changes on more meta-theoretical views on the status of scientific theories, logic, and semantics, and even as undercurrents to Carnap's incompatible official position of the *Aufbau*, only to resurface later on and become Carnap's final view, shows how resilient they are. So resilient, in fact, that they are at the core of the account of scientific theories by one of Carnap's staunchest opponents.

15.7. Acknowledgments

Parts of a very early version of this contribution were part of my dissertation at Utrecht University and presented under the title "Carnap's Unchanging Correspondence Rules" at the *Second* SIFA *Graduate Conference at the Cogito Research Centre*, Bologna, Italy, on 29 October 2009 and under the title "Two Constants in Carnap's Philosophy of Science" at the HOPOS 2010 at the Central European University in Budapest, Hungary, on 25 June 2010. I thank the audiences, Thomas Müller, Christopher Pincock, and Ádám Tuboly for helpful comments.

Notes

1. Note that his example of a correspondence rule does not, contrary to Carnap's conjecture about the definability of observational terms, provide an explicit definition of "warmer than" in theoretical terms. If the conjecture is right, the relation is at best *entailed* by the complete correspondence rule.
2. Here is an abbreviated list of further correspondences in content between the "Aufgabe", the monograph on concept formation, and the introductory textbook, given with the respective page numbers: The role of simplicity in choosing scientific laws and correspondence rules: §§IV-V/16–17, 27, 30–31/69, 84, 145, 168. The possibility of choosing what counts as rigid body: 91/25/91. The possibility of choosing what counts as periodic process: 91–92/39/80. Time dependence of counting: – /15/60. Five rules for measurement scales: – /22–23/63–65. Standardization of temperature and length through successive approximation: – /35–36/98–99.
3. George translates the title of the paragraph as "The World of Physics". The original uses the same phrase ("Die physikalische Welt") as the article on three-dimensionality and causality.

4. All (and only) these three works of Carnap's are listed in the references to §136 of the *Aufbau*, contrary to Carus (2019b, note h).
5. Jordi Cat describes in Section 2.3.4 of this volume the specific scientific theories that Carnap considers for axiomatization.
6. Thus, this description is missing the conceptual step from the formalism to the interpretation as a space, corresponding to the step from Liouville, Jacobi, and Boltzmann to Gibbs, Poincaré, and Paul Ehrenfest (Nolte 2010).
7. By treating time as another geometric dimension rather than a parameter, Carnap technically does not use phase space but rather the more general extended phase space (Scheck 2010, §1.20).
8. In this focus on field theories, Carnap is in the company of Schlick: See Jordi Cat's Section 2.3.1 in this volume.
9. This was first mentioned by Montague (1965, 143).
10. A minor difference is that Carnap, when discussing finite dimensional systems, assumes a formalization in extended phase space (see n. 7), while van Fraassen assumes normal phase space. I thank Bobby Vos for pointing this out to me.
11. The translation of the parenthetical remark ("späteren Naturgesetzen zuliebe, die bei dem Begriff der spezifischen Wärme auftreten") is changed from the translation by Dean and Richardson.
12. Carnap (1936, 168) referred to Tarski's article (1935) soon after its publication and so at least could have seen the relation then.
13. The translation of the last sentence ("Von der eindeutigen Voraussage physikalischer Zustände dagegen bleibt die Wissenschaft auch bei noch so kleinen Zeitabständen immer unendlich weit entfernt.") is modified from the one given by Friedman et al.
14. I thank Christopher French for extensive discussions of this passage, although we still disagree on both its content and its relevance.

References

Andrews, P. B. (2002), *An Introduction to Mathematical Logic and Type Theory: To Truth Through Proof*. 2nd ed. Dordrecht: Kluwer Academic Publishers.
Campbell, N. R. (1920), *Physics: The Elements*. Cambridge: Cambridge University Press.
Carnap, R. (1922), *Der Raum. Ein Beitrag zur Wissenschaftslehre*. Vol. 56. „Kant-Studien". Ergänzungshefte im Auftrag der Kant-Gesellschaft. Berlin: Verlag von Reuther und Reichard.
——— (1923/2019), 'On the Task of Physics and the Application of the Principle of Maximal Simplicity', in A. W. Carus et al. (eds.), *The Collected Works of Rudolf Carnap: Early Writings, Volume 1*. Translated by M. Friedman, A. W. Carus, J. Hafner, P. Mancuso, C. Pincock, T. Ryckman, H. Treuper, H. Wilson, and R. Zach. Oxford: Oxford University Press, pp. 209–241.
——— (1924/2019), 'Three-Dimensionality of Space and Causality: An Investigation of the Logical Connection between Two Fictions', in A. W. Carus et al. (eds.), *The Collected Works of Rudolf Carnap: Early Writings, Volume 1*. Translated by M. Friedman. Oxford: Oxford University Press, pp. 247–289.
——— (1926/2019), 'Physical Concept Formation', in A. W. Carus et al. (eds.), *The Collected Works of Rudolf Carnap: Early Writings, Volume 1*. Translated by E. Dean and A. Richardson. Oxford: Oxford University Press, pp. 339–427.

—— (1928/1967), *The Logical Structure of the World: Pseudoproblems of Philosophy*. Translated by R. A. George. Berkeley and Los Angeles: University of California Press.

—— (1934/1967), *The Logical Syntax of Language*. Translated by A. Smeaton. Reprinted with corrections. London: Routledge and Kegan Paul Ltd.

—— (1936), 'Testability and Meaning', *Philosophy of Science*, 3 (4): 420–468.

—— (1937), 'Testability and Meaning: Continued', *Philosophy of Science*, 4 (1): 2–35.

—— (1939), *Foundations of Logic and Mathematics*. Foundations of the Unity of Science. Toward an International Encyclopedia of Unified Science 3. Chicago and London: University of Chicago Press, pp. 139–213. References are to the two-volume edition.

—— (1950/1962), *Logical Foundations of Probability*. 2nd ed. Chicago: University of Chicago Press.

—— (1956), 'The Methodological Character of Theoretical Concepts', in H. Feigl and M. Scriven (eds.), *The Foundations of Science and the Concepts of Psychology and Psychoanalysis*. Minneapolis, MN: University of Minnesota Press, pp. 38–75.

—— (1958/1975), 'Observation Language and Theoretical Language', in J. Hintikka (ed.), *Rudolf Carnap, Logical Empiricist: Materials and Perspectives*. Translated by H. Bohnert. Dordrecht: D. Reidel, pp. 75–85.

—— (1963), 'Intellectual Autobiography', in P. A. Schilpp (ed.), *The Philosophy of Rudolf Carnap*. Chicago and LaSalle, IL: Open Court Publishing Company, pp. 1–84.

—— (1966), *Philosophical Foundations of Physics: An Introduction to the Philosophy of Science*. Edited by M. Gardner. New York and London: Basic Books, Inc.

Carus, A. W. (2019a), 'Editorial Notes on "On the Task of Physics"', in A. W. Carus et al. (eds.), *The Collected Works of Rudolf Carnap: Early Writings, Volume 1*. Oxford: Oxford University Press, pp. 242–245.

—— (2019b), 'Editorial Notes on "Three-Dimensionality of Space and Causality"', in A. W. Carus et al. (eds.), *The Collected Works of Rudolf Carnap: Early Writings, Volume 1*. Oxford: Oxford University Press, pp. 291–295.

Carus, A. W. and Friedman, M. (2019), 'Introduction', in A. W. Carus et al. (eds.), *The Collected Works of Rudolf Carnap: Early Writings, Volume 1*. Oxford: Oxford University Press, pp. xxiii–xli.

Creath, R. and Richardson, A. (2019), 'Editorial Notes on "Physical Concept Formation"', in A. W. Carus et al. (eds.), *The Collected Works of Rudolf Carnap: Early Writings, Volume 1*. Oxford: Oxford University Press, pp. 430–439.

Da Costa, N. C. A. and Chuaqui, R. (1988), 'On Suppes' Set Theoretical Predicates', *Erkenntnis* 29 (1): 95–112.

Feigl, H. (1970), 'The "Orthodox" View of Theories: Remarks in Defense as Well as Critique', in M. Radner and S. Winokur (eds.), *Analyses of Theories and Methods of Physics and Psychology*. Minneapolis, MN: University of Minnesota Press, pp. 3–16.

Friedman, M. (1992), 'Epistemology in the Aufbau', *Synthese*, 93 (1–2): 15–57.

—— (1999), *Reconsidering Logical Positivism*. Cambridge: Cambridge University Press.

Halvorson, H. (2013), 'The Semantic View, If Plausible, Is Syntactic', *Philosophy of Science*, 80 (3): 475–478.

Hempel, C. G. (1952), *Fundamentals of Concept Formation in Empirical Sciences*. Vol. 2, No. 7. Foundations of the Unity of Science. Toward an International Encyclopedia of Unified Science. Chicago and London: University of Chicago Press. References are to the two-volume edition.

—— (1969), 'On the Structure of Scientific Theories', in R. Suter (ed.), *The Isenberg Memorial Lecture Series 1965–1966*. East Lansing: Michigan State University Press, pp. 11–38.

—— (1970), 'On the "Standard Conception" of Scientific Theories', in M. Radner and S. Winokur (eds.), *Analyses of Theories and Methods of Physics and Psychology*. Minneapolis, MN: University of Minnesota Press, pp. 142–163.

Hudetz, L. (2017), 'The Semantic View of Theories and Higher-Order Languages', *Synthese*, 196: 1131–1149.

Leivant, D. (1994), 'Higher Order Logic', in D. M. Gabbay, C. Hogger, and J. Robinson (eds.), *Deduction Methodologies*. Oxford: Oxford University Press, pp. 229–321.

Lutz, S. (2012), 'On a Straw Man in the Philosophy of Science: A Defense of the Received View', *HOPOS: The Journal of the International Society for the History of Philosophy of Science*, 2 (1): 77–120.

—— (2014a), 'Empirical Adequacy in the Received View', *Philosophy of Science*, 81 (5): 1171–1183.

—— (2014b), 'What's Right with a Syntactic Approach to Theories and Models?', *Erkenntnis* 79: 1475–1492.

—— (2017), 'Carnap on Empirical Significance', *Synthese*, 194: 217–252.

Montague, R. (1965), 'Set Theory and Higher Order Logic', in J. N. Crossley and M. A. E. Dummett (eds.), *Formal Systems and Recursive Functions: Proceedings of the Eight Colloquium, Oxford, July 1963*. Amsterdam: North-Holland, pp. 131–148.

Mormann, T. (2007), 'The Structure of Scientific Theories in Logical Empiricism', in A. Richardson and T. Uebel (eds.), *The Cambridge Companion to Logical Empiricism*. Cambridge: Cambridge University Press, pp. 136–162.

Nolte, D. D. (2010), 'The Tangled Tale of Phase Space', *Physics Today*, 63 (4): 33–38.

Quine, W. V. O. (1969), 'Epistemology Naturalized', in *Ontological Relativity and Other Essays*. New York and London: Columbia University Press, pp. 69–90.

Reck, E. H. (2007), 'Carnap and Modern Logic', in M. Friedman and R. Creath (eds.), *The Cambridge Companion to Carnap*. Cambridge: Cambridge University Press, pp. 176–199.

Scheck, F. (2010), *Mechanics: From Newton's Laws to Deterministic Chaos*. 5th ed. Heidelberg: Springer.

Suppe, F. (1972), 'What's Wrong with the Received View on the Structure of Scientific Theories?', *Philosophy of Science*, 39 (1): 1–19.

—— (1974), 'The Search for Philosophic Understanding of Scientific Theories', in F. Suppe (ed.), *The Structure of Scientific Theories*. Urbana, IL: University of Illinois Press, pp. 3–241.

—— (1989), *The Semantic Conception of Theories and Scientific Realism*. Urbana, IL: University of Illinois Press.

Suppes, P. (1967), 'What Is a Scientific Theory?', in S. Morgenbesser (ed.), *Philosophy of Science Today*. New York: Basic Books, pp. 55–67.

Tarski, A. (1935), 'Einige methodologische Untersuchungen über die Definierbarkeit der Begriffe', *Erkenntnis*, 5 (1): 80–100.

van Fraassen, B. C. (1970), 'On the Extension of Beth's Semantics of Physical Theories', *Philosophy of Science*, 37 (3): 325–339.

—— (1980), *The Scientific Image*. Oxford: Clarendon Press.

—— (1989), *Laws and Symmetry*. Oxford: Clarendon Press.

16 From the Periphery to the Center
Nagel, Feigl, and Hempel

Matthias Neuber

[T]he quest for an ever more adequate statement and defense of some of the basic conceptions of empiricism has come to play the role of the treasure hunt in the tale of the old winegrower who on his death-bed enjoins his sons to dig for a treasure hidden in the family vineyard. In untiring search, his sons turn over the soil and thus stimulate the growth of the vines: the rich harvest they reap proves to be the true and only treasure in the vineyard.

(Hempel 1963, 707)

16.1. Introduction

Ernest Nagel (1901–1985), Herbert Feigl (1902–1988), and Carl Gustav Hempel (1905–1997) all came as immigrants to the United States, where they published the main part of their respective philosophical work. All three thinkers stood in close contact with the logical empiricism of the Vienna Circle, although Feigl was the only one who really descended from the Viennese tradition. Whereas Hempel was originally loosely allied with the so-called Berlin Group for Scientific Philosophy, Nagel (who came to the United States as early as 1911) stood from the very beginning under the influence of American pragmatism. It is not exaggerated to say that Nagel, Feigl, and Hempel – in their respective mature writings – represent three closely related points of view. However, there were also significant differences. One of these differences concerns the question of how to interpret the cognitive status of physical theories. It is this question on which I shall focus in the following.

16.2. Feigl's 'Realist Provocation'

To begin with, Herbert Feigl's reconstruction of the theories of physics no doubt had a minority status within the logical empiricist camp. As a matter of fact, during his student days, Feigl felt highly attracted by 'critical realism', as it was defended by thinkers such as Alois Riehl, Oswald Külpe, and the early Moritz Schlick. It was primarily the latter's

Allgemeine Erkenntnislehre from 1918 that motivated Feigl to become Schlick's student at the University of Vienna in 1922 (see Feigl 1963/1981, 39). However, Schlick's conception underwent significant development, and Feigl became somewhat perturbed. In his own words:

> Although I had been in the 'loyal opposition' in regard to the positiv-
> ism of the Vienna Circle, I had a hard time maintaining against them
> the sort of critical realism that I had originally learned to adopt from
> Schlick's own early work. . . . Under the influence of Carnap and
> the early Wittgenstein, Schlick and Waismann were converted to a
> sort of phenomenalistic positivism during the middle twenties. Their
> brilliant and powerful arguments overwhelmed me temporarily. But
> encouraged and buttressed by the support of Popper, Reichenbach,
> and Zilsel, I regained my confidence in my earlier realism and devel-
> oped it in my first book on *Theorie und Erfahrung in der Physik*
> (1929), and later in several articles written during my academic
> career in the United States.
>
> (Feigl 1974/1981, 9–11)

In the mentioned book from 1929, Feigl, among other things, raised questions like the following:

> Is theory a true picture . . . of reality? Or is it only a fictitious model,
> a working hypothesis which helps us to discover new states of
> affairs? . . . Do we obtain knowledge of an objective, transcendent
> reality in physical theories, or do they merely signify an economical,
> simplified description of our immediate experience?
>
> (1929/1981, 116)

After his emigration to the United States in 1931, Feigl recast these questions within the framework of what he called 'semantic realism' and thereby attempted to shield the logical empiricist program against extreme positivist restrictions.[1]

This is not the place to go into the details of Feigl's philosophical development (see, in this connection, Neuber 2018a). Instead, I shall concentrate on Feigl's paper "Existential Hypotheses" from 1950. Published in
the journal *Philosophy of Science*, this paper might be seen as Feigl's
major contribution to interpreting the cognitive status of scientific, especially physical, theories. Its principal aim is to argue that semantic realism
is the most promising way to go. Thus its methodological advice is to
take the 'linguistic turn'. Or, as Feigl himself puts it (alluding to a paper
by Wilfrid Sellars 1947a):

> A generation ago the bone of contention between realists and posi-
> tivists (or idealists) was the "independent existence" of the objects of

science. This issue has since been reformulated in "the new way of words": Those who hold the translatability-thesis may now be called "phenomenalists"; those who oppose it are "realists" in some new sense of this ambiguous word.

The glib and easy dismissal of the issue as a pseudo-problem will no longer do. No doubt there are ways of putting the issue that makes it into a question devoid of any specifiable significance. But the advance of modern syntactical and semantical techniques enables us not only to restate the problem in a new and sharpened fashion; it also offers some hope that the issue may now be more responsibly and more satisfactorily adjudicated.

(1950a, 36)

In short, by reflecting on certain aspects of scientific language, progress in the controversy over realism can reasonably be expected to be achieved.

Now the crucial feature of Feigl's particular approach toward this issue is that he primarily relies on insights from *semantics* to drive his point home. Accordingly, the concepts of 'factual reference' and 'surplus meaning' figure prominently in that approach. As it appears, these very concepts help effectively to overcome old fashioned 'styles of reasoning'. At any rate, Feigl's strategy to make the case for realism is quite straightforward: theoretical terms such as 'electron' or 'molecule' refer to the respective entities and thus are not exhaustively translatable into the language of pure observation. This is where phenomenalism as well as early logical empiricism got things wrong. In fact, Feigl maintains, there always remains a non-translatable "surplus meaning" (1950a, 48) which should be dealt with in the context of a "pure semantics" (1950a, 50), as can be found in the contemporary writings of Alfred Tarski and Rudolf Carnap.[2] Accordingly, Feigl claims that "we must distinguish between the radical empiricist's meaning of 'meaning' (i.e. epistemic reduction) and another, commonsensical meaning of 'meaning' (factual reference)" (1950a, 49). Furthermore, the strict criterion of direct verification (by immediate observational data) is supplanted by the weaker criterion of *confirmability*, so that theoretical statements as well as the states of affairs that render these statements true can be adequately represented in semantic terms. Feigl therefore declares: "The factual reference of not directly verifiable statements is to be construed in such a manner that it is semantically perfectly on a par with the factual reference of directly verifiable statements" (1950a, 49–50).

However, pure semantics does not suffice in Feigl's view. To be sure, the semantic approach to the language of science prevents us from falling back into the realm of transcendent – metaphysical – realism. Thus

[t]he semantic conception of reference . . . *explicates* what a cautious empirical realism can legitimately mean by "reference", "independent

existence", etc. If we handle our concepts responsibly, we can avoid metaphysical perplexities. . . . The essential requirement of empiricism is thus safeguarded. But the very adoption of the *confirmability* criterion (in preference to the narrower *verifiability* criterion) allows as much realism as we are ever likely to warrant.

(Feigl 1950a, 51–52)

Consequently, semantic realism, according to Feigl, circumvents the disadvantages of both phenomenalism (including positivism as well as early logical empiricism) and metaphysical realism.[3] But as already indicated, *semantics alone* does not suffice for Feigl. What is furthermore needed is a conceptual framework for integrating the observational evidence basis as such. In order to account for the criterion of confirmability, a link must be established between the theoretical terms (and statements) of science, on the one hand, and observable facts, on the other. As Feigl (1950a, 58) makes clear, the required conceptual framework is provided by Wilfrid Sellars' (1947b) idea of a "pure pragmatics". He points out:

Only when we impose the requirements of pure pragmatics do we attain the desired scope of genuinely designating terms. That is to say, that in the language of empirical science all those terms (and only those terms) have factual reference which are linked to each other and to the evidential base by nomological relationships. Concepts and constructs that designate observable items of the world and those which do not, but are required for the coherent spatio-temporal-causal account to which science aspires are thus properly related to each other by means of the metalanguage of pragmatics and semantics.

(Feigl 1950a, 50)

Unfortunately, Feigl does not elaborate much further on the idea of a pure pragmatics.[4] But so much should be clear: semantic realism is, as Stathis Psillos (1999, 12) has put it, "an anti-reductive position" (Psillos 1999, 12), and it has, as Psillos (2011, 303) has pointed out at another place, "a pragmatic ring to it".

Reasons of space prevent a more detailed discussion of Feigl's plea for realism (for an extended scrutiny, see Neuber 2011, cf. Psillos 2011). However, one should not wonder that "Existential Hypotheses" did not remain uncontested. Especially within the logical empiricist movement, Feigl's paper was perceived as a downright provocation.

16.3. The Symposium of 1950

Interestingly enough, immediate reactions to Feigl's 'realist provocation' came both from Hempel and from Nagel. To be more concrete,

"Existential Hypotheses" had appeared in the 1950 January issue of volume 17 of *Philosophy of Science*, and in the subsequent 1950 April issue of that volume, the journal published a "symposium" on Feigl's paper. Contributors to the symposium were, besides Hempel and Nagel, the logical empiricist Philipp Frank, C. West Churchman, A. G. Ramsperger, and Feigl himself. In our context, it is enough to shed some light on Hempel's and Nagel's critiques of Feigl's "Existential Hypotheses" and to briefly comment on Feigl's rejoinders to these critiques.

Hempel's critical survey is titled "A Note on Semantic Realism". As Hempel makes clear from the very beginning, his principal concern pertains to the notion of factual reference. In his view, this notion can actually be reformulated in *phenomenalistic* terms, and this without any substantive loss. By explicitly associating the notion of factual reference with the criterion of confirmability, Feigl, according to Hempel, in fact prepares such a reformulation. Hempel writes:

> [T]he basic import of that criterion [i.e., confirmability] consists in the interpretation of the factual reference of scientific constructs in terms of the logical relationships which connect scientific hypotheses involving those constructs, with the evidential base of science. But those relationships can be characterized in a "phenomenalistic" reconstruction; therefore I think the latter cannot be "found wanting" in anything essential that semantic realism has to offer for the analysis of the "surplus meaning" of existential or other scientific hypotheses. In particular, Feigl's assertion that hypotheses involving constructs do assert more than the relevant segment of the evidential base can be adequately interpreted and duly justified in terms of a phenomenalistic analysis which does not refer at all to the moot, and largely terminological, issue of the semantic referents of hypothetical constructs.
>
> (1950, 170)

Accordingly, towards the end of his critique, Hempel (1950, 173) concludes that the notion of the reference of theoretical constructs is "unnecessary ... and can be eliminated, in effect, by means of Feigl's own criterion of factual reference". That is to say, theoretical constructs are semantically on a par with other scientific terms in that they must satisfy the confirmability criterion. Yet nothing less is demanded by the phenomenalistic type of reconstruction.

It is primarily for this reason that Hempel is not convinced by Feigl's plea for realism. But what, then, is his alternative to it? This very question will be discussed in the following section. For the time being, it might suffice to point out that Hempel, in his 1950 contribution, commits himself to some sort of *meaning holism*. In this account, the meaning of theoretical terms forms part of a "complex relational network" (1950, 172), within which the theoretical statements of science are connected with the "evidence sentences" (*ibid.*) via the criteria of empirical confirmability and logical coherence. "[T]he meaning of a statement in the language of science", Hempel (1950, 172) thus maintains, "is reflected in the totality of its logical relationships (those of entailment as well as those of confirmation) to all other sentences of that language" (*ibid.*). Astonishingly enough, Hempel (1950, 173) draws from this the conclusion that "a purely syntactical account of scientific knowledge is . . . impossible".

Feigl's rejoinder to Hempel's critique is quite convincing. As he sees it, Hempel's variant of meaning holism still presumes "syntactical positivism" (Feigl 1950b, 193). To be sure, Hempel (1950, 173) explicitly concedes that science stands in need of "a semantical interpretation of at least some of its terms". But, as Feigl correctly observes, the project of a semantic interpretation in Hempel's hands gets confined to purely observational terms, after all. "I suspect", Feigl (1950b, 193) writes, "that Hempel has here in mind only the predicates whose designata are observable thing-properties and the proper names which designate the objects of direct acquaintance". In other words, Hempel does not succeed in providing a satisfactory semantics for theoretical terms and sentences. We will come back to this point in a moment (see the following section).

As for Nagel's critique of "Existential Hypotheses", it is interesting to note that Nagel (1950, 176), in contrast to Hempel, finds himself "in agreement with Feigl's rejection of phenomenalism". However, as Nagel immediately adds, phenomenalism is not the sole alternative to a realist approach to science. As he understands the matter, what he calls "operationism" (*ibid.*) should be seriously taken into consideration. The essential feature of an operationist reconstruction of scientific theories is, according to Nagel, that it focuses not on the alleged translatability of hypothetical constructs into the language of observation but rather on the *function* of theoretical terms and statements in concrete contexts of inquiry. Thus

> the emphasis of operationism is upon the *instrumental role* of hypothetical constructs in the conduct of inquiry; it is not (as is the case with phenomenalism and realism) upon their supposed status as symbolic *representations* of some realm of events or objects.
>
> (ibid.)

Nagel illustrates this point of view by the analogy of a hammer. As he points out, it would be absurd to ask whether a hammer adequately 'represents' the things producible with its help. Neither would it be plausible to raise the question whether there is any 'surplus' group of potential products 'designated' by the hammer. Now, like hammers, the hypothetical constructs of science should, in the operationist reading, be interpreted in purely instrumental terms. In Nagel's own words:

> [A]ccording to the operationist interpretation of science, as I under-
> stand it, hypothetical constructs and theories are comparable to
> hammers and other tools. They are symbolic constructs deliberately
> devised for bringing into mutual relations matters of direct observa-
> tion which might otherwise be taken as irrelevant to one another, and
> for directing efficiently the course of experimental procedure.
>
> (1950, 177)

It is reasonable to presume that American pragmatism, particularly John
Dewey's 'instrumentalist' interpretation of the theories of science (see, for
example, Dewey 1929, Ch. 8), is lurking in the background here. Further
considerations in this regard will be addressed in the section after the
following.

But let us now come to the main point of Nagel's critique. Like Hempel, Nagel is highly skeptical of the notion of factual reference. For, in his
opinion, the supplementation of pure semantics by pure *pragmatics* does
definitely not help to make things clearer. Feigl's account of the nature of
pure pragmatics is, Nagel (1950, 178) complains, "regrettable meager".
That semantic realism is supposed to provide a semantic meta-language
in which theoretical terms and sentences are semantically on a par with
observational terms and sentences is, according to Nagel (1950, 179),
"a trivial achievement, of which its rivals are no less capable". So the
question remains how pure pragmatics should bring us in the position
to *transcend* this "trivial achievement". Nagel sees no way to accomplish
this ambitious aim and therefore declares: "After all, 'pure pragmatics'
possesses magic powers no more than does 'pure semantics'" (*ibid.*). Even
worse, in *programmatic terms*, semantic realism amounts to just those
consequences that characterize the operationist account. Nagel writes:

> Feigl is pressing for a distinction without a difference. Precisely how
> does he distinguish himself from those operationists who, in refusing
> to characterize atoms as "fictions" for example, wish to maintain
> simply that statements in current science containing the term "atom"
> are well-supported by a variety of experimental data? And indeed,
> is he not thinking entirely within the framework of ideas of opera-
> tionism, in spite of his nominal espousal of semantic realism, when
> he declares that "No concrete existential hypothesis . . . is factually
> meaningful unless it is confirmable".[5]
>
> (1950, 179)

In brief, 'semantic realism' is no realism at all. Particularly the idea of a
pure pragmatics turns out to be obscure and all but a convincing contribution to the controversy at stake.

Feigl's rejoinder to Nagel's critique is rather disappointing. What one
would expect is a further elucidation of the idea of a pure pragmatics.

However, the only thing Feigl (1950b, 192) provides is the following concession: "[P]ure pragmatics has not been developed to the extent that its indispensability or fruitfulness is as obvious as is (to my mind at any rate) the value of pure syntax and pure semantics". To be sure, Feigl actually attempts a reply to Nagel's critique. But this reply has nothing to do with pure pragmatics. Rather, it is (at least in my view) a strong as well as sound argumentation based on probabilistic grounds. "The difference that makes a difference", Feigl (1950b, 194) states, "can be explicated by the differing inductive probabilities of concrete predictions". Yet the strategy of arguing in terms of probability is nothing specific of 'semantic' realism. It can already be found in much earlier writings by Feigl's logical empiricist fellow Hans Reichenbach (1938).[6]

16.4. Hempel's Disguised Syntacticism

Summing up thus far, it can be said that the question of how to interpret the cognitive status of physical theories was answered by Feigl in a way that obviously did not comply with Hempel's and Nagel's point of view. But what did they propose instead? We have already seen that Hempel defended phenomenalism and Nagel defended operationism against semantic realism. But were these their actual respective stances?

Let us begin with Hempel. According to Michael Friedman (2000, 40), "virtually all the seeds" of Hempel's later philosophical development can be traced back to his encounter with the Vienna Circle in 1929–1930. This is clearly exaggerated, all the more since the early phase of Hempel's philosophical development primarily took place in Berlin, where he studied under Reichenbach from 1926 to 1933 and eventually wrote a dissertation on probability (see Hempel 1935/36). As is well known, Reichenbach, until his emigration to Turkey in 1933, figured as a leading member of the 'Berlin Group' of scientific philosophers. Hempel himself also attended seminars of Walter Dubislav, another member of the Group. Thus, to associate Hempel's philosophical socialization with the Vienna Circle as tightly as Friedman does is not in accord with the historic facts.[7]

But be that as it may, if one takes a look at Hempel's writings from the 1940s on (i.e., the writings he published after his emigration to the United States in 1937), then it becomes obvious that what might be called 'syntacticism' is the prevailing point of view. As we have seen before, Feigl, in his rejoinder to Hempel's critique in the context of the 1950 symposium, diagnoses that the latter's commitment to a "semantical interpretation" is significantly delimited. In point of fact, semantics is kept away from theoretical terms and sentences. To be sure, Hempel's meaning holism seems to be intended as an expression of the insight that "a purely syntactical account of scientific knowledge" does not suffice, since it does not connect to the world at all. However, meaning holism, as Hempel actually conceives of it, is at best a *disguised* form of syntacticism. Or so it shall be argued.

To begin with, in his writings before the 1950 symposium, Hempel explicitly defends syntacticism. For example, in his paper "Studies in the Logic of Confirmation" from 1945, he states that "[c]onfirmation as here conceived is a logical relationship between sentences, just as logical consequence is" (Hempel 1945, 25). Accordingly, confirmation is not to be seen as "a semantical relation obtaining between certain extra-linguistic objects on the one hand and certain sentences on the other" (1945, 22). Rather, confirmation is purely inter-sentential and thus syntactical (see already Hempel [1943] on this). Given a theoretical sentence (hypothesis), this sentence is confirmed (or disconfirmed) not by observational events 'out there' but by sentences from the observational *vocabulary*. Sentences are confronted with sentences, and it should therefore be possible "to establish purely syntactical criteria of confirmation" (1945, 26).

Seven years later, in his influential *Fundamentals of Concept Formation in Empirical Science*, Hempel connects syntacticism with the meaning holistic approach as follows:

> A scientific theory might . . . be likened to a complex spatial network: Its terms are represented by the knots, while the threads connecting the latter correspond, in part, to the definitions and, in part, to the fundamental and derivative hypotheses included in the theory. The whole system floats, as it were, above the plane of observation and is anchored to it by rules of interpretation. . . . By virtue of those interpretative connections, the network can function as a scientific theory: From certain observational data, we may ascend, via an interpretative string, to some point in the theoretical network, thence proceed, via definitions and hypotheses, to other points, from which another interpretative string permits a descent to the plane of observation.
>
> (1952, 36)

This is probably Hempel's most articulate characterization of what a scientific theory is. The crucial message is that there is no need for semantics in the guise of Tarski's, Carnap's, or even Feigl's particular conception. Hypothetical constructs receive their meaning solely through being embedded in a system of sentences, and it is the job of 'rules of interpretation' to link these constructs with the observational evidence base (for a similar approach, though in the realm of mathematics, see Carnap 1939; for a critical assessment, see Feigl 1970, esp. 6–8).[8]

Making now a further time jump to 1958, we are faced with what is certainly Hempel's most pertinent contribution to the issue at hand, his oft-quoted paper "The Theoretician's Dilemma". At first glance, this paper documents a fundamental change in Hempel's view. Had he rejected the realist approach to science both (at least implicitly) in his "Studies in the Logic of Confirmation" and (explicitly) in his 1950 critique of Feigl, he now apparently endorses this approach. Thus, we find him pointing out that

> the greatest advances in scientific systematization have not been accomplished by means of laws referring explicitly to *observables*, i.e., to things and events which are ascertainable by direct observation, but rather by means of laws that speak of various *hypothetical, or theoretical, entities*, i.e., presumptive objects, events, and attributes which cannot be perceived or otherwise directly observed by us.
>
> (Hempel 1958, 41, original emphases)

Every scientific realist would agree. And it is interesting to see that Hempel himself has actually been interpreted as an advocate of scientific realism.[9] However, one must be careful here: Hempel no doubt concedes the plausibility of a realistic attitude to theoretical terms and entities, but, in the last analysis, he merely seeks to make plain that full-blown *anti*-realism is not defensible.

We shall not attempt a detailed reconstruction of Hempel's argumentation, for a discussion of the numerous issues it raises would take us far afield.[10] Rather, we shall restrict our focus to the very last section of "The Theoretician's Dilemma". The section is titled "On Meaning and Truth of Scientific Theories", and one might accordingly suspect that Hempel is providing a semantic examination here. But is that really the case? Obviously not. For Hempel frankly states:

> [S]emantics does not enable us to decide whether the theoretical terms in a given system T' do, or do not, have semantical, or factual, or ontological reference. . . . Let us note here that from a purely semantical point of view, it is possible to attribute semantical reference to any term of a language L that is taken to be understood: the referent can be specified in the same manner as the truth condition of a given sentence in L, namely by translation into a suitable meta-language. . . . Plainly, this kind of information is unilluminating for those who wish to use existential reference as a distinctive characteristic of a certain kind of theoretical terms; nor does it help those who want to know whether, or in what sense, the entities designated by theoretical terms can be said to exist. . . . [W]e have to look elsewhere for criteria of significance for theoretical terms and sentences.
>
> (1958, 82)

So where exactly do we have to look? Is Hempel perhaps eager to go one step further, namely in direction of something like Feigl's pure pragmatics? Not at all. Rather, he is going one step back, that is, back to the realm of pure syntax. This becomes particularly evident from Hempel's following comment in the context of his contribution to the 1963 Schilpp volume on Carnap:

> [T]he semantical criteria of truth and reference which can be given for the sentences and for the terms, or "constructs", of a partially interpreted theory offer little help towards an understanding of those expressions. . . . Fortunately, however, a partially interpreted theory may be understood even when full semantical criteria of truth and reference are not available in a language which we previously understand. For if we know how to use the terms of V_B we may then come to understand the expressions in terms of V_T by grasping the rules which govern their use and which, in particular, establish connections between the "new" theoretical vocabulary and the "familiar" basic one.
>
> (1963, 696)

In other words: by relying on the given observational (basic) vocabulary ("V_B") and by establishing respective rules (of 'interpretation') we are in a position to account for theoretical terms and sentences. Near the end of "The Theoretician's Dilemma", Hempel (1958, 85) puts the deflationist aspect of this consequence as follows:

> [T]he existence of hypothetical entities with specified characteristics and interrelations, as assumed by a given theory, can be examined inductively in the same sense in which the truth of the theory itself can be examined, namely, by an empirical investigation of its V_B-consequences.

In brief, an autonomous semantics for theoretical constructs is not needed after all.

By way of conclusion, it can be stated that Hempel ultimately remained within the framework of syntacticism. The meaning holism he advocated in his critique of Feigl is, as documented by the last three quotes, entirely reconstructable within this framework.[11]

16.5. Nagel's Plea for Neutralism

Given that Hempel's syntacticism did not significantly differ from what he himself, in his 1950 critique of Feigl, characterized as 'phenomenalism', the question remains whether Nagel was in fact an advocate of 'operationism'. Furthermore, the origins of Nagel's relationship to the logical empiricist movement should be somewhat clarified. In order to adequately answer these questions, some bio-bibliographical background information will be needed.[12]

Nagel was born in Nové Mesto nad Váhom (now Slovakia, then part of the Austro-Hungarian Empire) and came to the United States at the age of ten. In 1931, he received a Ph.D. in philosophy from Columbia University, the topic of his dissertation being the logic of measurement. His academic mentors were Morris R. Cohen and John Dewey. With Cohen, Nagel published *An Introduction to Logic and Scientific Method* (1934). A Guggenheim fellowship brought him to Europe in 1934–1935. He visited Vienna and attended several meetings of the members of the Vienna Circle. In 1936, his "Impressions and Appraisals of Analytic Philosophy in Europe" (Nagel 1936) were published in *The Journal of Philosophy*. Three years later, his "Principles of the Theory of Probability" were published in the *International Encyclopedia of Unified Science*, which was edited by Otto Neurath, Rudolf Carnap, and Charles W. Morris. Nagel, who taught at Columbia until his retirement in 1970, wrote plenty of other books and articles. However, undoubtedly his most influential work was *The Structure of Science*, which appeared in 1961. On the whole, Nagel was very influential. Levi (2000, 641) even considers him "arguably the pre-eminent American philosopher of science from the mid 1930s to the 1960s".

As can be seen from this short overview, Nagel's contact with the logical empiricist movement began quite early and was rather intense.[13] However, a further point should be mentioned in advance: Nagel's two teachers, Cohen and Dewey, represented two different philosophical approaches. Whereas the pragmatist Dewey, as already indicated, favored an instrumentalist understanding of science, Cohen (1931) defended a more or less encompassing form of realism. There can be little doubt that Nagel was to some degree inspired by this apparent contrast of outlooks (for a similar assessment, see Levi 2000, 642).

Which brings us back to our central question. Recall that Nagel, in his 1950 critique of Feigl, rejected realism in favor of what he called 'operationism'. In his seminal *The Structure of Science*, Nagel chooses another strategy. There, he distinguishes between three different views as to the "cognitive status of theories in physical science" (Nagel 1961, 117), namely the 'descriptive view', the 'instrumentalist view', and the 'realist view'. Concerning the first of these three views, Nagel is convinced: a purely descriptive account of physical theories does not work. Being more precise, such an account – as it was, according to Nagel (see 1961, 120, fn. 11), proposed, for instance, by Ernst Mach, Karl Pearson, and the early Bertrand Russell – would amount to the exhaustive translation of the theoretical language of physics into the language of "pure sense contents" (1961, 122).[14] However, this strategy is doomed to fail, since the assumption of an autonomous language of sensory data cannot convincingly be sustained. "[N]o one", Nagel (1961, 122) writes, "has yet succeeded in constructing such a language". Accordingly, "the translatability thesis remains . . . a highly debatable program for analyzing theoretical statements" (1961, 128).

But what about the other two views? As for instrumentalism, it is important to note that this view is identical to the one Nagel had labeled 'operationism' in 1950. This becomes obvious both from his illustration of the instrumentalist view by the hammer analogy (see Nagel 1961, 130) and from the following general characterization of the core assumptions of instrumentalism:

> Theories are intellectual tools, not physical ones. They are neverthe-less conceptual frameworks deliberately devised for effectively direct-ing experimental inquiry, and for exhibiting connections between matters of observation that would otherwise be regarded as unre-lated. . . . On this view, theories, like other instruments, do indeed have a "factual reference" – namely, to the subject matter for whose exploration they have been constructed and in which they have an effective role.
>
> (Nagel 1961, 131)

Main proponents of the instrumentalist view are, according to Nagel (1961, 129, fn. 22), John Dewey, Moritz Schlick, and Gilbert Ryle. Now, the instrumentalists' admission of 'factual reference' is severely qualified. It is *not* to be understood in Feigl's terms, that is, in terms of semantics. Rather, factual reference *reduces* to the instrumental function played by the respective theoretical terms. "From the perspective of the instrumen-talist standpoint", Nagel (1961, 130) thus states, "it is . . . gratuitous to ask whether a theory has a 'surplus meaning' and what its 'factual refer-ence' is, over and above its meaning and reference as revealed by its orga-nizing role in inquiry". Consequently, physical theories are conceived of as "inference tickets" (1961, 130).[15] They are, according to the instru-mentalist view, neither true nor false but rather to be evaluated under the criterion of their heuristic usefulness. Thus the essential question for the instrumentalist is whether theories are "effective techniques for rep-resenting and inferring experimental phenomena" (1961, 133). Nagel illustrates this point by the instrumentalist's stance toward the molecu-lar (kinetic) theory of gases. That this theory employs limiting (and thus theoretical) concepts such as 'point-particle', 'instantaneous velocity', or 'perfect elasticity' is, according to instrumentalism, not a drawback of the theory. Rather, by applying these very concepts, the physicist is in a better position to infer information concerning certain experimental properties of given volumes of gases. From the instrumentalist perspective, what matters is that the theory is not interpreted as "a faithful portrayal of what transpires within a gas" (ibid.) but rather as an effective means of calculating and inferring experimental gaseous data at the purely phe-nomenological level. Or, to take another of Nagel's examples: inquiries into the thermal properties of gases are performed under the assump-tion of gases as aggregations of discrete particles, whereas inquiries into acoustic phenomena in connection with gases are performed under the assumption of a continuous medium. However, for the instrumentalist, this apparent contradiction is definitely not embarrassing:

> Construed as statements that are either true or false, the two theories are on the face of it mutually incompatible. But construed as techniques or leading principles of inference, the theories are simply different though complementary instruments, each of which is an effective intellectual tool for dealing with a special range of questions.
>
> (ibid.)

So much for Nagel's presentation of the instrumentalist view. What is qualitatively new as compared to Nagel's 1950 discussion of 'operationism' is that he is now reflecting on possible objections to this view. Thus, he explicitly addresses "limitations of the instrumentalist standpoint" (1961, 137) and condenses the existing critical assessments to the following crucial aspect:

> Some of the most eminent scientists, both living and dead, certainly have viewed theories as statements about the constitution and structure of a given subject matter; and they have conducted their investigations on the assumption that a theory is a *projected map* of some domain of nature, rather than a set of *principles of mapping*. . . . [N]either logic nor the facts of scientific practice nor the frequently explicit testimony of practicing scientists supports the dictum that there is no valid alternative to construing theories simply as techniques of inference.
>
> (Nagel 1961, 139)

As can be easily guessed, the aforesaid 'valid alternative' is represented by the *realist* view of theories.

Due to limitations of space, a detailed reconstruction of Nagel's discussion of the realist view cannot be provided here. However, the most striking feature of this discussion can readily be grasped: in the realist view, physical theories are either true or false. Thus, they are definitely more than mere instruments for prediction. Being 'projected maps', they are constructed in order to record the 'blueprint of the universe'. They have, in other words, semantic and explanatory impact. But does this really make a difference? Recall that, in the instrumentalist view, mutually incompatible theories might turn out to be equally effective means of organizing experimental data. In contrast, in the realist view, such incompatible theories cannot be *true* at the same time. However, the realist has an apparently good answer to that:

> Indeed, inconsistencies between theories, each of which is nevertheless useful in some limited domain of inquiry, are often a powerful

incentive for the construction of a more inclusive but consistent theoretical structure. Accordingly, a proponent of the view that theories are true or false statements can escape any embarrassment for his position from the circumstance that incompatible theories are sometimes employed in the sciences; he can insist on the corrigible character of every theory and refuse to claim final truth for any theory.
(Nagel 1961, 143)

Thus in the long run, the succession of theories in, for example, physics is "a series of progressively better approximations to the unattainable but valid ideal of a finally true theory" (Nagel 1961, 144). On the other hand, the instrumentalist could reply to this by appealing to the literally undeniable *disunity* of the various theoretical approaches within physics.

Cases such as this are for Nagel (1961, 141) a clear indicator that the controversy between realists and instrumentalists "can be prolonged indefinitely". Therefore, his final advice is to consider this whole controversy as a mere dispute over words. Nagel declares:

It is . . . difficult to escape the conclusion that when the two apparently opposing views on the cognitive status of theories are each stated with some circumspection, each can assimilate into its formulations not only the facts concerning the primary subject matter explored by experimental inquiry but also all the relevant facts concerning the logic and procedure of science. In brief, the opposition between these views is a conflict over preferred modes of speech.
(1961, 152)

The position reached by this line of reasoning might be plausibly called *neutralism*: since both the realist and the instrumentalist view of theories can be made sufficiently strong to account for the facts, it is reasonable to abstain from taking sides and thus to remain neutral.

All of this sounds rather harmonious. Like Hempel, Nagel concedes the plausibility of realism but actually does not hook up with it. Instrumentalism is 'neutralized' as well. However, if one takes a closer look at *The Structure of Science*, then a somewhat different image emerges. To put it more concretely, by explicating his understanding of the processes of scientific theory construction, Nagel is – not surprisingly – *unable* to remain neutral. In fact, he vehemently *repudiates* the realist 'mode of speech'. Of particular interest here is Chapter 11 (which probably contains the most often consulted passages of the whole book). As is well known, the topic of the chapter is the *reduction* of physical and other theories in science.[16] As Nagel makes clear from the very beginning, "the phenomenon of a relatively autonomous theory becoming absorbed by, or reduced to, some other more inclusive theory is an undeniable and recurrent feature of the history of modern science" (1961, 336–337). His principal case studies are the putative reduction of phenomenological thermodynamics to statistical mechanics and that of classical genetics to molecular biology.[17]

Confining ourselves to the first of these two cases, a *realist* interpretation of the issue of reduction might be summed up as follows. Phenomenological thermodynamics is both less fundamental and less inclusive than statistical mechanics. By reducing the former to the latter, we achieve a deeper understanding of thermal phenomena and are thus in a position to account for what is going on at the micro level. In doing so, we causally explain, for example, the phenomenon of heat by reference to the kinetic energies of individual molecules. From the perspective of semantics, the *truth predicate* finds its primary application in the context of the reducing theory, that is, in our case, statistical mechanics. To be more precise, what counts is that the reducing theory is not merely 'empirically adequate' but rather that it *literally represents* the actual states of affairs.

Nagel's own conception of reduction is not that straightforward. For him,

> a reduction is effected when the experimental laws of the secondary science (and if it has an adequate theory, its theory as well) are shown to be logical consequences of the theoretical assumptions (inclusive of the coordinative definitions) of the primary science
>
> (1961, 353)

Thus, any direct reference to theoretical *entities* (and their respective features) would lead us astray. It is for this reason that Nagel (1961, 364) sharply distinguishes between 'reduction of statements' and 'reduction of properties'. In his view, the proper object of reduction is only statements. The conception of reduction in terms of properties is, as he states, "potentially misleading and generates spurious problems" (*ibid.*). Nagel explicates:

> [T]he conception ignores the crucial point that the "natures" of things, and in particular of the "elementary constituents" of things, are not accessible to direct inspection and that we cannot read off by simple inspection what it is they do or do not imply. *Such "natures" must be stated as a theory.*
>
> (1961, 364; emphasis added)

Given that, according to Nagel, theories are systems of statements, a reduction at the level of properties (or 'natures') is indeed impossible. Any realist attempt to establish a property-based conception of reduction must fail because, Nagel (1961, 365) maintains, it "converts what is eminently a logical and empirical question into a hopelessly speculative one". Moreover, *truth* does not play a substantive role in this connection. In Nagel's view, the crucial (and sufficient) point is that by reducing one branch (or theory) of science to another more inclusive one, the aim of *explanatory unification* can be achieved (Nagel 1961, 359–361). In short, reduction is an issue of logic rather than of semantics and ontology.[18]

Interestingly, Hempel reacted quite enthusiastically to Nagel's account of reduction (see Hempel 1969). This does not surprise, since Hempel (unlike Feigl) shared Nagel's distrust toward the application of semantic and ontological categories to questions arising from the philosophy of science.[19] It might be suspected that Nagel himself was closer to the instrumentalist view of scientific theories than his plea for neutralism actually suggests. At any rate, his explicit rejection of realist assumptions in the context of reduction warrants the conclusion that the controversy between realists and instrumentalists has more than "only terminological interest" (Nagel 1961, 141).

16.6. Three Varieties of Logical Empiricism?

As common wisdom has it, philosophy of science witnessed a significant shift in the early 1960s. Not only did Thomas Kuhn's *The Structure of Scientific Revolutions* (1962) lead to a change of perspective among philosophers of science, but specifically the advent of what became known as 'the semantic view of theories' forcefully contributed to the shift. Authors such as Patrick Suppes, Frederick Suppe, and Joseph D. Sneed challenged the prevailing account of theories as syntactic entities and proposed instead a non-linguistic, model-based account. The previously prevailing conception was labeled 'the received view of theories', and it is no accident that the works of Hempel and Nagel are still most frequently associated with that label. Furthermore, the received view is often identified with logical empiricism *in toto*.

The foregoing discussion of Feigl's, Hempel's, and Nagel's respective interpretations of the cognitive status of physical theories calls for a more differentiated picture. To be sure, both Hempel and Nagel gave sufficient occasion to their being regarded as proponents of the received, particularly statement-based, view. And there can be little doubt that they both stood very close to the center of the logical empiricist movement and its 'official' commitments and doctrines. However, in the case of Feigl, things are not that easy. Admittedly, his defense of semantic realism turned out to be unstable and in need of further elaboration. But in point of fact, his specific conception of theories is *not* prone to being subsumed under the received view. At least in this regard, Feigl ultimately remained at the periphery.

16.7. Acknowledgments

I am deeply indebted to Ádám Tuboly for his encouragement and assistance in working out this chapter. I further wish to thank two anonymous referees for their suggestions to improve the chapter.

Notes

1. As Feigl makes clear in one place, his collaboration paper with Albert E. Blumberg, "Logical Positivism: A New Movement in European Philosophy" (1931), merely had the status of a "fanfare article" (Feigl 1963/1981, 38), standing somewhat beside his actual point of view.
2. Presumably Feigl (without being explicit here) is referring to Tarski (1944) and Carnap (1942).
3. It should be noted that this is a point where logical empiricists divided into two camps, namely those who thought that semantics is dangerous to empiricism (most prominently Neurath) and those who thought the exact opposite (most prominently Carnap). Feigl no doubt stood on Carnap's side. For an instructive overview of the logical empiricists' debate over the legitimacy and allegedly metaphysical connotations of semantics, see Mormann (1999).
4. However, an attempt to figure out what could be implied by this (rather intricate) idea is to be found in Neuber (2017).
5. Nagel is referring here to Feigl (1950a, 51).
6. The connection between Feigl's and Reichenbach's respective views is too complex to permit brief summary. But see Neuber (2018b, 9–12).
7. According to Nikolay Milkov (2013, 295), Hempel "never felt obliged to identify himself with a particular philosophical school – not the Berlin Group, not the Vienna Circle, nor any other philosophical coterie". See further Nikolay Milkov's chapter in the present volume on the Berlin Circle.
8. Since hypothetical constructs (theoretical terms) are often couched in dispositional language ('solubility', 'chemical affinity', etc.), Hempel concedes that "surely only a partial interpretation of the theoretical terms is achieved" (1952, 38). Furthermore, he draws a link to the theory of probability and indicates that the interpretation of theoretical terms should be performed within this framework. However, as Maria Carla Galavotti (2007, 118) has correctly observed, "[p]robability was for Hempel somewhat of a side interest that he took up when writing his dissertation under Reichenbach but did not cultivate much in his later writings".
9. For instance, Wesley Salmon (1999, 337) contends that, for the Hempel of "The Theoretician's Dilemma", "theoretical language is essential" and that it is therefore "reasonable to conclude that theoretical terms denote unobservable entities".
10. In a nutshell, Hempel's main arguments against a full-blown anti-realist reconstruction of the theories of physics are directed against the various attempts to eliminate theoretical terms (and with them the respective entities) by application of such logical techniques as 'Craig's theorem' and the method of the 'Ramsey-sentence'. For an extended discussion of this issue, see Friedman (2008).
11. One of the referees of this chapter objects that Hempel *never* adopted a syntactic theory of theories, since the key issues of his account revolve around the meaning of theoretical terms, which is a semantic and epistemic affair, not a syntactical one (for a similar interpretation, see Fetzer 2017). To this it can be replied that Hempel typically relies on inferential relations between sentences of the object language, while the referee refers to the purported goal of Hempel's elucidations, that is, finding the meaning of theoretical terms. But finding the meaning of theoretical terms can, according to Hempel's meaning holism, be attempted in a purely syntactic fashion by determining the terms' inferential relations to other, directly understood terms. Consequently, my understanding of 'syntacticism' applies to the *formalism* underlying Hempel's 'solution' of the theoretician's dilemma. Semantics remains, on this account,

confined to observable terms, and it is only here that science connects to 'the world'. As Adam Tamas Tuboly has recently noted, Carnap, in his *Logische Syntax der Sprache* (1934), put forward a similar account of semantics like the one I ascribe to Hempel here. According to Tuboly, for Carnap (before his switch to 'Tarskianism' in 1937), "the meaning of (logical and non-logical) expressions is exhausted by their inferential roles" (Tuboly 2017, 73). Thus, just as with Hempel, for Carnap, finding meanings is associated with "inferentialism" (2017, 74). See, in this connection, further Peregrin (2020).

12. The following short sketch draws primarily on Levi (2000) and Suppes (2006).

13. In the preface to *The Structure of Science*, Nagel mentions Carnap and Frank in this connection. He writes: "Neither Rudolf Carnap nor Philipp Frank have been formally my teachers, but I have profited greatly from the numerous conversations I have had with them since 1934 on the logic of science" (Nagel 1961, x).

14. See Don Howard's contribution to this volume for a detailed argument on why this misinterprets Mach's views. See further Neuber (2002).

15. Nagel is obviously referring here to Ryle (1949, 121). Also, Schlick (1931/1979, 188) could be meant in this context. Among Nagel's immediate contemporaries Stephen Toulmin (1953, 84) and Otto Bird (1961, 534) suggested the 'inference-ticket view'.

16. That the discussion of this topic has a significant prehistory in Nagel's work is documented by Nagel (1935, 1949).

17. For a detailed discussion of Nagel's theory of reduction, see Schaffner (2012).

18. The previous reconstruction echoes to some extent a (still ongoing) discussion in the relevant literature: Raphael van Riel (2011) has made the point that Nagel's model of reduction should be (re)interpreted in ontological terms. This has been – convincingly – contested by Sahotra Sarkar (2015, 48) who insists that Nagel's model represents an "anti-ontological stance".

19. Nagel's respective position becomes particularly clear in his 1942 review of Carnap's *Introduction to Semantics*. There, he declares: "The worth of semantical distinctions in the study of the foundations of abstract logic and mathematics has been established beyond dispute by the investigations of Tarski and others. Nevertheless, the significance of semantics for general philosophy of science and empirical methodology can be seriously debated, especially if the study of semantics is pursued in the manner proposed by Professor Carnap. To the present reviewer, at any rate, Professor Carnap's general method raises as many problems as it may settle" (Nagel 1942, 470). And two pages later, he comments that "[i]t appears doubtful . . . whether a semantics which deliberately abstracts from all reference to the users of a language has much to contribute to the resolution of general philosophical issues" (1942, 472). There are (at least as far as I can see) no hints that Nagel had fundamentally changed this view by 1961.

References

Bird, O. (1961), 'The Re-Discovery of the Topics: Professor Toulmin's Inference-Warrants', *Mind*, 70: 534–539.

Carnap, R. (1939), *Foundations of Logic and Mathematics*. Chicago: University of Chicago Press.

———— (1942), *Introduction to Semantics*. Cambridge, MA: Harvard University Press.

Cohen, M. R. (1931), *Reason and Nature: An Essay on the Meaning of Scientific Methods*. London: Kegan Paul.

Dewey, J. (1929), *The Quest for Certainty: A Study of the Relation of Knowledge and Action*. New York: Minton, Balch & Company.

Feigl, H. (1929/1981), 'Meaning and Validity of Physical Theories', in R. S. Cohen (ed.), *Herbert Feigl: Inquiries and Provocations, Selected Writings, 1929–1974*. Dordrecht: Reidel, pp. 116–144.

—— (1950a), 'Existential Hypotheses: Realistic versus Phenomenalistic Interpretations', *Philosophy of Science*, 17: 192–223.

—— (1950b), 'Logical Reconstruction, Realism and Pure Semiotic', *Philosophy of Science*, 17: 224–236.

—— (1963/1981), 'The Power of Positivistic Thinking', in R. S. Cohen (ed.), *Herbert Feigl: Inquiries and Provocations, Selected Writings, 1929–1974*. Dordrecht: Reidel, pp. 38–56.

—— (1970), 'The "Orthodox" View of Theories: Remarks in Defense as Well as Critique', in M. Radner and S. Winokur (eds.), *Minnesota Studies in the Philosophy of Science, Volume IV: Analyses of Theories and Methods of Physics and Psychology*. Minneapolis, MN: University of Minnesota Press, pp. 3–16.

—— (1974/1981), 'No Pot of Message', in R. S. Cohen (ed.), *Herbert Feigl: Inquiries and Provocations, Selected Writings, 1929–1974*. Dordrecht: Reidel, pp. 1–20.

Feigl, H. and Blumberg, A. E. (1931), 'Logical Positivism: A New Movement in European Philosophy', *The Journal of Philosophy*, 17: 281–296.

Fetzer, J. (2017), 'Carl Hempel', *Stanford Encyclopedia of Philosophy*, https://plato.stanford.edu/entries/hempel/#ProbProv.

Friedman, M. (2000), 'Hempel and the Vienna Circle', in J. Fetzer (ed.), *Science, Explanation, and Rationality: Aspects of the Philosophy of Carl G. Hempel*. Oxford: Oxford University Press, pp. 39–64.

—— (2008), 'Wissenschaftslogik: The Role of Logic in the Philosophy of Science', *Synthese*, 164: 385–400.

Galavotti, M. C. (2007), 'Confirmation, Probability, and Logical Empiricism', in A. W. Richardson and Th. Uebel (eds.), *The Cambridge Companion to Logical Empiricism*. Cambridge: Cambridge University Press, pp. 117–135.

Hempel, C. G. (1935/1936), 'Über den Gehalt von Wahrscheinlichkeitsaussagen', *Erkenntnis*, 5: 228–260.

—— (1943), 'A Purely Syntactic Definition of Confirmation', *The Journal of Symbolic Logic*, 8: 122–143.

—— (1945), 'Studies in the Logic of Confirmation (I)', *Mind*, 54 (213): 1–26.

—— (1950), 'A Note on Semantic Realism', *Philosophy of Science*, 17: 169–173.

—— (1952), *Fundamentals of Concept Formation in Empirical Science*. Chicago: University of Chicago Press.

—— (1958), 'The Theoretician's Dilemma: A Study in the Logic of Theory Construction', in H. Feigl, M. Scriven, and G. Maxwell (eds.), *Minnesota Studies in the Philosophy of Science, Volume II: Concepts, Theories, and the Mind-Body Problem*. Minneapolis, MN: University of Minnesota Press, pp. 37–98.

—— (1963), 'Implications of Carnap's Work for the Philosophy of Science', in P. A. Schilpp (ed.), *The Philosophy of Rudolf Carnap*. La Salle, IL: Open Court, pp. 685–707.

———— (1969), 'Reduction: Linguistic and Ontological Issues', in S. Morgenbesser, P. Suppes, and M. White (eds.), *Philosophy, Science, and Method: Essays in Honor of Ernest Nagel.* New York: St. Martin's Press, pp. 179–199.

Kuhn, Th. (1962), *The Structure of Scientific Revolutions.* Chicago: University of Chicago Press.

Levi, I. (2000), 'Nagel, Ernest (1901–85)', in E. Craig (ed.), *Routledge Encyclopedia of Philosophy.* London: Routledge, pp. 641–643.

Milkov, N. (2013), 'Carl Hempel: Whose Philosopher?', in N. Milkov and V. Peckhaus (eds.), *The Berlin Group and the Philosophy of Logical Empiricism.* Dordrecht: Springer, pp. 293–307.

Mormann, Th. (1999), 'Neurath's Opposition to Tarskian Semantics', in J. Wolénski and E. Köhler (eds.), *Alfred Tarski and the Vienna Circle: Austro-Polish Connections in Logical Empiricism.* Dordrecht: Kluwer, pp. 165–178.

Nagel, E. (1935), 'The Logic of Reduction in the Sciences', *Erkenntnis*, 5: 46–52.

———— (1936), 'Impressions and Appraisals of Analytic Philosophy in Europe', *The Journal of Philosophy*, 33: 5–25 and 29–53.

———— (1942), 'Review of Carnap's Introduction to Semantic', *The Journal of Philosophy*, 39: 468–473.

———— (1949), 'The Meaning of Reduction in the Natural Sciences', in R. C. Stouffer (ed.), *Science and Civilization.* Madison: University of Wisconsin Press, pp. 99–135.

———— (1950), 'Science and Semantic Realism', *Philosophy of Science*, 17: 174–181.

———— (1961), *The Structure of Science: Problems in the Logic of Scientific Explanation.* New York: Harcourt, Brace and World.

Nagel, E. and Cohen, M. R. (1934), *An Introduction to Logic and Scientific Method.* New York: Harcourt, Brace and Company.

Neuber, M. (2002), 'Physics without Pictures? The Ostwald-Boltzmann Controversy, and Mach's (Unnoticed) Middle-Way', in M. Heidelberger and F. Stadler (eds.), *History of Philosophy of Science: New Trends and Perspectives.* Dordrecht: Kluwer, pp. 185–198.

———— (2011), 'Feigl's "Scientific Realism"', *Philosophy of Science*, 78: 165–183.

———— (2017), 'Feigl, Sellars, and the Idea of a "Pure Pragmatics"', in S. Pihlström, F. Stadler, and N. Weidtmann (eds.), *Logical Empiricism and Pragmatism.* Cham: Springer, pp. 125–137.

———— (2018a), 'Herbert Feigl', *Stanford Encyclopedia of Philosophy*, http://plato.stanford.edu/entries/feigl/.

———— (2018b), 'Realism and Logical Empiricism', in J. Saatsi (ed.), *The Routledge Handbook of Scientific Realism.* London: Routledge, pp. 7–19.

Peregrin, J. (2020), 'Rudolf Carnap's Inferentialism', in R. Schuster (ed.), *The Vienna Circle in Czechoslovakia.* Cham: Springer, pp. 97–109.

Psillos, S. (1999), *Scientific Realism: How Science Tracks Truth.* London: Routledge.

———— (2011), 'Choosing the Realist Framework', *Synthese*, 180: 301–316.

Reichenbach, H. (1938), *Experience and Prediction: An Analysis of the Foundations and the Structure of Knowledge.* Chicago: University of Chicago Press.

Ryle, G. (1949), *The Concept of Mind.* Chicago: University of Chicago Press.

Salmon, W. (1999), 'The Spirit of Logical Empiricism: Carl G. Hempel's Role in Twentieth-Century Philosophy of Science', *Philosophy of Science*, 66: 333–350.

Sarkar, S. (2015), 'Nagel on Reduction', *Studies in History and Philosophy of Science*, 53: 43–56.

Schaffner, K. (2012), 'Ernest Nagel and Reduction', *The Journal of Philosophy*, 109: 534–565.

Schlick, M. (1918), *Allgemeine Erkenntnislehre*. Berlin: Springer.

—— (1931/1979), 'Causality in Contemporary Physics', in H. Mulder and Barbara F. B. van de Velde-Schlick (eds.), *Moritz Schlick: Philosophical Papers, Volume II (1925–1936)*. Dordrecht: D. Reidel, pp. 176–209.

Sellars, W. (1947a), 'Epistemology and the New Way of Words', *The Journal of Philosophy*, 44: 645–660.

—— (1947b), 'Pure Pragmatics and Epistemology', *Philosophy of Science*, 14: 181–202.

Suppes, P. (2006), 'Ernest Nagel', in S. Sarkar and J. Pfeifer (eds.), *The Philosophy of Science: An Encyclopedia*. Vol. 2. New York: Routledge, pp. 491–496.

Tarski, A. (1944), 'The Semantic Conception of Truth and the Foundations of Semantics', *Philosophy and Phenomenological Research*, 4: 341–376.

Toulmin, S. (1953), *The Philosophy of Science: An Introduction*. London: Hutchinson's University Library.

Tuboly, A. T. (2017), 'From "Syntax" to "Semantik": Carnap's Inferentialism and Its Prospects', *Polish Journal of Philosophy*, 11: 57–78.

van Riel, R. (2011), 'Nagelian Reduction beyond the Nagel Model', *Philosophy of Science*, 78: 353–375.

17 Understanding Metaphysics and Understanding Through Metaphysics

Philipp Frank on Scientific Theories and Their Domestication

Adam Tamas Tuboly

17.1. Introduction

Though most logical empiricists studied physics, and some of them even wrote doctoral dissertations in physics or on related topics, only one of them comes to mind when discussing actual physical research: Philipp Frank. Frank not only studied physics and mathematics in Vienna and later in Göttingen, but he actually worked as a theoretical physicist and played a significant role in the institutional structures of physics for decades. His most important works (some of which caught the attention of Albert Einstein) appeared in the 1910s, but he continued to follow the latest achievements in the field from the frontline, so to speak, as the director of the Theoretical Physics Department of the German University of Prague between 1912 and 1938.[1] In the United States, during and after the Second World War, he taught physics and philosophy of science at Harvard until his 1954 retirement.

While Section 17.2 provides a bird's-eye view of the relations between Frank, physics and the philosophy of the physical sciences, Sections 17.3 and 17.4 will both restrict and widen the focus of this chapter by discussing the relation between metaphysics and the natural sciences. Since, for Frank, metaphysics was closely connected to the humanities, the discussion will also touch upon questions of the relation between the humanities and the natural sciences.

In his 1949 review of Frank's *Modern Science and Its Philosophy*, Percy Black claimed that Frank tried to use the same physicalist and logical method across the humanities that he had so fruitfully applied in the physical sciences. He structurally compared Frank's work to the supposedly typical positivist (or behaviorist) conception of "dementalized psychology", which concerns bodily outputs, quantifiable data and natural law-like generalities. In Black's (1949, 424) eyes, such a

> dementalized psychology is not psychology at all; it may be good physiology or good physics, but as such leads us not further into an understanding of ourselves, qua human beings, than a mentalized

physics or astronomy gives us an understanding of what happens in the physics lab or in the heavens.

Frank thought otherwise, at least regarding the second half of Black's opinion.

Frank, as a true believer in Neurath's unified science project and the director of the Institute for the Unity of Science at Boston from 1947 on, always sought connections between the sciences. He was thus interested in a "mentalized physics" or, as he put it, "in the humanistic background of science". Since *"human values [are] intrinsic in science itself"*, the most faithful approach to science cannot ignore the human aspect, that is, the perspective and approach of the humanities, since an *"interest in humanities is the natural result of a thorough interest in science"* (1946/1949, 261; original emphases).

In a recent book, James Ladyman, Don Ross, David Spurett and John Collier (2007, esp. Ch. 1) have argued for their version of "naturalistic metaphysics", where metaphysics is a bundle of ideas aiming to "unify hypotheses and theories that are taken seriously by contemporary science" (2007, 1). Metaphysics will not revise contemporary scientific theories, but the two should work together, reinforcing each other's results, hence the "naturalism" of the title. In the authors' opinion, the distinctive feature of their conception is that metaphysical theories provide *true explanations*. To better explain what they mean, the authors contrast this with a more traditional conception:

> it is a tradition which aims at domesticating scientific discoveries so as to render them compatible with intuitive or "folk" pictures of structural composition and causation. Such domestication is typically presented as providing "understanding". This usage may be appropriate given one everyday sense of "understanding" as "rendering more familiar".
>
> (Ladyman, Ross, Spurett and Collier 2007, 1)

"Domesticating scientific discoveries" amounts partially to knowledge dissemination, which is traditionally interlinked with metaphysical approaches and ideas. In this process, metaphysics serves as a bridge that connects the field of abstract scientific views and the common-sense or "folk" view of the world. Whether these views are separated by a gap or fall on the same continuous line is a matter of further discussion. Ladyman, Ross, Spurett and Collier (2007), along with Wilfrid Sellars (1962), believe that the "scientific image" and the "manifest image" of the world amount to a "synoptic view", wherein these two images or pictures reinforce each other.

While this problem horizon was associated with Sellars from the mid-20th century on, and it has recently been taken up in great detail by

Ladyman and others, it was Philipp Frank who investigated how the dissemination of scientific knowledge via metaphysics connects to the domestication of the sciences from a philosopher's point of view.[2] As I will argue, the investigation of Frank's work on the topic provides compelling reasons for re-evaluating the relation between logical empiricism and the rejection of metaphysics (an area where the oversimplified view of a clear-cut rejection still prevails); for Frank, metaphysical interpretations are a means of providing laypeople with an understanding of the physical sciences.

17.2. Frank's Career as a Physicist and the Themes in His Philosophy of Physics

In order to get a broader picture of Frank's oeuvre and to understand his basic stance and background, this section will briefly review some of the main topics of Frank's life and work. While almost all of these would deserve an article on their own, later in this chapter, I will focus on just one, namely the understanding of scientific theories via metaphysics.

Frank was trained in physics, with a strong mathematical background, and, as he told Thomas Kuhn in their famous interview, he was "mostly a student of Boltzmann" (Frank 1962). Among his classmates, teachers and colleagues, we find Erwin Schrödinger, Hans Thirring, Paul Ehrenfest, Felix Ehrenhaft, Friedrich Hasenöhrl, Karl Herzfeld and many other young scholars who would become prominent physicists and mathematicians in the following years. Although Frank briefly moved to Göttingen after defending his dissertation (on the Principle of Least Action) in order to study with David Hilbert and Felix Klein (see Stöltzner 2002), he soon went back to Vienna to teach physics as a *Privatdozent*.

As a young and promising physicist who was intensely interested in contemporary philosophy (of science), Frank started his work as an ardent defender of Einstein's theory of relativity (both special and general). While it is often noted that Hans Reichenbach was one of the few students and scholars who attended Einstein's groundbreaking lectures on relativity in Berlin during the 1918/1919 winter semester, and that it was Schlick who wrote one of the first and best philosophical appreciations of Einstein's theory (with the latter's approval and recognition), Frank's role in this general scientific story is often neglected. He was among the first generation of physicists who grasped the *meaning* and *significance* of Einstein's theory right away and tried to contribute to it themselves. Not long after gaining his Ph.D., Frank published papers, on his own and with the Austrian mathematician and engineer Hermann Rothe, which were recognized as important contributions to Einstein's ideas. To give one prominent example, Wolfgang Pauli (1958) discussed Frank's ideas in his textbook-length article on the theory of relativity, published originally in the *Encyklopädie der mathematischen*

Wissenschaften in 1921 (though he rejected them, along with Wladimir Ignatowski's approach). Even more importantly, Einstein himself wrote quite positively about Frank's work in his "Report to the Philosophical Faculty of the German University on a Successor to the Chair of Theoretical Physics".

Einstein left Prague after one and a half years to take up his new chair in Zurich. From among the candidates, Einstein – with the other committee members Anton Lampa and Georg Pick – suggested Frank *primo loco*. In the recommendation (written in May 1912), he described Frank as an "excellent lecturer", who is "a capable mathematician and extraordinarily knowledgeable in theoretical physics" and, also taking into account Frank's philosophical works, he emphasized that "his talents are singularly versatile" (Einstein 1912/1995, 303).[3]

Frank took up his position in Prague in 1912 and stayed on as director of the Department of Theoretical Physics until 1938, when he emigrated to the United States. Somewhat like Vienna, the German University in Prague also had a history of philosophically oriented physical studies. As Frank (1962) himself stated, "it was just a branch of Vienna", partly due to the influence of Ernst Mach, who held the chair in experimental physics for almost 30 years in the late 19th century. Mach had a lasting effect on the Prague community, and thanks to Lampa – an ardent follower of Mach's positivistic *Weltanschauung* – his legacy persisted, even at the time of Frank's arrival.

During his nearly 30 years in Prague, Frank built up a significant department with its own library, providing an institutional foundation for many students and researchers. Already in the 1930s, he tried to maintain good relations with the university's other departments and disciplines, for instance with philosophy of science (led by Carnap), experimental physics (headed by Reinhard Heinrich Fürth) and mathematics (represented by Karl Löwner and Arthur Winternitz). In his homage to Frank, the physicist Peter Bergmann (1965, ix), whom Frank recommended to Einstein, noted that "there were frequent joint seminars and other common undertakings. But the spirit of the unity of science and of scientific inquiry was to taken . . . for granted". Nonetheless, as Fürth (1965, xiv) remembered, "[Frank] preferred to work on his own and never had a 'research school'".

Frank was widely known as a teacher of physics, philosophy and philosophy of physics. All the obituaries about Frank (Holton 1966; Cohen and Wartofsky 1965) emphasized the depth and breadth of his knowledge of the history of science and the humanities and the way he disseminated this knowledge in the classroom. While he frequently gave courses on Maxwell's theory of electromagnetism, probability theory, statistical mechanics and kinetic theory, he also taught relativity theory and quantum mechanics in Prague (Frank 1962). Later, he also taught thermodynamics in the Physics Department at Harvard (where Thomas Kuhn was possibly one of his students).[4]

For many, Frank's name is synonymous with the philosophy of relativity theory. He was not just a steadfast and outspoken defender of Einstein and relativity in general but also a well-known popularizer of the theory and its implications.[5] In the 1940s, Frank, as the only physicist, participated in the "Science, Philosophy, and Religion" series of symposia for almost ten years. At the end, he collected and rearranged his talks into a book, entitled *Relativity: A Richer Truth* (1951a), in which he aimed to show that "relativity" or "relativism" is entirely harmless from a broader cultural and political point of view. He argued that instead of eliminating any possibility of discussing values and rationality, relativism actually opens up a refined space for democracy, pluralism and tolerance by being a "significant representation of the enrichment of human expression which is inseparably connected with our gradually increasing experience" (Frank 1951a, 12).[6]

According to another story that is told about Frank (see e.g. Wolters 1987, 163–165), in 1910, Ernst Mach, years after his retirement, sent out a circular to young physicists asking them to explain Einstein's new theory of relativity to him. Among those physicists was the freshly habilitated Frank, who visited Mach personally in May 1910 to discuss relativity and Minkowski's four-dimensional geometry, which did not gain Mach's approval at first.[7] Nonetheless, Frank – as the young authority on Einstein – repeatedly acknowledged his Machian background and always tried to find a way to integrate Mach, positivism and Einstein's physics (see Frank 1949, 2021; cf. Mormann 2017).

Finally, long before the quantum revolution, Viennese physicists prepared the ground for and accepted an indeterministic, statistical and probabilistic worldview of physics. Michael Stöltzner (1999, 2003) shows clearly that Frank was indoctrinated in this milieu and forged his own work from this viewpoint. Frank already engaged with probability and statistics during his formative years, and he later connected probability and statistics to both quantum mechanics and causality (Stöltzner 2009). Besides his former teacher Franz Exner, Frank's good friend Richard von Mises also developed an interest in these topics.[8] Discussing these issues had a lasting effect on Frank. While he defended a purely conventionalist account of the law of causality (Frank 1907/1949) at first, after the emergence of quantum mechanics, he tried to develop a new empirical account (Frank 1932/1998) in one of his early major works. This long and detailed book (titled *The Law of Causality and its Limits*) is not just a simple philosophical account of some basic physical concepts (such as causality, inertia, mass, laws, etc.) but also forms part of a much broader project of Frank's. From the 1920s on, he tried to identify the socio-cultural and political factors behind scientific investigations (usually called "external conditions") that determined how concepts and theories were formed, structured, accepted and denied.

Though this was a huge project and Frank never finished it, some pieces of the big picture are preserved throughout his scattered writings. By studying these articles, we may get a better sense of how logical empiricists *could* have understood the relation between science and society. For Frank, perhaps somewhat surprisingly, one central aspect was metaphysics.

17.3. Fighting Metaphysics in the United States

When Frank started to teach at Harvard, his new, unexpected experiences regarding the attitude of students and the general intellectual atmosphere in the United States opened up new territories and new theoretical possibilities (Reisch 2017). He presented the following observations in a letter to his life-long friend Otto Neurath:

> From the viewpoint of empiricism, one thing is striking. Students who have had very little scientific training are, unconsciously, influenced by a sort of vulgarized scholastic philosophy. They may have put up from [sic] the church or from "philosophical introductions" to textbooks or what not. Before the science teaching got a certain grip on these boys, they were genuine Aristotelians.
>
> (Philipp Frank to Otto Neurath,
> December 10, 1943. ONN.)

In order to account for this peculiarity of his students' metaphysical engagement, Frank developed some new ideas in the United States. What exactly was this change in Frank's views, and how did it happen? According to Uebel (2011, 53), during Frank's first philosophical phase (1932–1938), his views were close to Carnap's, in that for him, "metaphysical assertions were cognitively meaningless (for they presupposed an untestable existential hypothesis)". Frank was keenly interested in the logic of science (though he did not write anything about this topic, he often referred to its importance) and argued that metaphysics is indeed just plainly meaningless. Nonetheless, when he moved to the United States, he experienced a new atmosphere and attitude towards metaphysics in the context of scientific investigations. He recognized that metaphysical interpretations of science are often identical to earlier scientific principles. In the same letter to Neurath, he wrote – with reference to one of his latest papers – that "[m]y point is that every 'philosophical system' is the petrification of a former scientific doctrine. Therefore, necessarily, such a system will be contradicted by the advance of science".

Said paper was published in 1941 on the pages of *Reviews of Modern Physics* as "Why Do Scientists and Philosophers so Often Disagree about the Merits of a New Theory?" Frank claimed that often, when a new scientific theory is formulated, philosophers disagree with it. With time, this disagreement "weakens and finally the philosophers come to agree too completely. Frequently just at this moment the physical theory in question turns out to be doubtful to the physicist" (1941/1949, 207). Frank thought that certain scientific principles, for example, the law of gravitation and the law of inertia, which philosophers found absurd before the end of the 18th century, later became self evident or directly derivable from pure reason.[9] That is exactly what happened with Newton's principles in Kant's philosophy, at least according to Frank. Newton's scientific principles were proved by philosophical means, and thus they became elements of a general philosophical system with the claim of having eternal and necessary validity.

After a while, however, scientific principles were subjected to internal criticism, and thus they were either dropped or simply modified to some extent. When this happened, scientists no longer accepted the principles that had been approved by philosophers. As scientists and philosophers start to disagree, a rupture thus once more threatens the relation of science to philosophy. In this situation, a certain compromise can ease the tension: scholars (both scientists and philosophers) often admit that a certain theory is scientifically true; that is, it is mathematically consistent and provides accurate predictions, but philosophically false; that is, it is not in agreement with the established principles of philosophy. By distinguishing two types of truth, Frank joined a long list of philosophers from Thomas Aquinas to Francis Bacon and George Berkeley. All of them admitted that the new scientific theories they faced could be thought of as mathematically correct but false from the viewpoint of philosophy. They were thus not *intelligible*; that is, they were not *understandable*.

In Frank's view – and this is a rather strong, underdiscussed and underargued claim – we are dealing with a certain form of necessity here:

> Why do philosophers and scientists so often disagree about the merit of a new theory? They mostly disagree because the new theory seems to be in contradiction to established philosophic principles. Moreover – and this is my chief point here – this disagreement arises from necessity, for the established philosophic principles are mostly petrifications of physical theories that are no longer appropriate to embrace the facts of our actual physical experience.
>
> (Frank 1941/1949, 215.)

If philosophers hold, for example, that the statements of Newtonian physics are eternal, necessary, universal truths (as some did), then changes in the physical sciences may, in fact, cause new troubles of how to account for the truth of their philosophical counterparts. As a result, the philosophers and physicists of the old era often claimed that while relativity theory and quantum mechanics may be mathematically true (and useful, for example, for certain calculations), they are philosophically false, since they do not describe the furniture of the world as it is.

Philosophical truths about the nature and structure of the world that were previously considered scientific statements are "metaphysical truths", and Frank thought that if we want to engage in any critical discussion of science and metaphysics, we first need to understand metaphysics. That is, we need to know what a metaphysician does or wants to do in order to show why and how the project fails. A more detailed and interpretative *understanding* of metaphysics thus became a prerequisite for the *critique* of metaphysics. Although both Carnap and Schlick had their own general account of what metaphysicians do, Frank spent much more time working on this problem from various perspectives, perhaps more than anyone else in the Circle – except, of course, for Neurath.

17.4. The Nature of Metaphysics

In 1950, Frank published a highly interesting but underrated paper in the opening volume of *The British Journal for the Philosophy of Science*, titled "Metaphysical Interpretations of Science, Part I and Part II". Frank starts by saying that it has been generally accepted (echoing the old Viennese tune) that the "tenets of metaphysics . . . are neither true nor false, but meaningless" (1950a, 60). But he quickly leaves this idea behind, mainly because he is interested in something else: he notes that, "[a]lthough this argument can hardly be refuted, the interest in metaphysics has abated very little" (ibid.). Thus, despite its alleged meaninglessness, philosophers, and even many scientists, continue to pursue metaphysics. Why is that? Of course, this question about the genealogy of metaphysics is not new; even the logical empiricists posed it from time to time. Take, for example, Carnap's famous dictum from his metaphysics paper of 1932:

> [H]ow could it be explained that so many men in all ages and nations, among them eminent minds, spent so much energy, nay veritable fervor, on metaphysics if the latter consisted of nothing but mere words, nonsensically juxtaposed? And how could one account for the fact that metaphysical books have exerted such a strong influence on readers up to the present day, if they contained not even errors, but nothing at all? [Metaphysical statements] serve for the *expression of the general attitude of a person towards life* ('Lebenseinstellung, Lebensgefühl').
>
> (1932/1959, 78, original emphasis.)

Frank also admits this role of metaphysics. He claims that it is possible to argue that scientific truth "does not provide the only valuable kind of truth. Metaphysics might be meaningless for the scientist 'as a scientist', but may be of the highest value for human life" (ibid.). In that case, metaphysics is "a trait of human behavior", and its aim is to orient our behavior and relations towards others by means of certain values and determinate pictures of the world. "To agree on this point", Frank continues (ibid.), "we need only to consider the 'cold war' which is now going on and is threatening our civilization with destruction". That is, besides Carnap's doctrine that metaphysics expresses our life feelings, Frank claims that metaphysics is "a method of education intended to influence the behavior of people, a kind of high-level propaganda talk" (ibid.).[10]

From this perspective, Frank, in his unpublished late book manuscript, *The Humanistic Background of Science*, interpreted the logical and philosophical claims of Carnap's famous *Überwindung* paper (Carnap 1932/1959) in order to integrate logical empiricism into the American atmosphere (which was characterized by various forms of pragmatism and the influence and authority of James, Peirce and Dewey):

> Since an "attitude towards life" certainly influences human actions, metaphysical propositions have [according to Carnap] a meaning if we understand "meaning" in the sense of Neurath. . . . The salient point is that metaphysical propositions about the physical universe are actually meaningful propositions about human behavior or, in other terms, they are propositions of sociology. . . . The point is that, again, according to the Vienna Circle metaphysical propositions about the physical world are meaningless within the system of physical concepts. But they have meaning within the wider "universe of discourse" that embraces physical and sociological concepts.
>
> (2021, Part 2, Ch. 5, Sect. 6)

Frank, of course, could not deny the fact that according to Carnap, metaphysics is just plainly and strictly *meaningless*. But he also had a positive story to tell: the claim that "metaphysics is meaningless" is *elliptic* and needs to be *contextualized*. When reading metaphysical claims *as physical claims* about observable data, they are simply meaningless, but they can be meaningful *as sociological claims* about the place of humanity and its values in the physical universe.

This points towards a new, more interpretative understanding of metaphysics. Frank is thus not so much interested in whether metaphysics is *cognitively meaningless* based on some highly questionable and hardly determinate logical criterion but in the *purpose and role* of metaphysics in light of its regular recurrence in life, as well as in scientific fields (Frank not only accepted Quine's and Hempel's famous arguments against the one-and-final formal criterion of cognitive meaningfulness, but his views even predate them; see Uebel 2011).

Frank aimed at finding out "what statements of metaphysics mean if we regard them as statements about objective facts, and we shall investigate typical metaphysical statements in order to find out what testable facts they are asserting" (1950a, 61). Frank's answer is not entirely unambiguous – perhaps we should talk about multiple answers rather than a single one. And, for some guidance on this matter, we first have to consider the notion of "understanding".

17.5. Understanding Through Metaphysics

Scientists want to *understand* the natural world, while non-professionals want to *understand* both the natural world *and* the science behind it. Consequently, questions of understanding with regard to science are ambiguous, as domestication of science often means domestication of the world that surrounds us. But this is a temporal process: while the new quantum mechanics domesticated certain segments of the world for physicists, for laymen, the latest theories were entirely unintelligible and unintuitive. "The impression . . . that the language of quantum physics is artificial and contradictory to common sense", Frank argues (1939/1949, 157), "arises from the fact that this language is unnecessarily exact for use in the affairs of daily experience". But if some form of mathematical precision and cohesion, and the success of predictions, were sufficient for understanding the world, the latest theories would be more than beneficial. Alternatively, another possibility would be if our *daily experiences* were different: in fact, Frank imagines and describes a parallel world in which "tennis balls behave in the fashion of the atoms and their nuclei of our actual world" (ibid.). But as laymen require something other than abstract rational reasoning based on equations, and as our daily experiences are in tension with some of the statements of scientific theories, some further domestication is needed.

Frank is not entirely explicit, however, about what precisely he is talking about at any given moment: whether he is referring to science, scientific theories, the practices of scientists, or the dissemination of scientists' knowledge. He claims, for example, that science formulates general statements or principles that consist of certain symbols. These are symbols such as "electromagnetic field strength", "gravitational tensor potential", "coordinate", "force" and the like, usually those that appear in the core equations of theories. Science is always related to experiences (either by structuring them or by explaining them, as the source or goal of contemplation), but the symbols, Frank writes (1950a, 61), "do not designate any observable properties of bodies". In order to connect experience (i.e. observation) and theory, we have to derive statements that are directly about observational events from statements that involve the symbols and descriptions of certain accepted sense experiences in a manner that is logical and mathematical.

This is, however, a quite complex issue, which requires much logical and mathematical work, as well as a huge amount of knowledge of statements involving abstract symbols. Frank provides the following example:

> From the principle of conservation of energy we can derive, by using the equations of thermodynamics, relations between measurable quantities (temperature, heat capacity, mass, velocity, etc.) and can check these relations by actual measurements which can be described in commonsense statements as, for example, the statement that the top of the mercury column in a thermometer coincides with the mark "100" of a scale.
>
> (1950b, 88)

As Frank (ibid.) concludes, "[t]his is the long, scientific path between scientific principles and commonsense statements". But, in a certain sense, this makes the principles of science and scientific theories understandable: they are related to operational moves, and because we know the operational definitions of scientific concepts, we know what they *mean* and can design and conduct experiments with their help. Understanding in this case is gained through an experiential, practice-oriented operational approach that goes from abstract formulas to common-sensical or observational statements. Understanding thus becomes *practicalized*, or *operationalized*.

This logico-empirical path is a long and difficult one to tread for layman and non-scientific philosophers, but fortunately there is another way of understanding science and the world, based on *metaphysical interpretations of science*. By metaphysical interpretation, Frank means, according to his 1950 paper, "a search for the reality behind the physical phenomena" or "a system of statements about [the] third realm" (1950a, 62–63), beyond symbols and experiences. The problem for metaphysicians is often whether this third realm can be *described*, and if yes, in what *language* this can be done. Frank discusses the solution of Thomist philosophers who claimed that this realm could be described and investigated through *analogies* and *metaphors* that are related to our *common sense* and *everyday language*.[11]

These analogies and metaphors establish another route, which Frank called "short circuits" or "short cuts". Let us take the following example: according to Einstein's theory, mass can be converted into energy. This is not the language of science; in scientific language, this should be expressed with certain formulas, equations, abstract symbols and operational definitions in order to arrive at sentences that can be coordinated with experimental scenarios. As "mass" and "energy" seem to be two entirely different things in our common-sense language, "conversion" stands for a totally obscure process. When scientists try to explain what is happening – that is, they try to make scientific principles *understandable* to laypeople in common-sensical terms – they have two choices:

> If forced to introduce common-sense language, [they] would have to say either that energy is mass and one kind of mass is converted into

another kind of mass, or else that mass is basically energy and that there is no "mass" in the common-sense meaning of the world.

(Frank 1958, 62)

Let us take another example. When Arthur Eddington (1928, 328) claimed that, in light of the new atomic theory of matter, the plank on which we are standing "has no solidity of substance [and that to] step on it is like stepping on a swarm of flies", Susan Stebbing (1937/1944, 45) replied that "there is no common usage of language that provides a meaning for the word 'solid' that would make sense to say that the plank on which I stand is not *solid*" (original emphasis). If the expression "solid" is meaningful, then what could be solid if not a plank? Our ordinary language and our common-sense field of experience prevent such statements as "the plank has no solidity". The point is that common-sense language has its own rules (though, as we have seen, some of these may just be petrified ideas of former scientific statements), and if scientists, for some reason, formulate their theories in the language of common sense, then decisions have to be made whether to overrule common-sense language or to adjust to its established usage one way or another.

The "conversation of mass" formula is therefore a special common-sense translation of scientists for laymen. One might argue that physicists should "just shut up and calculate", as the infamous slogan would have it that summarized the general attitude of many working physicists about the danger of interpreting quantum mechanics. And many physicists did just that, claiming that they are *just* physicists. However, and this may be a rather general and broadly accepted consideration among Marxists and sociologists, there is no such thing as *just* being a physicist, rejecting every ideology, worldview and philosophy. As Frank (1946/1971, 428) claimed, "a physicist who dodges all logical analysis and tries to be a 'physicist and only a physicist' will imbue the presentation of his subject with some 'chance philosophy', usually a very obsolete one". You can thus only be silent about the philosophical interpretation of your scientific theory if you *already accept a special interpretation* of theories in general (e.g. that they do not describe anything but merely structure certain experiences; thus they are in no need of interpretation) and thus commit yourself to additional values and their implications.

Frank goes on to say that the interpretation which claims that energy is mass after all, and that one type of mass is being converted to another, "would favor materialism", while that which claims that mass is basically energy "would favor idealism". But how is it possible to interpret simple principles, such as the "conversation of mass", so differently? Is there any force that could provide an idealist or a materialist metaphysical interpretation of scientific theories? Common-sense language is vague and inexact by its nature, and there is thus no exact, one-dimensional or direct, logical route from abstract principles to particular common-sense formulations. In a sense, they are "arbitrary", but as Frank (1958, 62) notes, "'arbitrary' does not mean that these common-sense interpretations are the result of 'free choice'". Henri Poincaré (1905, xxii–xxiv) similarly remarked, in the introduction to his famous work *Science and Hypothesis*, that conventions can, in one sense, be arbitrary but that they are nevertheless restricted by certain empirical considerations. Frank, however, emphasized not empirical observations, but something more interesting. "The choice is, as a matter of fact, determined by our 'values', and our predilection for a certain philosophy of life, since different common-sense interpretations could and would support different philosophies of life" (1958, 62).

Since materialism and idealism are philosophical schools, Frank sees an opening for connecting science and philosophy through philosophy of science. How so? According to Frank (2021, 1.3.3), "all experimental observations are described in the language of common sense", as are metaphysical analogies, metaphors and descriptions of scientific matters. Therefore, both ways share certain fundamental features of the common-sense language, namely that they are highly ambiguous, dubious and value-laden. On the other hand, "metaphysical interpretations also arise when we look in the domain of common sense for analogies to the general principles of science" (ibid.). That is, irrespective of whether we interpret the abstract principles of science through observations or through metaphysics, we will quickly arrive at metaphysically loaded, common-sensical descriptions. We could restrain ourselves, of course, by focusing only on the mathematical apparatus. "The positivistic attitude attempts to present science as an instrument for derivation of observable facts from general principles", Frank writes (2021, 1.7.4). In other words, by highlighting the mathematical, logical and physical parts of theories, we could focus on those parts that are relevant for predictions and the systematizing of experiences. However, "the metaphysical attitude attempts also an 'understanding' of the principles by analogies taken from experience on the common-sense level" (ibid.).[12] If our aim is to understand science, then restricting ourselves to the purely mathematical part is insufficient, and we need to take a broader look at the human phenomenon of science – and that aspect is studied by philosophy.[13]

It would require a huge amount of historical, sociological and perhaps even anthropological work to show how external socio-cultural factors influenced particular scientists or groups of scientists in their pursuit of a definite way of science and their acceptance of certain physical theories. The famous Forman thesis (1971) about the external influence on German physicists regarding the notion of causality in forming and accepting quantum mechanics could serve as a good example of this type of work. Nonetheless, even though Frank had an enormous knowledge of history of science and a knack for sociological considerations, he never produced any detailed work on the matter. Instead, he set the agenda for others and inspired generations of scientists and philosophers (through the work of Robert S. Cohen and Marx Wartofsky) to do historical philosophy of science. As he himself noted, "[a]ll this exploitation of stalemates" – that is, all the discussions of contradictory (but from an empirical point of view equally well supported) metaphysical interpretations of abstract physical principles – "makes a fascinating topic in research in the sociology of science which has been explored very little as yet" (1951b, 26). That is why, in the context of Frank, we should not ask how exactly these external conditions affected the history of science.

Metaphysical interpretations of scientific theories are thus *interpretations that give common-sensical interpretation to the abstract formulations of science.* As we appear to be able to understand our common-sense vocabulary (or at least we are more familiar with this type of vocabulary), relating science to common sense indeed seems to be a possible route towards understanding. In Frank's own words,

> [i]f we want to make a system of abstract principles (like relativity or quantum theory) intelligible to a layman we have to explain it to him by some analogy which is expressed in commonsense language and which has, if possible, some application to human behaviour.
>
> (1950a, 66)

Of course, we might also ask how metaphysical interpretations have come into existence, and on this topic Frank also had a generic story to tell.

He starts with a very general socio-historical introduction from a bird's-eye view. I have already mentioned the necessary disagreement between science and traditional philosophy. But according to Frank, this was not always the case. Before the scientific revolution – or, to be more precise, what is usually called the scientific revolution – there was another separation, namely between the practical work of engineers and the theoretical work of scientists, a.k.a. philosophers. As engineers and technicians worked according to "traditions which had been transmitted in the profession[,] they were little co-ordinated with the scientific theories which had been developed, e.g. in the 'physics' of Aristotle" (1950a, 67). Frank continues with this idea, which became known as the "Zilsel thesis":

> Because these philosophical systems did not claim to give technical advice to the engineer or artisan, they could devote themselves to the task of "understanding" the phenomena of nature, leaving the technical application to a socially "lower" type of men. "Science" and "Philosophy" were nearly one and the same thing or, to use the language of this paper: science and its metaphysical interpretation were nearly one and the same thing.
>
> (Frank 1950a, 67.)

According to Edgar Zilsel (1942), another old friend of Frank from Vienna, *sociological* and *economic* grounds made the scientific revolution possible. When capitalism was about to be rigidified as the major economic form, three different social strata became mingled in the newly developing European cities, and, as previous social barriers came down, mixed forms of theoretical and experimental knowledge and interests appeared. The three social classes were the scholastic intelligentsia interested in reason and abstract philosophical principles; humanists concerned with language and antiquity and finally the superior artisans, that is, practitioners who had access to experimental methods and the relevant practical skills to measure and change nature. As these classes grew closer to each other in the cities, modern experimental science was born.

However, the more the precision in scientific observations has advanced, "the more the emphasis has been shifted towards 'yielding observable facts' as the criterion for the validity of the principles" (1950a, 71). As philosophy and metaphysics are not particularly concerned with observable facts, the separation between science and philosophy became stricter with time. Due to this separation, the question of understanding is even more pressing for science. "Science is not interested in whether its principles are 'intrinsically knowable and intelligible'" (1950a, 74). As scientists are able to do their work without any detailed communication of their processes and ideas to laypeople, they do not have to be concerned with the intuitive intelligibility of their ideas. On the other hand, as Susan Stebbing (1937/1944, 11) noted, physical discoveries and results often end up in the daily headlines. Scientists thus seemingly have a certain *responsibility* towards society to explain their results or, better yet, to make them understandable, intelligible.

Since science does not properly respond to laypeople's demand for making itself understandable, "this task is left to metaphysics, which becomes now an interpretation of science in an 'intelligible language'". The question is, of course, what this *intelligible language* might be. "It will turn out more and more", Frank concludes (1950a, 74), "that there is no other 'intelligible language' than the 'language of common sense' in which we have learned to express – since our childhood – the experience of our everyday life".

17.6. In Place of a Summary

Suppose that we do not want to use natural language – with all its traps – to make science intelligible. It seems, then, that we are left with two choices. First, we could say that we do not want to understand science anymore and that scientists can just calculate and work on their own. While Frank did not offer a detailed narrative about the origins and sources of scientific curiosity, he often noted that science, especially in the Cold War context (where scientific and technological advances played a crucial role in determining the conflicting cultural and political values and ideologies), is not done in a social vacuum. Consequently, the demand to make science intelligible, that is, understandable, will always be present. Being silent about the delicate points is thus not an option.

The other option, in order to satisfy our curiosity, is thus simply to fall back on an operationalized understanding of science and leave behind the metaphysical discussion. But, as Frank noted, leaving behind the metaphysical discussion and relying on purely operational-logical issues is already a certain worldview and thus a value commitment. On the other hand, Frank also remarked that due to the insights of Einstein, and presumably those of Poincaré and many others, we recognized that the structure of science is not only being built up by strict and abstract logical moves. While principles of science, Frank notes, "are not dehydrated abstractions", at certain points (be they either contingent gaps or necessary splits), the "creative imagination of the scientist steps in" (1958, 59). Furthermore, this creative imagination is backed up by "operational definitions by which a connection between the symbols at the top and the measurements at the bottom was established" (1958, 60).

It therefore seems to be the case that science is always in need of understanding, of domestications, across many various levels and by different means. For Frank, any explanation of understanding is entangled with metaphysics, which enables us to connect the dots of the history of logical empiricism. In other words, we can see how, in the concrete case of the physical sciences, metaphysics became utilized in the old-Carnapian sense of "expression of life-feelings".

17.7. Acknowledgments

I am indebted to Sebastian Lutz and an anonymous referee for their most valuable comments. Any mistakes and generic statements are my own responsibility. I am grateful to George Reisch for the numerous discussions in the last few years about Philipp Frank's philosophy. I am also grateful to the audience members of "Physicalism and reduction: a Jerusalem-Budapest twin workshop" and to Christoph Gottstein for correcting my English. This research was supported by the MTA BTK *Lendulet Morals and Science* Research Group and by the MTA Premium Postdoctoral Scholarship.

Notes

1. As Richard von Mises noted in a letter to Otto Neurath, Frank's most important and novel contributions to physics consisted in his remarks on relativity theory (especially on the relation between the principle of relativity and mechanics); Frank also did important work on geometrical optics and dynamics, which he published in the famous Frank-Mises book on differential and integral equations; Frank initially wrote about analytic mechanics,

but in the second edition, he also to contributed to quantum mechanics (by means of a chapter on "Classical Mechanics and Ray Optics"). Finally, during his Prague years, Frank wrote several significant papers together with function-theory mathematicians. For von Mises' letter to Neurath, see Siegmund-Schultze (2007, 52–53).

2. C.P. Snow's crusade about the "two cultures" comes to mind, "between [physical scientists and the representatives of the humanities there is] a gulf of mutual incomprehension – sometimes (particularly among the young) hostility and dislike, but most of all a lack of understanding. They have a curious distorted image of each other" (Snow 1961, 4). Nonetheless, while Snow's lectures and his subsequent book provoked heated and persistent discussions that allowed him and his "two cultures" narrative to enter the mainstream of intellectual history, Frank organized his Institute for the Unity of Science so as to inquire into and "work for an integration between the sciences and the humanities" (Frank 1951c, 6). This is yet another issue that points toward the need of re-introducing Frank into the general narratives of 20th-century philosophy of science.

3. The combination of physics and mathematics – especially in their advanced, 20th-century forms – is a quite delicate matter. For some of the complexities (in the context of Frank), see Stöltzner (2002). On Einstein's relation to Frank, see Don Howard's chapter in the present volume, while the role of mathematics in physics and its relation to Einstein and logical empiricism will be further discussed in Thomas Ryckman's chapter.

4. Frank published a lesser-known textbook on thermodynamics; see Frank (1945). There are occasional references to a textbook on relativity by Frank as "Relativitätstheorie. Leipzig: Teubner, 1920", but presumably, this work was only announced then and Frank never finished writing it – he later earned a certain reputation for not finishing his works (Reisch 2005, Ch. 15). Nonetheless, it is telling that Frank at least considered such a textbook.

5. As a popularizer, Frank (1947/1972) wrote a long socio-cultural and philosophical biography of Einstein. Though it was criticized for being too much about Frank himself, the book became a standard reference text on (the context of) Einstein's life and work for years. See further Don Howard's chapter in the volume.

6. It should be noted, however, that the difference between "relativism" and "the theory of relativity", the former being a philosophical approximation and the latter a technical-physical theory, is often diluted in the secondary literature, especially in critiques of Einstein. Frank seems to admit (or take for granted) a certain relativistic implication of Einstein's theory and argues only that since physics has been harmlessly "relativized", other areas (such as ethics, politics and cultural issues) could be relativized in the same manner following the enrichment of our experiences provided and ensured by Einstein's work in physics. See Nemeth (2003) and Siegetsleitner (2017).

7. The only letters between Frank and Mach that seem to survive are published in Blackmore, Itagaki and Tanaka (2001, 77–78) and concern a manuscript of Paul Gerber and the translation of Pierre Duhem's *L'evolution de la mechanique*. After asking Mach for his help, Frank translated Duhem's book into German in 1912. See Duhem (1903/1912).

8. Together with von Mises, Frank edited the well-known textbook on *Differential- und Integralgleichungen der Mechanik und Physik*, which appeared in 1925 (the mathematical part edited by von Mises) and in 1927 (the physical part edited by Frank). On the significance and history of this volume, see Siegmund-Schultze (2007), and on von Mises, see Maria Carla Galavotti's chapter in this volume.

9. Reichenbach (1951, 141) also criticized "self-evidence" as a guide to philo- sophical insights, claiming that "what [the philosopher] regarded as laws of reason [in the context of geometry] was actually a conditioning of human imagination by the physical structure of the environment in which human beings live".

10. Although I will not discuss the matter here, Frank, like many others, pointed out the metaphysically biased argumentation strategy of the Soviets, but, and this is the interesting aspect, he pointed out the same regarding the Ameri- cans. That is, he highlighted that democracy and the Western world order have their own *socio-political base* and *historically developed structure*. The fact that he, in a certain sense, equated Americans (who saw themselves as neutral) and Soviets (who were obviously biased from an American point of view) also contributed to the neglect of Frank in the United States. On Frank's general, scholarly and political situation in the United States, see Richardson (2012, 10–13), Reisch (2005), Tuboly (2017).

11. Frank's discussion of Thomist philosophers was a well-motivated decision. He was unable to get a job in Chicago (next to Carnap) due to the concerns of some Thomist philosophers about the negative effects of logical empiri- cism (Reisch 2017). On the other hand, Thomist philosophers were keen on discussing the latest scientific results (mainly of physics and of biology) and often criticized actual scientific approaches on their own philosophi- cal basis: Thomist philosophy consisted of supposedly intuitive, rational and intelligible principles. In fact, Thomist philosophers and pragmatists equally determined and influenced American philosophy during the mid-20th cen- tury. Consequently, Frank devoted several chapters to Thomist philosophy in his last major and systematic undertaking (Frank 2021).

12. In this context, one might think of the following passage of Carnap (1939, Ch. 25): "When abstract, nonintuitive formulas, as, e.g., Maxwell's equations of electromagnetism, were proposed as new axioms, physicists endeavored to make them 'intuitive' by constructing a 'model', i.e., a way of representing electromagnetic micro-processes by an analogy to known macro-processes, e.g., movements of visible things." This passage is taken from Carnap's *Foun- dations of Logic and Mathematics*, namely the section titled "'Understand- ing' in Physics". Nonetheless, Carnap's move points towards a more technical way (representation with models), while Frank emphasized the common- sense language immanent in the scientific language of representations.

13. Therefore, "the influence of social factors upon the validity of scientific (even physical) theories is a necessary part of pragmatic epistemology" (Frank 2021, 1.6.1.). By "pragmatic epistemology", Frank meant those investiga- tions of science that take into considerations the socio-cultural background and context of science in theory-building and theory validation. The sources of Frank's "pragmatic epistemology" go back to the semiotics of his Ameri- can brother-in-arms Charles Morris, who was responsible, with Neurath and Carnap, for the publication of the *International Encyclopedia of Unified Science*.

References

Bergmann, P. (1965), 'Homage to Professor Philipp G. Frank', in R. S. Cohen and M. W. Wartofsky (eds.), *Proceedings of the Boston Colloquium for the Philoso- phy of Science, 1962–1964*. New York: Humanities Press, pp. ix–x.

Black, P. (1949), 'Review of Frank's Modern Science and Its Philosophy', *The Scientific Monthly*, 69 (6): 423–424.

Blackmore, J., Itagaki, R., and Tanaka, S. (eds.) (2001), *Ernst Mach's Vienna 1895–1930: Or Phenomenalism as Philosophy of Science*. Dordrecht: Springer.

Carnap, R. (1932/1959), 'Elimination of Metaphysics through Logical Analysis of Language', in A. J. Ayer (ed.), *Logical Positivism*. New York: The Free Press, pp. 60–81.

—— (1939), *Foundations of Logic and Mathematics*. Chicago: University of Chicago Press.

Cohen, R. S., and Wartofsky, M. (eds.) (1965), *In Honor of Philipp Frank: Proceedings of the Boston Colloquium for the Philosophy of Science, 1962–1964*. New York: Humanities Press.

Duhem, P. (1903/1912), *Die wandlungen der mechanik und die mechanische Naturerklärung*. Translated by Philipp Frank. Leipzig: Verlag von J. A. Barth.

Eddington, A. (1928), *The Nature of the Physical World*. Cambridge: Cambridge University Press.

Einstein, A. (1912/1995), 'Report to the Philosophical Faculty of the German University on a Successor to the Chair of Theoretical Physics', in M. J. Klein et al. (eds.), *The Collected Papers of Albert Einstein: Volume 5: The Swiss Years: Correspondence, 1902–1914*. Princeton: University of Princeton Press, pp. 300–303.

Forman, P. (1971), 'Weimar Culture, Causality, and Quantum Theory: Adaptation by German Physicists and Mathematicians to a Hostile Environment', *Historical Studies in the Physical Sciences*, 3: 1–115.

Frank, Ph. (1907/1949), 'Experience and the Law of Causality', in *Modern Science and Its Philosophy*. Cambridge, MA: Harvard University Press, pp. 53–60.

—— (1932/1998), *The Law of Causality and Its Limits*. Translated by M. Neurath and Robert S. Cohen. Dordrecht: Springer.

—— (1939/1949), 'Modern Physics and Common Sense', in *Modern Science and Its Philosophy*. Cambridge, MA: Harvard University Press, pp. 144–157.

—— (1941/1949), 'Why Do Scientists and Philosophers So Often Disagree about the Merits of a New Theory?', in *Modern Science and Its Philosophy*. Cambridge, MA: Harvard University Press, pp. 207–215.

—— (1945), *Thermodynamics*. Providence, RI: Brown University.

—— (1946/1949), 'Science Teaching and the Humanities', in *Modern Science and Its Philosophy*. Cambridge, MA: Harvard University Press, pp. 260–285.

—— (1946/1971), 'Foundations of Physics', in O. Neurath et al. (eds.), *Foundations of the Unity of Science*. Chicago: University of Chicago Press, pp. 423–504.

—— (1947/1972), *Einstein: His Life and Times*. New York: Alfred A. Knopf.

—— (1949), 'Einstein, Mach, and Logical Positivism', in P. A. Schilpp (ed.), *Albert Einstein: Philosopher-Scientist*. New York: Tudor Publishing Company, pp. 269–286.

—— (1950a), 'Metaphysical Interpretations of Science, Part I', *British Journal for the Philosophy of Science*, 1 (1): 60–74.

—— (1950b), 'Metaphysical Interpretations of Science, Part II', *British Journal for the Philosophy of Science*, 1 (2): 77–91.

—— (1951a), *Relativity: A Richer Truth*. London: Jonathan Cape.

—— (1951b), 'The Logical and Sociological Aspects of Science', *Proceedings of the American Academy of Arts and Sciences*, 80 (1): 16–30.

—— (1951c), 'Introductory Remarks: Contributions to the Analysis and Synthesis of Knowledge', *Proceedings of the American Academy of Arts and Sciences*, 80 (1): 5–8.

——— (1958), 'Contemporary Science and the Contemporary World View', *Daedalus*, 87 (1): 57–66.

——— (1962), 'Interview of Philipp Frank by Thomas S. Kuhn on 1962 July 16', Niels Bohr Library & Archives, American Institute of Physics, College Park, MD USA, www.aip.org/history-programs/niels-bohr-library/oral-histories/4610.

——— (2021), *The Humanistic Background of Science*. Edited by G. Reisch and A. T. Tuboly. New York: SUNY Press.

Fürth, R. (1965), 'Reminiscences of Philipp Frank at Prague', in R. S. Cohen and M. W. Wartofsky (eds.), *Proceedings of the Boston Colloquium for the Philosophy of Science, 1962–1964*. New York: Humanities Press, pp. xiii–xvi.

Holton, G. (ed.) (1966), *Philipp Frank 1884–1966: Expressions of Appreciation as Arranged in the Order Given at the Memorial Meeting for Philipp Frank, October 25, 1966*. Cambridge: Harvard University Press.

Ladyman, J., Ross, D., Spurrett, D., and Collier, J. (2007), *Every Thing Must Go: Metaphysics Naturalized*. Oxford: Oxford University Press.

Mormann, Th. (2017), 'Philipp Frank's Austro-American Logical Empiricism', *HOPOS*, 7 (1): 56–87.

Nemeth, E. (2003), 'Philosophy of Science and Democracy: Some Reflections on Philipp Frank's *Relativity: A Richer Truth*', in M. Heidelberger and F. Stadler (eds.), *Wissenschaftsphilosophie und Politik/Philosophy of Science and Politics*. Wien and New York: Springer, pp. 119–138.

Pauli, W. (1958), *The Theory of Relativity*. New York: Pergamon Press.

Poincaré, H. (1905), *Science and Hypothesis*. London and Newcastle: The Walter Scott Publishing Co., Ltd.

Reichenbach, H. (1951), *The Rise of Scientific Philosophy*. Berkeley: University of California Press.

Reisch, G. (2005), *How Cold War Transformed Philosophy of Science: To the Icy Slopes of Logic*. New York: Cambridge University Press.

——— (2017), 'Pragmatic Engagements: Philipp Frank and James Bryant Conant on Science, Education, and Democracy', *Studies in East European Thought*, 69 (3): 227–244.

Richardson, A. W. (2012), 'Occasions for an Empirical History of Philosophy of Science: American Philosophers of Science at Work in the 1950s and 1960s', *HOPOS*, 2 (1): 1–20.

Sellars, W. (1962), 'Philosophy and the Scientific Image of Man', in R. Colodny (ed.), *Frontiers of Science and Philosophy*. Pittsburgh: University of Pittsburgh Press, pp. 35–78.

Siegetsleitner, A. (2017), 'Philipp Frank on Relativity in Science and Morality', *Studies in East European Thought*, 69 (3): 215–225.

Sigmund-Schultze, R. (2007), 'Philipp Frank, Richard von Mises, and the Frank-Mises', *Physics in Perspective*, 9: 26–57.

Snow, C. P. (1961), *The Two Cultures and the Scientific Revolution*. Cambridge: Cambridge University Press.

Stebbing, S. (1937/1944), *Philosophy and the Physicists*. London: Penguin Books.

Stöltzner, M. (1999), 'Vienna Indeterminism: Mach, Boltzmann, Exner', *Synthèse*, 119: 85–111.

——— (2002), 'How Metaphysical Is "Deepening the Foundations"?: Hahn and Frank on Hilbert's Axiomatic Method', in M. Heidelberger and F. Stadler (eds.), *History of Philosophy of Science: New Trends and Perspectives*. Dordrecht: Springer, pp. 245–262.

———— (2003), 'Vienna Indeterminism II: From Exner to Frank and von Mises', in P. Parrini, W. Salmon, and M. Salmon (eds.), *Logical Empiricism: Historical and Contemporary Perspectives*. Pittsburgh: Pittsburgh University Press, pp. 194–229.

———— (2009), 'The Logical Empiricists', in H. Beebee et al. (ed.), *The Oxford Handbook of Causation*. Oxford: Oxford University Press, pp. 108–127.

Tuboly, A. T. (2017), 'Philipp Frank's Decline and the Crisis of Logical Empiricism', *Studies in East European Thought*, 69 (3): 257–276.

Uebel, Th. (2011), 'Beyond the Formalist Meaning Criterion: Philipp Frank's Later Metaphysics', *HOPOS*, 1 (1): 47–72.

Wolters, G. (1987), *Mach I, Mach II, Einstein und die Relativitätstheorie. Eine Fälschung und ihre Folgen*. Berlin and New York: Walter de Gruyter.

Zilsel, E. (1942), 'The Sociological Roots of Science', *The American Journal of Sociology*, 47: 544–562.

About the Editors and Authors

Editors

SEBASTIAN LUTZ is Senior Lecturer of theoretical philosophy at Uppsala University. He received a master's degree in physics from the University of Hamburg and a Ph.D. in philosophy from Utrecht University. He works on philosophy of science, the history of logical empiricism, and philosophical methodologies. He has published on the history and philosophy of logical empiricism in *Philosophy and Phenomenological Research*, *HOPOS*, *Synthese*, *Philosophy of Science*, and *Erkenntnis*.

ADAM TAMAS TUBOLY is postdoctoral researcher at the Institute of Philosophy, Eötvös Loránd Research Network, Budapest, MTA BTK Lendület Morals and Science Research Group, and a research fellow at the Institute of Transdisciplinary Discoveries, Medical School, University of Pécs. He works on the history of logical empiricism and edited numerous volumes on it at Bloomsbury, Palgrave Macmillan, Routledge, Springer, and SUNY Press.

Contributors

JORDI CAT is Associate Professor at the Department of History and Philosophy of Science and Medicine, Indiana University Bloomington. His research interests include philosophy of science, history of science, and history of philosophy of science. He is the author of *Fuzzy Pictures as Philosophical Problem and Scientific Practice* (Springer 2016) and *Maxwell, Sutton and the Birth of Color Photography* (Palgrave 2013) and co-author of *Otto Neurath: Philosophy between Science and Philosophy* (Cambridge Univ. Press 1996) with Nancy Cartwright, Lola Fleck, and Thomas Uebel; co-editor of Neurath Reconsidered: New Sources and Perspectives (with Adam Tamas Tuboly, Springer, 2019).

RICHARD DAWID is Professor of philosophy of science at Stockholm University. His work focuses on philosophical aspects of contemporary theories in high-energy physics and cosmology. He research areas include Everettian quantum mechanics, issues of data analysis in high energy physics, the philosophical impact of string dualities, the cosmological multiverse, and anthropic reasoning. He published numerous articles on the philosophy of physics, authored *String Theory and the Scientific Method* (Cambridge University Press, 2013), and edited *Why Trust a Theory? Epistemology of Fundamental Physics* (Cambridge University Press 2019, with Radin Dardashti and Karim Thébault).

ROBERT DiSALLE is Professor of philosophy at the University of Western Ontario. He works on the history and philosophy of science, especially the history and philosophy of physics from Newton to the present, philosophical problems of space and time, and the history of philosophy of science. He authored *Understanding Spacetime: The Philosophical Development of Physics from Newton to Einstein* (Cambridge University Press, 2006) and edited *Henri Poincaré, Philosopher of Science: Problems and Perspectives* (with M. DePaz, Springer, 2014) and *Analysis and Interpretation in the Exact Sciences* (with M. Frappier and D. Brown, Springer, 2012).

KATHERINE DUNLOP is Associate Professor at the University of Texas at Austin. She received her Ph.D. from the University of California, Los Angeles, and has published numerous articles on the philosophy of mathematics and geometry, including on Henri Poincaré, in *Synthese*, *HOPOS*, and *Philosophy and Phenomenological Research*.

JAN FAYE is Associate Professor at the Department of Communication, University of Copenhagen. His research areas cover metaphysics, philosophy of science and the humanities, epistemology, and philosophy of mind. He has authored and edited numerous books on philosophy of science, especially on Niels Bohr.

MARIA CARLA GALAVOTTI is Professor Emerita at the Department of Philosophy and Communication Studies at the University of Bologna. She authored *Philosophical Introduction to Probability* (CSLI, 2005) and edited *Stochastic Causality* (with P. Suppes and D. Costantini, CSLI, 2001), *Cambridge and Vienna. Frank P. Ramsey and the Vienna Circle* (Springer, 2006), and *New Directions in the Philosophy of Science* (with D. Dieks, W.J. Gonzalez, S. Hartmann, T. Uebel, and M. Weber, Springer, 2014). She has also published numerous articles on the foundations and the history of probability, scientific explanation, prediction, and causality in the *British Journal for the Philosophy of Science*, *Erkenntnis*, and *Synthese*.

MARCO GIOVANELLI is a research fellow at the University of Tübingen. He received his Ph.D. from the Universities of Zurich and Turin, and he works on the history and/of philosophy of physics, especially on

Kant, neo-Kantianism, logical empiricism, Weyl, and Einstein. He has published the book *Reality and Negation: Kant's Principle of Anticipations of Perception* (Springer, 2011) and various articles on the history of physics in *Perspectives on Science, Studies in History and Philosophy of Science, HOPOS, Synthese,* and *Erkenntnis.*

CLARK GLYMOUR is Alumni University Professor at the department of philosophy, Carnegie Mellon University. He has published numerous books and articles on the philosophy and history of science, machine learning, reasoning, and recently on Bayesian probability.

DON HOWARD is Professor of philosophy and former Director of the Reilly Center for Science, Technology, and Values at the University of Notre Dame. A former assistant and contributing editor for the *Collected Papers of Albert Einstein*, he has written extensively on a wide array of topics in the history and philosophy of late-nineteenth- and early-twentieth-century physics, as well as the history of the philosophy of science.

RASMUS JAKSLAND is Ph.D. Candidate at the Department of Philosophy and Religious Studies at the Norwegian University of Science and Technology. His working project focuses on the prospects of naturalized metaphysics, but he works on foundational issues in high-energy physics, holographic gravity, and the nature of space-time in the light of the AdS/CFT correspondence. He has published papers in *Synthese, Journal of General Philosophy of Science,* and *Studies in History and Philosophy of Science Part B.*

NIKOLAY MILKOV is Professor at Fach Philosophie, Universität Paderborn. His research areas include the history of analytic philosophy, history of philosophy of science, metaphysics, and philosophy of language. His latest monograph is *Early Analytic Philosophy and the German Philosophical Tradition* (London: Bloomsbury, 2020).

MATTHIAS NEUBER is Privatdozent at the University of Mainz. He works on the philosophy of Kant, neo-Kantianism, logical empiricism (especially on realism) and philosophy of nature. He has published two books (*Die Grenzen des Revisionismus: Schlick, Cassirer und das „Raumproblem"*, Springer, 2012; *Der Realismus im logischen Empirismus: Eine Studie zur Geschichte der Wissenschaftsphilosophie*, Springer, 2017) and edited volumes on logical empiricism and related themes.

FLAVIA PADOVANI is Associate Professor of philosophy at the Department of English and Philosophy, Drexel University. Her research addresses issues in both history and philosophy of science and general philosophy of science. She published several papers in *Synthese, Studies in History and Philosophy of Modern Physics, Studies in History and*

Philosophy of Science, and *European Studies in Philosophy of Science* and edited, with Alan Richardson and Jonathan Y. Tsou, the volume *Objectivity in Science: New Perspectives from Science and Technology Studies* (Dordrecht: Springer, 2015).

ALAN RICHARDSON is Professor of philosophy at the University of British Columbia. He works mainly in the history of philosophy of science in the early twentieth century. He published the book *Carnap's Construction of the World* (Cambridge University Press, 1998) and edited various volumes, such as *Objectivity in Science: New Perspectives from Science and Technology Studies* (Springer, 2015), *The Cambridge Companion to Logical Empiricism* (Cambridge University Press, 2007), *Logical Empiricism in North America* (University of Minnesota Press, 2003), and *The Origins of Logical Empiricism* (University of Minnesota Press, 1996). He has also published numerous papers on logical empiricism and general philosophy of science.

THOMAS RYCKMAN is Professor at Stanford University. He works on the philosophy of science, especially on the history and philosophy of physics. He published *The Reign of Relativity: Philosophy in Physics 1915–1925* (Oxford University Press, 2007) and *Einstein* (Routledge, 2017).

Index